Developments in
Integrated Circuit Testing

Editors:

Werner Rheinboldt
University of Pittsburgh
Pittsburgh, Pennsylvania

Daniel Siewlorek
Carnegie-Mellon University
Pittsburgh, Pennsylvania

Editorial Advisory Board:

Kazuhiro Fuchi, Director
Institute for New Generation Computer Technology (ICOT)
Tokyo, Japan

Makoto Nagao
Kyoto University
Kyoto, Japan

PERSPECTIVES IN COMPUTING, Vol. 18
(Formerly "Notes and Reports in Computer Science and Applied Mathematics")

Developments in Integrated Circuit Testing

D. M. Miller, editor
Department of Computer Science
University of Victoria
Victoria, Canada

ACADEMIC PRESS LIMITED
Harcourt Brace Jovanovich, Publishers

London San Diego New York Berkeley
Boston Sydney Tokyo Toronto

TK
7874
.D49
1987

ACADEMIC PRESS LIMITED
24–28 Oval Road, London NW1 7DX

United States Edition published by
ACADEMIC PRESS, INC.
San Diego, CA 92101

Copyright © 1987 by
Academic Press Limited

All rights reserved. No part of this book may be reproduced or transmitted in any form or by any means, electronic or mechanical, including photocopy, recording, or any information storage and retrieval system without permission in writing from the publisher.

British Library Cataloguing in Publication Data

Developments in integrated circuit testing,
—(Perspectives in computing; v. 18).
1. Digital integrated circuits—Testing
I. Miller, D. M.   II. Series
621.381'73   TK7874

ISBN 0-12-496735-3

Printed in Great Britain by
St Edmundsbury Press Limited, Bury St Edmunds, Suffolk

Contents

Preface vii
Contributors ix

Introduction 1
   D. M. Miller
Comparing Causes of IC Failure 13
   E. J. McCluskey and S. Mourad
Matrix Methods in the Detection and Elimination of Redundancy 47
in Combinational Circuits
   A. R. Fleming, G. E. Taylor and A. C. Pugh
The Interrelationships Between Fault Signatures Based Upon 83
Counting Techniques
   S. L. Hurst
The Spectral Testing of Multiple-Output Circuits 115
   M. Serra and J. C. Muzio
Signatures and Syndromes are Almost Orthogonal 147
   J. P. Robinson and N. R. Saxena
Aliasing Probabilities of Some Data Compression Techniques 169
   J. C. Muzio, F. Ruskey, R. C. Aitken and M. Serra
An Introduction to an Output Data Modification Scheme 219
   V. K. Agarwal and Y. Zorian
Effectiveness Measures for Data Compression Techniques 257
Under Unequally Likely Errors
   N. R. Saxena
The Exhaustive Testing of CMOS Stuck-Open Faults 279
   J. A. Bate and D. M. Miller
Testability of Combinational Networks of CMOS Cells 315
   J. A. Brzozowski
Concurrent Checking Techniques – a DFT Alternative 359
   I. L. Sayers, G. Russell and D. J. Kinniment
The Testing of Integrated Circuits Incorporating Analogue Sections 391
   A. P. Dorey, P. J. Silvester and R. J. Ball
Selected Bibliography 407
   P. G. Bessler and D. M. Miller

Index 433

Preface

This volume is concerned with a number of recent developments in integrated circuit testing. Both theoretical and practical aspects are considered. The material is intended for graduate students, engineers and computer scientists working in the areas of logic design, testing of digital systems, built-in self-test and design for testability. It is a suitable foundation for those pursuing new research directions as well as those whose interest is in new techniques for dealing with the testing of their current and future designs.

Each chapter in this book is an extended version of a presentation made at the international workshop on New Directions in IC Testing held in Victoria, British Columbia, Canada, in April 1986. Sufficient background and tutorial material has been included so that each chapter is a self-contained presentation.

Administrative support for the workshop was provided by the University of Victoria and the University of Manitoba, Canada, and the University of Bath and the Open University, U.K. Facilities for the preparation of the camera-ready manuscript were provided by a number of the authors' institutions and in particular by the Department of Computer Science, University of Manitoba.

J. C. Muzio and S. L. Hurst were instrumental in the planning and execution of the workshop. I also acknowledge their invaluable assistance in the preparation of this work as well a number of reviewers whose comments have considerably improved this book. Finally I thank the authors for the opportunity to act as editor for this volume.

D. M. Miller
Winnipeg, Canada
April 10, 1987

Contributors

V. K. Agarwal, *VLSI Design Laboratory, Department of Electrical Engineering, McGill University, Montreal, Canada*
R. C. Aitken, *Department of Computer Science, University of Victoria, Victoria, B.C. V8W 2Y2, Canada*
R. J. Ball, *Department of Engineering, University of Lancaster, Bailrigg, Lancaster LA1 4YR, U.K.*
J. A. Bate, *Department of Computer Science, University of Manitoba, Winnipeg R3T 2N2, Canada*
P. G. Bessler, *Department of Computer Science, University of Manitoba, Winnipeg, Manitoba R3T 2N2, Canada*
J. A. Brzozowski, *Department of Computer Science, University of Waterloo, Waterloo, Ontario N2L 3G1, Canada*
A. P. Dorey, *Department of Engineering, University of Lancaster, Bailrigg, Lancaster LA1 4YR, U.K.*
A. R. Fleming, *Department of Electronic Engineering, University of Hull, Hull, U.K.*
S. L. Hurst, *The Open University, Faculty of Technology, Walton Hall, Milton Keynes MK7 6AA, U.K.*
D. J. Kinniment, *Department of Electrical and Electronic Engineering, University of Newcastle upon Tyne NE1 7RU, U.K.*
E. J. McCluskey, *Center for Reliable Computing, Computer Systems Laboratory, Stanford University, Stanford, CA 94305, U.S.A.*
D. M. Miller, *Department of Computer Science, University of Victoria, Victoria, B.C. V8W 2Y2, Canada*
S. Mourad, *Center for Reliable Computing, Computer Systems Laboratory, Stanford University, Stanford, CA 94305, U.S.A. and Computer Engineering Department, San Jose State University, San Jose, CA 95192, U.S.A.*
J. C. Muzio, *Department of Computer Science, University of Victoria, Victoria, B.C. V8W 2Y2, Canada*
A. C. Pugh, *Department of Mathematics, Loughborough University of Technology, Loughborough, U.K.*
J. P. Robinson, *Department of Electrical and Computer Engineering, University of Iowa, Iowa City, Iowa 52242, U.S.A.*

F. Ruskey, *Department of Computer Science, University of Victoria, Victoria, B.C. V8W 2Y2, Canada*

G. Russell, *Department of Electrical and Electronic Engineering, University of Newcastle upon Tyne NE1 7RU, U.K.*

N. R. Saxena, *Hewlett Packard, MS44UX, 19111 Prune Ridge Avenue, Cupertino, CA 95014, U.S.A.*

I. L. Sayers, *Department of Electrical and Electronic Engineering, University of Newcastle upon Tyne NE1 7RU, U.K.*

M. Serra, *Department of Computer Science, University of Victoria, Victoria, B.C. V8W 2Y2, Canada*

P. J. Silvester, *Department of Engineering, University of Lancaster, Bailrigg, Lancaster LA1 4YR, U.K.*

G. E. Taylor, *Department of Electronic Engineering, University of Hull, Hull, U.K.*

Y. Zorian, *VLSI Design Laboratory, Department of Electrical Engineering, McGill University, Montreal, Canada.*

INTRODUCTION

D. M. Miller

*Department of Computer Science
University of Manitoba
Winnipeg, MB
Canada R3T 2N2*

I. AN OVERVIEW OF FUNCTIONAL TESTING.

This chapter is intended to define the context within which the research reported in this volume was performed. Readers familiar with the area of digital testing may wish to skip this preparatory material.

Digital systems are subject to a great variety of physical failures. These failures can arise during the manufacturing process or can appear during the lifetime of the system. Depending on the nature of the failure the functional behaviour of the system may be affected or the effect may alter the timing or operating range parameters of the system. In this book, we are primarily concerned with techniques for the detection of functional failures.

By a functional failure we specifically mean one which can be detected by applying either one or a sequence of input patterns and observing the resulting output behaviour. Any deviation in response from the known good response of the system under test indicates a failure is present.

Our concern is with integrated circuits and in particular with complex VLSI devices. Since there is currently no practical means to repair a faulty chip, we consider the problem of fault detection. Isolating the cause of the fault is of little importance. Note that we are not directly concerned with design verification where the identification of a design flaw is important.

Failures can be classified as permanent or intermittent. As the name implies a permanent failure, once it has occured, persists. In this case, the failure will be detected the next time the system is tested. An off-line test, which exercises the system outside its normal operating mode, will detect such a fault. It is less than certain that such a test will detect an intermittent failure which by its very nature may or may not be present when the test is applied. The only certain way to detect an intermittent failure is to construct a test which is applied concurrently to the normal operation of the system. The majority of this book is concerned with off-line testing. Concurrent testing is the subject of the chapter by Sayers *et al*.

An off-line test can be accomplished in two ways. The traditional approach is to put the chip or system into a tester which applies the test stimulus and verifies the observed response. This has a significant disadvantage in the fact the unit under test is not being tested in its normal operating environment. For example, if a chip must be removed from a board and placed in a tester, this action can generate a failure or can fail to detect a failure in the board's interface to the chip.

The second approach which has been the subject of much recent research is *built-in self test* (1). In this method, the test stimulus and the response verification are part of the integrated circuit or the board itself. The chip or system has a normal/test mode control input. Putting the chip into test mode causes it to apply its built-in stimulus and to verify the response against a built-in 'correct' answer. The result of the verification is then reported as a single 'pass/fail' output. In the VLSI environment pins are scarce and such an approach has obvious attraction.

The possible disadvantage to built-in self test is the fact that it uses up valuable chip area. The current research thrust in this area thus concerns the development of built-in self-test techniques which are both time and area efficient while still offering reasonable failure coverage. A number of the results presented in this volume concern built-in self-test.

A central theme throughout the book is *design for testability* (1) which in general refers to techniques for designing chips and systems that are easier to test than those which are designed with no particular concern for the testing process. It is now clear that testability is crucial to successful VLSI design. A chip design, no matter how clever, is of little use if it can not be tested both at the design phase and during its lifetime.

The primary concern though most of this volume is the testing of digital circuits. In the chapter by Dorey *et al*, the scope is broadened with the introduction of the testing of circuits containg analogue sections.

II. FAULT MODELS.

In order to develop a test for a functional failure, the failure must be cast in functional terms. A *fault model* is a logical abstraction which attempts to represent the effect of physical failures. The degree to which a fault model truly does represent the potential physical failures in a particular technology is central to the success of any test method based on that model. A number of models have been proposed.

A. Stuck-at Faults.

The *single stuck-at* fault model (2) assumes that a physical failure is equivalent to one line in the circuit being permanently fixed at logic 0 or logic 1. This model was developed for discrete component circuits where a broken wire typically leads to the behaviour noted.

It is well known that many of the physical failures possible in a VLSI circuit are not properly represented by the single stuck-at model. However, due to its simplicity as well as the high investment in techniques and CAD tools oriented to the single stuck-at model, it remains the dominant model used in the testing industry and in much of the fault detection literature. Often a fault detection technique not specifically tied to the single stuck-at model is evaluated in terms of the percentage of single stuck-at faults which the method detects.

The single stuck-at model can be extended by allowing for multiple stuck lines. The high density of VLSI circuits suggests that failures equivalent to such *multiple stuck-at* (3) faults should be quite probable. However, while a circuit with t lines has 2t potential single stuck-at faults, there are $3^t - 2t - 1$ potential multiple stuck-at faults. This number prohibits explicit consideration of each multiple fault for large circuits.

B. Bridging Faults.

Many physical failures are clearly not adequately represented by the stuck-at model (4). A circuit line shorted to power (ground) will usually lead to that line being stuck-at 1 (0). However, two signal lines shorted will have a quite different effect.

A *bridging fault* (5) assumes that two signal lines shorted together results in both lines being replaced by a logical combination of those lines. Depending on the technology, the short is assumed to introduce an AND bridge or an OR bridge.

A circuit with t lines has on the order of t^2 possible bridging pairs. If the actual layout of the circuit is known the number that need be considered can be greatly reduced. Muzio and Serra (6) have introduced the term *stuck-at neighbour* to denote a bridging fault between two physically adjacent circuit lines.

A further consideration with bridging faults is the fact that such a fault can introduce a feedback path and can change a normally combinational circuit into a sequential circuit thereby greatly compounding the testing problem.

C. MOS Transistor Faults.

Stuck-at faults and usually bridging faults are used in the analysis of gate level circuit descriptions. Wadsack (7) has identified types of physical failures in MOS transistors which are not adequately modeled by the stuck-at and bridging models. A *stuck-open* fault assumes the physical failure results in a transistor which can never conduct current and is thus in switch-level terms permanently open.

A transistor which cannot conduct in an MOS combinational cell results in input conditions where the output of the cell is floating, i.e., it is not driven to a definite logic value. In this case the capacitance associated with the output tends to hold the previously driven output value. The failure in the transistor has thus introduced memory into the combinational cell and, as in the bridging case, the testing process is greatly complicated. Stuck-open faults are considered in the chapters by Bate and Miller and by Brzozowski.

Wadsack (7) also introduced the notion of a stuck-on transistor. Renovell and Cambon (8) have empirically examined floating gate failures in MOS circuits and found they are not adequately represented by any of the aforementioned models.

III. TEST METHODS.

A. Test Sets.

The traditional approach to testing a combinational circuit for functional faults involves the identification of a *test set* which is a subset of the possible input assignments to the circuit. These assignments called *test vectors*, are selected so that any fault within the target fault class, e.g. single stuck-at faults, will give different responses in the faulty and the fault-free circuits for at least one vector. It is thus common to refer to a single stuck-at fault test set, an input bridging fault test set, etc.

Many algorithms for determining test sets, particularly for single stuck-at faults, are known (9). The determination of test vectors is computationally complex and very time-consuming. However, it is a one-time design cost. The necessity to store the test set and the associated expected responses and the time required to apply the tests and to verify the responses are the serious drawbacks to the test set approach. All these factors frequently mean a suboptimal test set offering less than 100% single stuck-at fault coverage is used.

The generation of test sets for faults in sequential circuits has also been studied (9) and has been found to be considerably more complex than in the combinational case. Since most practical circuits are sequential or contain a substantial sequential component, this is a serious problem.

B. Scan Path Testing.

Numerous schemes (1) have been proposed whereby a sequential circuit can be treated as combinational at test time. Combinational test vectors are then sufficient to test the circuit. The difficulty of identifying a truly sequential test is thus bypassed but at the expense of the hardware required to reconfigure the circuit at test time, and with the possibility that certain types of faults may be go undetected since the circuit is being tested outside its normal operating mode. There is also the possibility the chip might be declared bad due to a fault in the test circuitry even though the chip performs its normal function correctly.

While the details of the various scan path schemes vary, the basic scenario is the same. The state latches associated with the sequential circuit are extended so that they have two operational modes. During 'normal' operation, they perform as simple latches. In 'test' mode, the normal feedback path of the sequential circuit is broken and the latches form a shift register. A test vector can be 'scanned' into this register from one chip pin. Subsequently, the result of the test vector can be latched and then 'scanned' out via a second chip pin. This result is then externally compared to that expected. Suitable use of this 'scan path' together with the primary inputs and outputs of the sequential circuit allow the combinational logic to be fully tested for whatever class of faults the test set used has been constructed. The latches are themselves tested by virtue of the fact they form the scan path used to apply the test set.

In IBM's level sensitive scan design (LSSD) (10), control of the function of the scan path is accomplished by using three clocks. Various sequences of clock pulses result in normal latch, shift in or shift out operation. In LSSD the normal operating speed of the circuit is not affected by the extra functionality associated with the latches. The chip area overhead varies from 15 to 30% (1)

The obvious advantages of scan path techniques are the allowance for combinational testing of a sequential circuit with little or no degradation of normal performance, and the simplification of the test generation process. The disdvantages include the additional chip area required, the addition of up to four chip pins, the rather elaborate external control required to set up and monitor the tests and the length of time to scan tests into an out of the chip.

C. Built-In Self-Test.

By *built-in self-test* (BIST) we mean any scheme where all circuitry associated with testing the chip is part of the chip itself. An input pin is required to select between normal and self-test modes. An output pin is required to deliver the pass/fail result. We assume the circuit to be tested is combinational. As in the scan path approach a sequential circuit is reconfigured with the latches forming a part of the test circuitry.

BIST requires an on-chip input source. This is typically accomplished by configuring certain latches into a linear feedback shift register (LFSR) (11). For small combinational blocks, say up to 20 inputs, the LFSR can be used to apply all possible input patterns to the circuit under test[†]. For larger blocks the LFSR is used to apply a pseudo-random subset of the input patterns.

Considerable response data is generated. For a 20-input, 10-output combinational block, on the order of 10^7 bits are produced. Some form of data compaction is thus used to map the response data into a *signature*.

[†] The LFSRs used can generate $2^n - 1$ of the 2^n input patterns required for an n-input combinational circuit. Additional gating must be added to achieve the input pattern 00....0.

Typical signatures proposed are on the order of 16 bits for single-output circuits.

The signature accumulated is compared on-chip with a built-in reference signature which is predetermined at design time. Depending on the compaction method used, the correct signature may be found by simulation or by analytic means. A deviation in the good and the observed signatures identifies a fault and test output 'fail' is indicated.

BIST has obvious advantages. No test generation is required although this is to some degree offset by the necessity of finding the good signature. Only two chip pins are required and the external test control required is minimal. Perhaps most important, a BIST can be applied at close to the normal operating speed of the chip and should thus detect a broader class of faults than techniques centred upon test vectors.

It is also significant that a chip incorporating BIST carries its test with it. Thus the investment of time and money required to establish the test is averaged across the life of the chip. Field testing is much simpler since no external test hardware is required. In a properly designed system, the chips can be self-tested without removal. This both quickens and simplifies the testing process.

BIST requires extra chip area. The actual cost is highly design dependent. Further, as in the scan path approach, there is the possibility that a fault isolated to the test circuitry will cause an otherwise good chip to be declared bad. For most applications, the advantages of BIST outweigh these concerns.

Since data compaction implies a loss of infomation, there is of course the possibility that a fault will be *masked* in BIST. This phenomenon, often termed *aliasing*, and the inherent difficulty in assessing the actual fault coverage afforded by a BIST, are the most serious problems associated with this approach.

D. Data Compaction Schemes

A number of data compaction[†] schemes have been proposed, several of which are considered in this book. In *syndrome* testing (10), the signature for a single-output combinational circuit is taken to be the number of ones in the response data when all possible input patterns are applied. Other counts such as the number of transitions in the data have also been proposed (13).

In *spectral coefficient* testing (14), the syndrome concept is generalized so that the count is the number of ones in the exclusive-OR of the circuit output and some subset of the inputs. A spectral signature typically contains two or more such counts relative to different input subsets. Spectral testing extends the fault coverage of syndrome testing at the expense of the area required for the exclusive-OR generators and the need for multiple counts. While spectral testing is therefore of questionable practical use as a testing method on its own, spectral techniques form a valuable mathematical basis for the construction and analysis of other test approaches as demonstrated in the chapter by Hurst and the chapter by Serra and Muzio.

The fault coverage afforded by syndrome and spectral coefficient testing can be assessed for the common fault models in a straightforward, if a somewhat time-consuming, analytical fashion.

In the single-output case, an LFSR can be used to compress the data. This is essentially the signature analysis approach proposed by Hewlett-Packard (14) for microprocessor board testing. LFSR compression has a straightforward mathematical interpretation as polynomial division over a Gallois field (16). Unfortunately, there is no easy characterization of the effect which the common fault models have on the circuit output when a pseudo-random sequence of input patterns is applied. Hence in the LFSR

[†] "Data compression" is often used rather than "data compaction". However, the latter more correctly reflects the inherent loss of information.

case, fault coverage must be assessed by simulation or by probabilistic arguments.

The LFSR compactor can be extended to the multiple-output case using a multiple-input shift register (17). In this case, the signature for an m-output circuit is m bits. The technique is thus not applicable for small m. The analysis problems remain and are indeed more complex.

The LFSR technique is of considerable interest due to the low chip area overhead required compared with other BIST techniques. LFSR based BIST is now capturing much attention in the IC industry. The results presented in several chapters of this volume concerning its effectiveness are thus quite timely.

IV. ORGANIZATION OF THE BOOK.

The body of this text begins with an overview of the causes of failures in integrated circuits and some empirical results regarding test sets. The test set theme is continued with a discussion of redundancy in combinational circuits and its effect on testing. Matrix methods for the detection and elimination of redundancy are presented.

A variety of counting techniques are presented including syndrome and spectral coefficient testing. The latter is extended to the multiple-output case and to PLAs.

The next four chapters consider various data compaction schemes with particular emphasis on comparative analysis of these schemes and on aliasing problems.

Faults and fault detection in CMOS circuits are considered in the next two chapters. The first examines the effect of stuck-open faults on exhaustive testing methods. The second presents a general theory of CMOS testability.

The notion of concurrent checking with particular reference to data path chips is considered in the penultimate chapter. Finally, the testing of circuits with analogue sections is introduced.

A reference list is included at the end of each chapter. A comprehensive bibliography appears in the appendix and will be of assistance to those wishing to pursue the research areas presented in this volume.

REFERENCES

1. Williams, T. W., and K. P. Parker (1982), "Design for testability -- a survey", *IEEE Trans. Comput.*, C-31, pp. 2-15.

2. Eldred, R. D. (1959), "Test routines based on symbolic logic statements", *J. Assoc. Comput. Mach.*, 6, pp. 33-36.

3. Breuer, M. A,. and A. D. Friedman (1975), Diagnosis and Reliable Design of Digital Systems, Computer Science Press, Potomac, MD.

4. Galiay, J., Y. Crouzet, and M. Vergniault (1980), "Physical versus logical fault models for MOS LSI circuits: impact on their testability", *IEEE Trans. Comput.*, C-29, pp. 527-531.

5. Karpovsky, M. G., and S. Y. H. Su (1980), "Detection and location of input and feedback bridging faults among input and output lines", *IEEE Trans. Comput.*, C-29, pp. 523-527.

6. Muzio, J. C., and M. Serra (1985), "Built-in testing and design for testability of stuck-at neighbour faults", *Canadian Conf. on VLSI*, Toronto, pp. 229-233.

7. Wadsack, R. L. (1978), "Fault Modelling and Logic Simulation of CMOS and MOS Integrated Circuits", *Bell System Technical Journal*, 57, pp. 1449-1473.

8. Renovell, M, and G. Cambon (1986), "Topology dependence of floating gate faults in MOS integrated circuits", *Electronics Letters*, 22, pp. 152-157.

9. Fujiwara, H. (1985), Logic Testing and Design for Testability, MIT Press, Boston, MA.

10. Eichelberger, E. B. and T. W. Williams (1978), "A logic design structure for LSI testability", *J. Design Automat. Fault Tolerant Computing*, 2, pp. 165-178.

11. Williams, T. W. (editor) (1986), VLSI Testing, North-Holland, Amsterdam.

12. Savir, J. (1980), "Syndrome-testable design of combinational circuits", *IEEE Trans. Comput.*, C-29, pp. 442-451, correction: pp. 1012-1013.

13. Hayes, J. P. (1976), "Transition count testing of combinational logic circuits", *IEEE Trans. Comput.*, C-25, pp. 613-620.

14. Hurst, S. L., D. M. Miller, and J. C. Muzio (1985), Spectral Techniques in Digital Logic, Academic Press, London.

15. Frohwerk, R. A. (1977), "Signature analysis: A new digital field service method", *Hewlett-Packard J.*, pp. 2-8.

16. Golomb, S. W. (1967), Shift Register Sequences, Holden-Day, San Francisco, CA.

17. Carter, J. L. (1982), "The theory of signature testing for VLSI", *Proc. 14th ACM Symp. Theory Comput.*, pp. 289-296.

COMPARING CAUSES OF IC FAILURE

Edward J. McCluskey and Samiha Mourad*

CENTER FOR RELIABLE COMPUTING
Computer Systems Laboratory
Stanford University
Stanford, CA 94305

* *Also with the Computer Engineering Department*
San Jose State University
San Jose, CA 95192

I. INTRODUCTION

Digital systems fail for various reasons. This chapter is an attempt to identify the different causes of failure in the digital logic portion of a system and to assess their relative importance. Software errors are also an important source of system failure (34). They are not discussed here in order to confine this presentation to a manageable scope rather than because they are less important than hardware failures. Digital logic failures can be distinguished according to whether the failing component was defective when the system was first assembled or the component malfunctioned during the system lifetime. Defects can be present initially because of incorrect design or because the test procedure used to eliminate bad devices was incomplete. Devices can fail during system operation either permanently or they can malfunction temporarily. The temporary malfunctions can be due to external causes - transient failures - or they can be due to parameter degradation - intermittent failures. Table 1 lists these failure categories. This chapter deals only with permanent failures. After briefly pointing out the problem areas related to improper design, the chapter examines in greater detail the reasons for digital testing being less than perfect.

Definitions of several testing related parameters are given in section III. Factors that determine the fault coverage required to keep the number of defective chips low is discussed in the following section. Then, the reasons for each type of testing to yield less than 100% fault coverage will be explained and the necessity for a fault model demonstrated. Even when perfect coverage is attained using the popular fault model, single stuck-at, testing is still imperfect because of 1) failure modes that are not adequately represented by the single stuck-at model, and 2) compaction techniques cause some faults to escape detection.

Table 1.

Causes of Digital Logic Failures.

Component Initially Defective
    Incorrect Design
    Incomplete Test of Element
Component Failure during Operation
    Permanent Failure
    Temporary Failure
        Transient
        Intermittent

II. DESIGN PROBLEMS

A system can fail to operate properly if there is something wrong in its design. There are many causes of design errors. Table 2 lists some of them. The most significant contributing factor to design errors is imperfect documentation. There is at present no formal method for specifying system requirements. Thus, they are typically documented in lengthy English prose. This means that it is not possible to verify formally that the design meets the specifications.

There is an interesting and sad parallel between documentation and testing. Taken together they are probably the major manufacturing cost of many systems. In spite of this, it is rare that the most experienced or best designers have any involvement with either testing or documentation. There is a tradition of not worrying about either activity until after the system design has been completed. This attitude is slowly changing with respect to test. The very large capital cost of modern test equipment has produced a recognition of the necessity to prepare for test during the very first design stages. Documentation costs are less concentrated: thus there is less recognition of their importance.

Table 2.

Some Causes of Design Errors

Incorrect Specification
Imprecise Specification
Unclear Specification
Incomplete Design Verification
Unclear Documentation
Incorrect Modifications

One very promising development for logic circuit documentation is the adoption by the IEEE of a Standard for Graphic Symbols for Logic Functions, ANSI/IEEE Std 91-1984. This standard is based on the International Electrotechnical Commission standard specified in IEC Publication 617-12: Binary Logic Elements and discussed in Kampel (22). The IEEE standard is now being used by a number of IC manufacturers, notably Texas Instruments. Unfortunately the work station manufacturers and ASIC chip companies still seem to be inventing their own notations. There is hope that this will change as the IEEE standard symbols start to appear in university logic design courses (31).

The logic aspects of a design are typically checked by simulating its operation for a set of input signals chosen by the designer. The input signals are called *simulation vectors* or *design verification vectors*. These inputs are supposed to be representative of the required system operation and their simulation is assumed to demonstrate the correctness of the design. Since it is only possible to simulate a subset of the possible input sequences, the effectiveness of this technique is very dependent on the skill of the designer.

During the operation of a system design, errors may be discovered or the system specifications may be modified. As a consequence the system hardware must be changed. These "engineering changes" are themselves another source of failures if they are not carried out perfectly. The problems are further aggravated by improper documentation.

In summary, design errors are a significant source of system failure. Improvements in documentation and formal verification are necessary to reduce their contribution to system failure.

III. IMPERFECT TESTING

One of the most important causes of system failure is the presence of faulty ICs. This is due to the lack of 100% fault coverage in the testing procedure used. Even if 100% coverage is possible, faults other than the stuck-at type remain undetectable because of the lack of adequate test patterns to provoke and propagate these faults. This section compares the various causes of less-than-perfect testing.

A. Boolean Testing

The words test and testing as used here refer only to "Boolean testing" of digital integrated circuits. A Boolean Test is defined as a test in which only logical signal values are considered. Parametric tests of digital circuits and testing analog circuits are specifically excluded from the discussion.
Based on when testing is performed two types of Boolean testing that are important for digital systems are distinguished:

1) *Implicit or concurrent testing.* This is sometimes called checking or built-in test. It refers to on-line testing to detect errors that occur during normal system operation. The most common checking techniques use parity codes. Computers often add a parity bit to each byte of a bus or register. The most important characteristics of implicit testing are the ability to detect errors caused by temporary failures and the fact that the circuit inputs are the normal system inputs rather than special test inputs.

2) *Explicit testing.* This is the testing that is carried out while the tested circuit is not in use. It includes the tests done on chips while still on a wafer, production tests of packaged chips and of chips on boards, acceptance tests, maintenance tests, and repair tests. This type of test is aimed at discovering chips with permanent failures. An important characteristic of this type of test is the necessity of controlling the test inputs to the chip.   Explicit Boolean testing is the only type of testing considered in this section. The object of this type of test is to determine whether or not a chip has failed. There are many physical phenomena that can cause chip failures, and production techniques are being continually improved to reduce their occurrence. In spite of this the increased densities and reduced geometries of VLSI cause process yields (the fraction of die on a wafer that

are free of defects) to remain below 50%. Such low yields combined with increasing requirements for improved reliability have created a need for higher quality testing (52). In particular, it is necessary to reduce the defect level (the number of faulty chips that pass the test procedure). This level is a function of the chip fabrication yield and the fault coverage. The determination of the *fault coverage* for a specific circuit and test set is called *fault grading*.

B. Fault Grading

Fault grading is almost always carried out by simulating the circuit activity caused by the test set inputs. This simulation is, in effect, carried out once for the fault-free circuit and then once for the circuit with each possible single stuck-at fault present. The fraction of the faulty circuits for which at least one test input produces an output different from that of the fault-free circuit is called the fault coverage. The various causes for the fault coverage to be less than 100% are listed in Table 3 and will be discussed subsequently. If the coverage is not 100% then a decision must be made whether to add more test vectors and repeat the simulation. The issues involved in deciding how close to 100% fault coverage is required are given in Williams (52). The reason for accepting a test set that doesn't have 100% fault coverage is the expense and time required for simulation. In fact, fault grading is often not done because of its high cost. As shown below this can result in low fault coverage with the consequence of faulty chips not being detected by the test.

Table 3.

Causes of Bad Chips Passing Test.

Single stuck-at fault coverage less than 100%
    Use of design verification vectors
    Use of toggle vectors
      Limited running time of the automatic test pattern generator
    Redundancy present in logic
    Use of incorrect implementation information
    Statistical sampling rather than full simulation
Presence of multiple stuck-at faults
Presence of non-stuck-at faults - bridges, stuck-open, stuck-on
Aliasing or masking in compaction technique

The cost of simulation is being reduced through the development of special hardware and more efficient simulation programs. In principle it should be possible to obtain test sets with 100% fault coverage except for circuits with untestable faults (discussed in Sec. VI.C). This does not guarantee that the test will detect all faulty chips. Failures that don't correspond to single stuck-at faults can be present. Also, compaction techniques may cause some faults (including single stuck-at type) to escape detection.

C. Fault Simulation

Fault simulation is usually carried out to compute fault coverage, evaluate test patterns, and provide diagnostic capability. Early simulators employed a gate level model, used 2-valued logic and simulated one fault at a time. The increase in circuit complexity and the emergence of new circuit technologies have resulted in more diversified and efficient simulators. At present, fault simulation can be performed on the functional and circuit level. Mixed levels may also be used within the same simulator. The number of logic values used has been increased. In order to improve performance, fault processing has evolved first from serial to parallel, then to deductive. Subsequently, concurrent fault simulation was developed and proven to be faster than the other techniques.

Because of the importance of simulation to circuit design and testing and the repetitive nature of simulation, hardware simulators were constructed. Among such engines are the Yorktown Simulation Engine (39), Zycad (49) and Daisy's Megalogician (8). A survey of hardware accelerators was given by Blank (5).

One novel approach reduces the cost of single-stuck fault simulation by deriving a statistical estimate rather than the exact coverage. A random sample of all the faults is simulated. The method requires a minimal sample size thus making it more efficient for very large circuits. The drawback of this method is that it gives no information about the faults not simulated and it is not possible to bound the inaccuracy of the statistical coverage. As higher fault coverages are demanded, the statistical estimates will be less and less useful.

An alternative to fault simulation was proposed by Jain (20). This technique, STAFAN, consists of calculating probabilities of detection for stuck-at faults on all nodes of the circuit. These detection measures are estimated by collecting statistics of the activities of the nodes. For this, fault free simulation is required. This technique suffers from the inability to bound the inaccuracy of the estimated fault coverage.

Besides speed, the main problems with fault simulators, is the fault model used by the simulator. Only single stuck-at faults are modeled. Although this fault representation was well suited for some technologies, it is not adequate for some failure modes in MOS technology. Hayes (16) and Bryant (7) have developed techniques to represent MOS failures.

The remainder of this chapter explores the various factors that permit a bad chip to escape detection. A more effective focus of testing research should result from ranking the items that reduce fault coverage. However, we will first address the issue of how much fault coverage is enough to obtain an acceptable defect level.

IV. FAULT COVERAGE AND YIELD

A number of studies (2), (43), (52) related the 3 parameters: chip yield (Y), fault coverage (T) and defect level (DL). The expression given by Eq. 1, was developed by Williams (51).

$$DL = 1 - Y^{(1-T)} \tag{1}$$

This equation shows that for a certain DL, the higher the yield the less demand is on the fault coverage to be close to 100%. However, a yield in the range 10% to 50% is more realistic. In order to determine the fault coverage for such a range, DL will be expressed in terms of $M = 1-T$, the percentage of missed faults, as shown in Eq. 2. This expression can be expanded in Taylor's series in terms of $(M.\ln Y)$ resulting in the expression given by Eq. 3.

$$DL = 1 - e^{M.\ln Y} \tag{2}$$

$$DL = 1 - \{ 1 + (M.\ln Y) + (M.\ln Y)^2/2! + (M.\ln Y)^3/3! + ... \}. \tag{3}$$

For a value of $(M.\ln Y) < 0.1$, second and higher order terms can be ignored and the expansion in Eq. 3 can be reduced to

$$DL = - M.\ln Y. \tag{4}$$

The minus sign indicates an inverse relationship between DL and either M or Y. This relationship is linear in M and logarithmic in Y. Thus, a decrease in DL is more sensitive to the change in M than that in Y. The defect level for typical values of Y and M are listed in Table 4. The last column of the table gives the defect level in a most often used unit, DPM (Defects Per Million). For Y=50%, and 0.1% faults missing detection, the DPM is equal to 690. If the DPM is to be reduced to say 10 (keeping the same yield), the percentage of missed faults will have to be reduced to 0.002. This corresponds to a fault coverage of 99.998%. Using the same parameters, the fault coverage would be 99.988 according to Seth's (43) calculations.

Table 4.

Defect level for Typical values of Y and M

Y	DL M = 10%	DL M = 1%	DL M = .1%	DPM M = .1%
50%	6.9%	.69%	.069%	690
25%	13.9%	1.39%	.139%	1390
10%	23 %	2.3 %	.23 %	2300

V. FUNCTIONAL TESTING

The term *Boolean testing* is used here to describe the test procedure that is called "functional testing" by the manufacturers of automatic chip test equipment. Functional testing is used differently in other contexts, hence the new term Boolean testing is introduced here to eliminate any possible confusion.

Only integrated circuit chip testing is discussed here -- board test is closely related and is also very important. The expression "functional testing" is used in connection with board test to specify that the test signals are applied through the board I/O edge connectors. The other major type of board test is called "in-circuit testing" or "bed-of-nails testing." This refers to the use of a special fixture that allows access to each node of the board and thus allows each component to be tested individually. Automatic test equipment for board test is discussed in Stover (45).

There is another use (or misuse) of the expression "functional testing". It is sometimes used to describe a test that detects failures by "verifying correct operation" rather than by verifying the absence of specific faults. This usage is based on the desire to generate test patterns without reference to a specific implementation or to a model of which faults are detected by the tests. It is not uncommon to use the design verification vectors for such a "functional test."

A. Design Verification Vectors

Definition: *Design verification* establishes that a design correctly implements a behavioral specification. The customary technique for design verification is a fault-free simulation of the circuit operation. This simulation uses a set of "design verification" or "simulation" vectors as inputs. The design verification vectors are typically generated manually by the designer. If the correct output values are present for each of the design verification inputs, it is assumed that the design is correct.

The design verification vectors for the dual 4-line to 1- line multiplexer of Fig. 1. are shown in Table 5. Part b of the table lists a set of 8 vectors obtained from Franz (11). These vectors are derived by assuming that it is sufficient to verify that both a 0-or a 1-signal on the addressed input are passed correctly to the output. He assumed that it was necessary to make sure that a 1-signal on a nonaddressed line wasn't propagated to the circuit output.

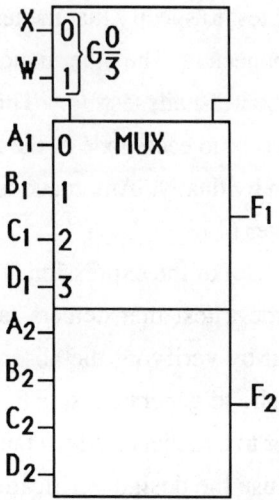

Fig. 1 Dual 4-line to 1-line multiplexer.

It's not surprising that different approaches lead to different design verification vectors since there isn't any standard approach to generating such vectors. In fact, the only way to truly guarantee the correctness of a design would be:

(1) to have a complete specification of the desired output for every possible input, including an identification of those inputs for which the output was not specified, and

(2) to do a detailed simulation of the circuit response to all possible inputs.

It is usually not considered feasible to do such a thorough verification of a design. Instead the designer makes some assumptions about the implementation and selects design verification vectors based on those assumptions. Design verification vectors are discussed here only because they are often used for Boolean testing as well as design verification.

The effectiveness of using the design verification vectors as test vectors

Table 5.

Test Sets for a Dual 4-line to 1-line multiplexer.
(a) 20 design verification vectors, (b) 8 design verification vectors,
(c) Single stuck-at fault test set.

(a)

W	0 0 1 1	0 0 1 1	0 0 1 1	0 0 1 1	0 0 1 1
X	0 1 0 1	0 1 0 1	0 1 0 1	0 1 0 1	0 1 0 1
A	0 0 0 0	1 1 1 1	0 0 0 0	1 1 1 1	0 0 0 0
B	0 0 0 0	0 0 0 0	1 1 1 1	0 0 0 0	0 0 0 0
C	0 0 0 0	0 0 0 0	0 0 0 0	1 1 1 1	0 0 0 0
D	0 0 0 0	0 0 0 0	0 0 0 0	0 0 0 0	1 1 1 1
F1	0 0 0 0	1 0 0 0	0 1 0 0	0 0 1 0	0 0 0 1
F2	0 0 0 0	1 0 0 0	0 1 0 0	0 0 1 0	0 0 0 1

(b)

W	0 0 1 1	0 0 1 1
X	0 1 0 1	0 1 0 1
A	0 0 0 0	1 0 0 0
B	0 0 0 0	0 1 0 0
C	0 0 0 0	0 0 1 0
D	0 0 0 0	0 0 0 1
F1	0 0 0 0	1 1 1 1
F2	0 0 0 0	1 1 1 1

(c)

W	0 0 1 1	0 0 1 1
X	0 1 0 1	0 1 0 1
A	0 1 1 d	1 d d d
B	1 0 d 1	d 1 d d
C	1 d 0 1	d d 1 d
D	d 1 1 0	d d d 1
F1	0 0 0 0	1 1 1 1
F2	0 0 0 0	1 1 1 1

can be illustrated with the implementation of the Fig. 1 multiplexer shown in Fig. 2. The 20-vector set of design verification vectors, Table 5(a), detects all single-stuck faults, but is a rather uneconomical test since it is possible to detect all single stuck-at faults with only the 8 vectors of Table 5(c).

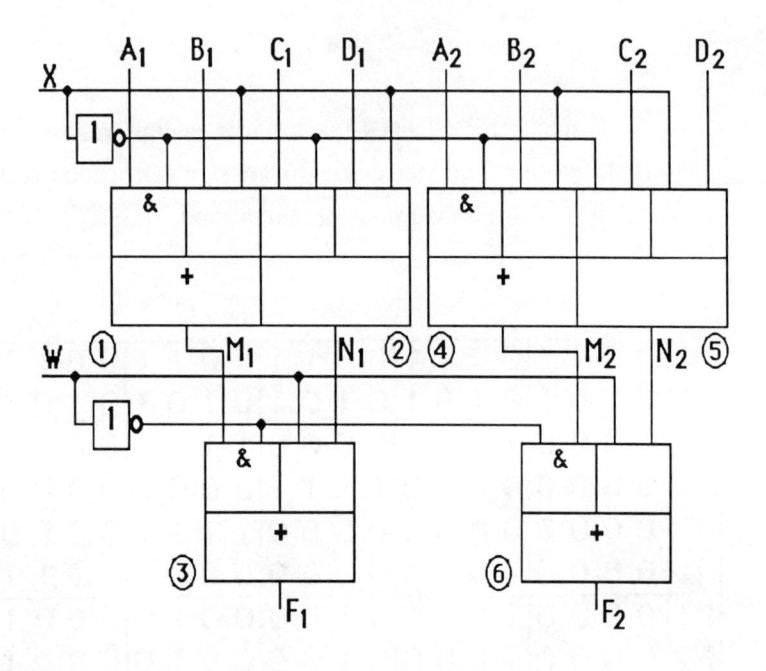

Fig. 2. One implementation of the multiplexer in Fig. 1.

The set of 8 design verification vectors, Table 5(b), detects only 68% of the single stuck node faults. Whether or not a specific fault coverage is acceptable can be approximated in terms of the process yield and the desired maximum defect level of the parts after testing. This determination was discussed in section III of this chapter. For a 68% fault coverage and a process yield of 30%, the defect level is predicted to be 32%. A lower defect level can be obtained by increasing the fault coverage. Using fault simulation, the design verification test set may be augmented to obtain an acceptable fault coverage. With 4 additional vectors it is possible to get 100% single-stuck fault coverage. Thus the augmented verification vector set has 12 vectors instead of the 8 required for design verification.

B. Toggle Test Vectors

The fault coverage (68%) of the 8 design verification vectors was determined by a complete fault simulation. Since fault simulation can be very expensive (in computer time) for large circuits, simpler methods are often used to estimate the quality of a test set. One popular technique checks that each node of the circuit is "toggled", that is, it receives both a 0-signal and a 1-signal. The advantage of this approach is that only a fault-free rather than a fault simulation is necessary. Such *Toggle Test Sets* are discussed in Hughes (19). Here, a toggle test set for the 74LS181 ALU was generated and simulated for all single-stuck faults. The fault coverage was found to be 70.5%.

Even simpler are the *Pin Test Sets* that check only that the primary inputs and primary outputs are toggled. All minimal test sets that give full coverage for single stuck faults on I/O lines of all 3- input functions were developed (42). The fault coverage for these test sets for all nodes (I/O lines and internal nodes) turns out to be a rapidly decreasing function of the number of nodes. Also, for implementations of these functions which differ only by inverters at the inputs and/or outputs, the fault coverages differ. This is because adding the inverters on the I/O lines increases the number of faults detected by the test set. Pin test sets may lead to bad judgement. For example, 2- and 3-input XOR circuits require 4 patterns to be tested exhaustively. However, the lengths of the pin test sets are respectively 3 and 2 This means that the 3-input circuit requires fewer patterns.

The 8 design verification vectors of Table 5(b) satisfy the requirements of both the Toggle Test Set and the Pin Test Set. For the multiplexer example considered here this shows that use of design verification vectors and either a Toggle Test Set or Pin Test Set validation could lead to the use of a test set with only 68% fault coverage. Toggle and pin test sets may be augmented to obtain an acceptable fault coverage as was explained for design verification test sets at the end of the previous section.

VI. STRUCTURAL TESTING

This discussion of using design verification vectors for Boolean testing demonstrates some of the problems that can arise when test sets are generated without attention to the structure of the circuit being tested. In fact it was the lack of success with the generation of tests based on the functional operation of circuits that led to the development of the stuck-at fault model and the use of structural information to generate test sets. The switch from "functional" tests to "structural" tests was signalled in a paper presented by R. Eldred (10) at the August 1958 meeting of the ACM. The opening sentence of this paper is:

In order for the successful operation of a test routine to guarantee that a computing system has no faulty components, the test conditions imposed by the routine should be devised at the level of the components themselves, rather than at the level of programmed orders.

In a study on test generation for circuits for which no information is provided (42), it has been shown that functional testing is ineffective. For example, 5 patterns are sufficient for 100% fault detection of the OR/AND realization of the function $f(x,y,z) = (x'+y)(y'+z)(z'+x)$. The same patterns yield only 83.3% coverage for the dual AND/OR implementation, $f(x,y,z) = xyz + x'y'z'$, of the same function. It is very likely that dependence on the implementation would cause greater discrepencies for larger circuits.

Because of the complexity and density of modern digital circuits Boolean testing is now recognized as an important, perhaps limiting, factor in the continued progress of integrated circuit manufacture. Two aspects of Boolean testing are currently being reexamined: (1) The adequacy of the single stuck-at fault model for VSLI and ULSI is being studied because other types of faults may become more important with the new circuits, and

(2) techniques to generate tests without using the detailed structure of the circuit are being sought either because detailed test pattern generation is becoming too expensive or because detailed structural information is not available for many catalog parts.

The following section discusses the necessity of a fault model for test pattern generation. Choices among various fault models are considered. A very restrictive fault model like single stuck-at faults permits small test sets but can fail to detect many actual failures. Weaker fault models that represent more realistic failure modes are also considered.

A. Fault Models

It is, in general, impossible to determine the quality of a test set for a circuit without making some assumption about the faults that are to be detected by the test. As an example to illustrate this statement consider the 2-line to 1-line multiplexer of Fig. 3(a). Since this is a 3-input combinational circuit, the exhaustive test set of Table 5(c) that contains all 8 valid input vectors should test the circuit completely. However if the test set is applied in the order given in the table, it fails to detect the bridging fault shown in Fig. 4(b). The bridging fault causes the topmost input to the multiplexer to be equal to Ay, the AND function of the A- input and the y-output. (Such a fault could not occur for all circuit realizations, but there are circuit configurations for which it is an accurate representation.)

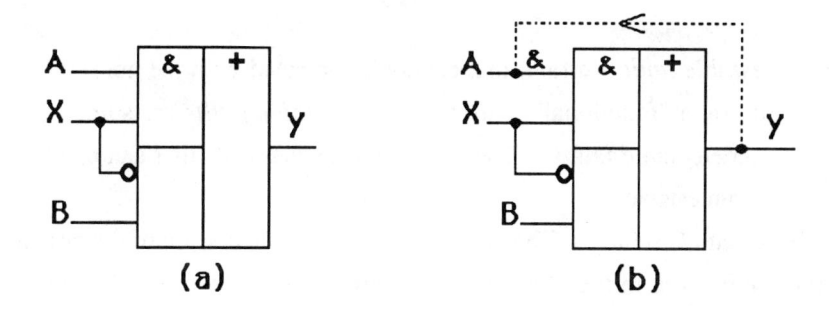

Fig. 3. Two-line to one-line multiplexer.
(a) Fault free, (b) With bridging fault.

The effectiveness of test sets developed for the single stuck-at fault model in detecting failures that are represented by other fault models will be discussed in section VII of this chapter. The necessity of a fault model is becoming more evident with increased use of new circuit technologies. In an atttempt to better understand the failure modes and to represent them at the logical level, modeling on the device level has been pursued recently by several researchers among them Bryant (7) and Maly (17). Also, new models have been developed by Hayes (17).

The largest fault class for which it is practical to generate a test set is the class of *combinational faults*. they are faults that change the circuit function but do not increase the number of states in the circuit. If no limit is imposed on the number of extra states in the faulty circuit, then it is impossible to construct a complete test set, a test set that detects all faults that change the circuit function (33). Test set generation for "bounded sequential faults," faults that can increase the number of circuit states by no more than a fixed number of additional states, is possible in principle but not in practice. The qualifier -- "that changes the circuit function" -- is used above in order to allow for the presence of untestable faults in the circuit under test.

C. Untestable Faults

An *untestable fault* is a fault that cannot be detected with the normal test procedure on a "functional" chip tester. *Untestable faults* are sometimes called undetectable faults . There are four mechanisms that can cause a fault to be untestable:
1) Redundant Circuitry -- The fault can occur in some portion of the circuit that is *redundant* in the sense that it has no effect on the circuit function. The usual reason for the presence of redundancy is a design mistake. This situation is demonstrated in Fig. 4(a). Stuck- at-1 faults on leads 1,2,3, or 4 are untestable as are stuck-at-0 faults on leads 5 or 6. Since either M or N

is always 0, the output of the bottom OR gate (lead 2) is always 1. Neither the bottom OR gate nor the output AND gate are required and, in fact, the output function could be taken from lead G rather than F.

2) Internal Signal Dependencies -- There may be one or more internal gates at whose input leads some pattern or patterns cannot occur. For example, in Fig. 4(b) it is not possible to have a 1-signal on both inputs to the OR gate. In such a situation, any fault that changes the gate function for only the input combinations that never occur will, of course, have no effect on the entire circuit operation. In Fig. 4(b) an example of such a fault is one which changes the OR gate to an XOR gate. The presence of such faults cannot be detected by a Boolean test.

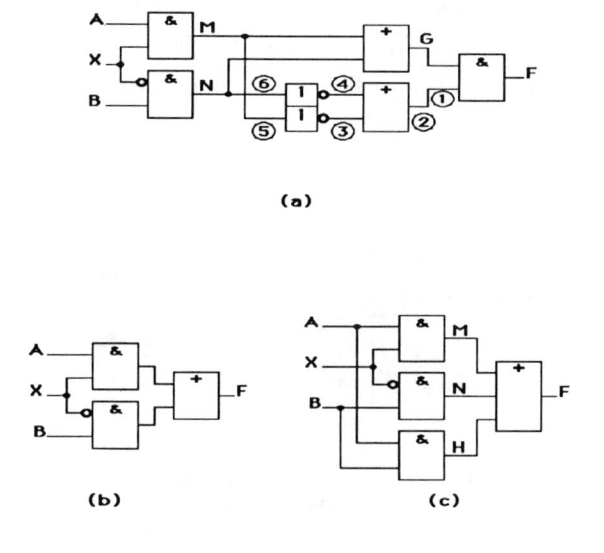

Fig. 4. Three implementations of a 2-line to 1-line multiplexer illustrating untestable faults.(a) Redundant circuitry, (b) Signal dependency, (c) Hazard control circuitry.

3) Hazard control circuitry -- If part of the circuit is present to eliminate a hazard, faults that disable the hazard control will not be detectable since they have no effect on the static circuit operation. In Fig. 4(c) the output H of the bottom AND gate is only 1 when both A and B are 1, in which case one of the other AND gate outputs must be 1 since X or X' must be 1. The bottom AND gate is present only to prevent a static logic hazard and cannot be detected by a Boolean test. For a discussion of hazards, see Chapter 3 in McCluskey (31).

4) Error Control Circuitry -- Circuitry is sometimes added to a design in order to detect failures while the system is in use. The circuit in Fig. 5 is a 2-line to 4-line decoder with added error detection circuitry. Since exactly one of the decoder outputs is equal to 1, one of the OR gate outputs is always 1 and the XOR gate output (E) is always 1. Any decoder failure that causes all outputs to be 0 will change E to 0 since both inputs to the XOR gate will be 0. Similarly any failure that makes both inputs to the XOR gate equal to 1 will change E to 0. When the decoder circuitry is fault-free, the E output is always 1 so that the fault E stuck-at-1 is undetectable.

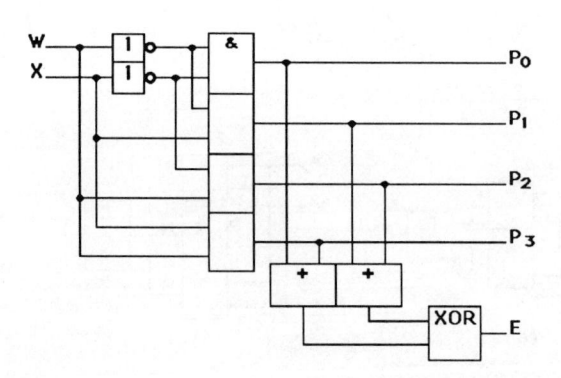

Fig. 5. Two line to four line decoder with error detection circuits.

VII. TYPES OF TEST SETS

It was shown above how "functional" and design verification testing may not guarantee the detection necessary to obtain an acceptable defect level. Also, the necessity of a fault model for proper testing of a circuit was demonstrated. Even with a fault model available, the method of test generation can affect the fault coverage. In this section, different test generation techniques will be presented and their limitations in providing 100% fault coverage will be discussed.

A. Exhaustive Test Sets

In order to test a circuit for the presence of combinational faults an exhaustive test set is required.

Definition. An Exhaustive Test Set for an n-input combinational circuit is the set of all 2^n possible input vectors. For a sequential circuit, an exhaustive test set is a checking experiment (13).

Only combinational circuits are considered here. Sequential circuit testing is important, but it appears uneconomic to test large sequential circuits without using a scan path technique to permit the use of combinational circuit tests (31).

Theorem. The smallest test set that detects any combinational fault in an n-input combinational circuit is the 2^n pattern exhaustive test set.

Proof. Suppose a shorter test set is used. Then at least one input pattern is not applied. The combinational fault that changes the circuit function for only this input pattern will not be detected. Suppose all input patterns are applied, then any combinational fault must change the circuit output for at least one pattern causing the fault to be detected.

There are two difficulties with exhaustive test sets:
1) Some physical failures can cause non-combinational faults which are not detectable by an exhaustive test. There are two classes of these failures,

bridging faults and stuck-open faults. Failures in which two points in a circuit are incorrectly connected together are not uncommon. Such failures can be modelled as bridging faults (32). Some bridging faults can introduce additional states into a circuit. The bridging fault of Fig. 3(b) is an example of one such fault class.

The other important non-combinational faults, *stuck-open faults*, occur in MOS circuits. They are caused by charge storage on nodes that have some of their connections removed due to open-circuit failures, Wadsack (78).

A great deal of attention is now directed towards detecting MOS stuck-open faults. One approach derives tests for specific MOS stuck-open faults and adds these to a stuck-at fault test set (21), (40). This process is expensive in the amount of required computation. Another method adds circuitry to the MOS circuit to facilitate testing for MOS stuck-open faults (25). The cost penalty for this technique is performance and the area overhead of the additional circuits. The most promising method when the layout can be controlled is to design the gate geometry so as to reduce the possibility of a MOS stuck-open fault to an acceptably low value (53). A recent study to determine whether stuck-open faults occurred in significant numbers in production CMOS chips did not detect any stuck-open faults (47). This suggests that stuck-open faults may not be a concern for well-designed CMOS chips.

Bridging faults have received much less attention than MOS stuck-open faults, but there seems to be no justification for this. The bridging faults appear to be, if anything, more common than the MOS stuck-open faults. Most efforts concentrated on very simple circuits and special cases (4), (14). Also, only bipolar technology was used. Bridging faults have been represented as wired AND and wired OR. For CMOS technology, the preliminary results (12) have shown that bridging faults are mostly wired AND. It is possible to generate tests for specific bridging faults and add these to a stuck-at fault test set (1). Another approach would be to control the layout to reduce the probability of non-combinational bridging faults.

2) The other difficulty with exhaustive test sets is that they can be very long. An exhaustive test for an n-input circuit requires 2^n input patterns. For circuits with a large number of primary inputs (more than 20), the length of an exhaustive test set is prohibitively long. In such cases, it is necessary to use a more restrictive fault model or, if possible, to resort to pseudo-exhaustive testing techniques which are discussed next.

B. Pseudo-exhaustive Test Sets

For circuits in which each circuit output depends on a subset of all the inputs, it may be feasible to test each output exhaustively even though the entire circuit is not tested exhaustively. This technique is based on a fault model, the *constrained dependence fault model*, which assumes that the set of inputs on which a circuit output depends is not increased by the presence of the fault. The corresponding test set that applies all input combinations to each set of inputs on which a circuit output is dependent is called a pseudo-exhaustive output function verification test set (30). Since each of the outputs in the Fig. 1 multiplexer depends on 6 inputs, each output can be tested exhaustively with 64 patterns. It is possible to test both outputs at the same time using only 64 patterns by using the pseudo-exhaustive test method.

For some circuits this output function verification test may be too lengthy. A further restriction on the fault class is then necessary. The circuit must be then segmented into subcircuits, each of which is tested exhaustively. This process is called *pseudo-exhaustive segment function verification* and is described in McCluskey (29). For the Fig. 1 multiplexer circuit this type of test can be done with 16 (rather than 64) patterns. Each of the six AND-OR circuits labelled in Fig. 2 is tested exhaustively.

The advantage of pseudo-exhaustive testing is the possibility of obtaining 100% single stuck fault coverage with a much reduced length test set. Also, fault simulation need not be performed. The optimal circuit segmentation for pseudo-exhaustive testing has been shown to be NP-complete (3). Techniques to partition a circuit were developed (3), (38), (41), (48). Preliminary results on the use of simulated annealing for circuit segmentation are promising (44). There is still much to be done in the area of pseudoexhaustive testing.

C. Stuck-at Fault Test Sets

There are a number of techniques for generating test sets that are shorter than exhaustive test sets. The most popular assume that only one stuck-at fault can occur and require the detailed logic diagram of the circuit to be tested. Path tracing methods such as the D algorithm can generate minimum-length single stuck-at fault tests, but are very computationally expensive. These methods are not very useful for Built-in Self-Test since they require that the test vectors be stored rather than computed when needed.

Pseudo-random patterns are used for test vectors in the other approach based on single stuck-at faults and a detailed logic diagram. The advantage of this approach is that the test patterns can be easily generated on-line by a Linear Feedback Shift Register. The disadvantage is that, at present, a full fault simulation is required to determine the number of patterns that must be used to achieve a specified fault coverage. This is particularly troublesome since the pseudo-random test sets tend to be much longer than a minimum-length single-stuck fault test. Current research is aimed at discovering some way to accurately estimate the fault coverage without a detailed fault simulation (9), (26), (50).

The expected fault coverage for a test set of length L is given in Wagner (50) as:

$$E(C) = 1 - \sum_{k=1}^{N-L} (1 - L/N)^k \, h_k/M \qquad \text{for} \quad (N - L) \ll k \qquad (5)$$

where

h_k is the number of faults that are detected by exactly k patterns,

M is the total number of faults ($M = \sum_{k=1}^{N} h_k$) in the circuit, and

N is the length of the exhaustive test set for the circuit, $N=2^n$.

Applying Eq. (5) to the 2-bit multiplexer of Fig. 4 yields a value of 77.14% for a pseudo-random test of length 8. The curve in Fig. 6 shows the expected fault coverage as function of the test length for the multiplexer of Fig. 2.

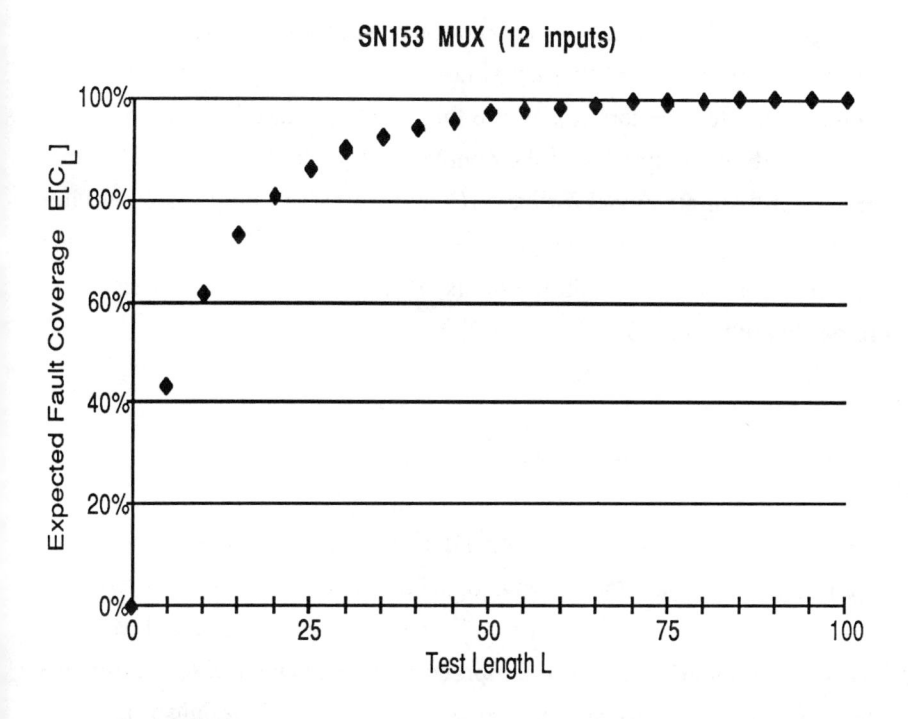

Fig. 6. Expected Fault Coverage for the Multiplexer of Fig. 2.

For the same circuit, the loss in expected fault coverage due to omitting the detection of a fault with detectability k is given approximately by

$$\Delta f_{Lk} = 1/M \cdot (1-L/N)^k \qquad \text{for} \quad (N-L) \gg k \qquad (6)$$

From this expresssion, faults causing the largest loss of coverage are those with the smallest detectabilities. They are the hardest to detect. However, the fault coverage need not be estimated on the basis of these faults solely but the whole fault set should be considered.

D. Effectiveness of Single Stuck-at Test Sets

All methods based on the single stuck-at fault assumption can fail to produce a satisfactory test when the actual failure modes are not detected by a single-stuck fault test. As an example of such a situation, consider the Parity tree of Fig. 7. Such a network can be tested for all single stuck-at faults with 4 test vectors (6). While this is a 100% test for single stuck-at faults, it detects only 89% of the double-stuck faults (35). Single failures are not an accurate model for most IC processes. The single stuck-at fault model is useful only because test sets based on this model usually also detect almost all of the multiple faults (18), as well as other faults such as bridging faults (1), (32).

E. Functional Fault Test Sets

There are many situations in which a detailed logic diagram is not available for test pattern generation. Often a manufacturer will be unwilling to release the logic diagram to a customer for use in incoming inspection or board test. This is particularly true for microprocessors or memory chips. Attempts have been made to do test pattern generation for such designs using only the functional specification and making some broad assumptions about

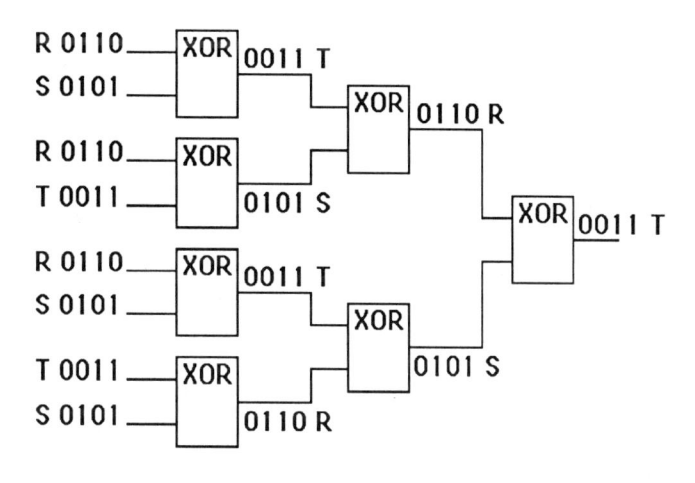

```
R 0110 ──[XOR] 0011 T
S 0101 ──┘         └──[XOR] 0110 R
R 0110 ──[XOR]         │
T 0011 ──┘  0101 S     │
                       ├──[XOR] 0011 T
R 0110 ──[XOR] 0011 T  │
S 0101 ──┘         └──[XOR]
                        0101 S
T 0011 ──[XOR]
S 0101 ──┘  0110 R
```

Fig. 7. Eight input parity tree, with single- stuck fault test pattern.

implementation. For example, in Knaizuk (23) a general Stuck-at Fault RAM test is derived by assuming that the Address Decoder is implemented by what they call a "noncreative decoder," a decoder circuit containing only AND gates and Inverters with no reconverging fanout paths. A similar approach is taken in Thatte (46) in which faulty decoder functions are characterized based on the assumption that the decoder is realized without any reconvergent fanout. The author uses the term functional fault model to specify this approach. This is a useful technique for those situations where it is applicable. Of course, it doesn't work when the assumptions are not valid as would be the case for the decoder example if reconvergent fanout was present. For examples of decoders with reconvergent fanout see pp. 57-59 of Maley (27) or Secs. 2-1.1 and 2- 1.2 of Lenk (24). Again, single-stuck faults are assumed in these "functional fault" models with all the drawbacks previously discussed for this model.

VIII. RESPONSE COMPACTION

In actually testing a physical chip it is necessary to analyze the output response to the test inputs in order to determine whether a fault has been detected. By carrying out a fault-free simulation the correct outputs for the test inputs can be determined. The actual test output can then be compared with a stored record of the correct output. This technique is sometimes called a *fault dictionary* method. The difficulty with use of a fault dictionary is the large volume of memory required. The required storage can be reduced by not recording the entire test response but instead some characteristic of the output response. This technique is called *response compaction* (37), or sometimes response compression. Compaction is preferred to emphasize the fact that information is lost in this process. Response compaction is particularly important for built-in self test applications. With the fault dictionary method all failures detected by the test set will be discovered by the test procedure. This is not guaranteed with response compaction: it is possible for a faulty circuit response to produce the same compacted response as a fault-free circuit. Thus the *effective fault coverage* can be less than the fault coverage of the test set.

The only compaction technique used in current designs is signature analysis (15). In order to study the impact of using this compaction method, the effective fault coverage of a number of complete single stuck-at fault sets for the 74LS181 was determined by simulation. The results are taken from a forthcoming paper (36) and are shown in Table 6. The test sets are those discussed in Hughes (18) where the results of simulating double stuck-at faults was reported. These results are included in Table 6 to provide a perspective on the compaction results. In all cases there are more double stuck faults missed than single stuck faults after compaction.

Table 6.

Results of Simulating Test Sets for 74LS181 ALU.

Test Set	Length	P_1	P_2	2-stuck	1-stuck
Krish.	12	0	0	9	0
Bry1	14	1	1	4	0
Bry2	14	2	0	14	0
Bry3	14	0	6	11	0
Mic1	17	0	0	3	0
Mic2	17	0	0	30	0
Goel	35	0	0	0	0
Pran	135	7	0	0	0
Pexh1	125	2	0	0	0
Pexh2	325	0	0	0	0
Toggle	7	-	-	6969	119

$P_1 = x^8 + x^4 + x^3 + x^2 + 1$
$P_2 = x^8 + x^6 + x^5 + x^2 + 1$

Examination of Table 6 shows that compaction is not a serious problem for the 181 ALU. Since it is necessary to simulate all single stuck faults for fault grading, the effect of the signature analyzer can be included in the simulation. When less than 100% fault coverage is obtained a different polynomial can be used to eliminate the aliasing effect of the signature analyzer. Multiple stuck-at faults are a more serious problem because it is not generally economically feasible to do a multiple stuck-at fault simulation due to the very large number of such faults.

X. CONCLUSIONS

Faulty chips can fail to be detected by a test for a variety of reasons. The most important factor causing the acceptance of faulty chips in practice is probably the use of test sets with less than 100% single stuck-at fault coverage. The issue involved in deciding how close to 100% fault coverage is required was discussed. Fault coverage is determined by fault grading which is almost always carried out by simulation. Problems preventing simulators to meet the challenge of the increase in circuit size and complexity were presented. The importance of simulation to design and testing of digital circuits has resulted in improvements of simulators and attempts to develop alternative techniques.

The necessity of a fault model assumption for assuring the quality of a test set has been demonstrated. A number of fault models and test pattern generation techniques have been presented and evaluated. The most general fault model is the combinational fault that does not increase the number of internal states. Such a fault model requires an exhaustive test set. However, this can be impractical for large number of primary inputs. Pseudoexhaustive testing is more realistic but this requires partitioning of the circuit. The single stuck-at fault model leads to the shortest test sets, but can miss failures that cannot be detected by single-stuck fault test sets. Pseudorandom test sets are used as alternative to the single stuck-at test sets. They have a low generation cost but still require simulation for fault grading.

Even when single stuck-at fault test sets guarantee 100% fault coverage, they are not necessary effective in detecting other types of fault models -- multiple stuck-at and bridging faults. Also, if compaction techniques are used for response verification, the fault coverage can drop lower than 100%. The effect of signature analysis on effective fault coverage was examined by simulating the 74LS181 ALU for a number of single stuck-at fault test sets. The loss of coverage due to signature analysis is shown to be minimal particularly when compared to double stuck-at faults.

In the future it will be necessary to direct attention to other factors causing the acceptance of faulty chips. It seems clear that the most important factors are the occurrence of non single stuck-at faults and the need for developing appropriate models for these faults.

X. ACKNOWLEDGEMENTS

This work was supported in part by the Innovative Science and Technology Office of the Strategic Defense Initiative Organization and administered through the Office of Naval Research under contract no. N00014-85-K-0600. The authors wish to thank David McCluskey his help in preparing this paper.

XI. REFERENCES

1. Abramovici, M and Menon, P. (1983). *Proc., Int'l Test Conference 138-144.*

2. Agrawal, V.D. and Seth S. (1982). *J. of Solid-State. Circuits,* 17, 57-61.

3. Archambeau, E.C. (1985). *Center for Reliable Computing, TR 85-10,* Stanford University, Stanford, CA..

4. Bhattacharya, B.B., Gupta B., Sarkar, S., Choudhury, A.K. (1985). *IEE PROCEEDINGS,* 132,155-161.

5. Blank, T. (1984). *Design and Test Computers.* 3, 21-39.

6. Bossen, D.C., D.L. Ostapko, and A.M. Patel. (1970). *Proc. FJCC* , 37, 63-68.

7. Bryant, R.E. (1984). *IEEE Transactions on Computers,* 33,160-177.

8. Catlin, G. and Paseman, B. (1985). *IEEE Int'l Conf. on Computer-Aided Design* . ICCAD-85, 131-132.

9. Chin, C.K., and McCluskey, E.J.(1985). *Proc., Int'l Test Conf.* 94-99.

10. Eldred, R.D. (1959). *J. ACM,* 6, 33-36.

11. Franz, M. (1985). *EDN* . June 13, 153-158.

12. Freeman, G., and McCluskey, E. J. (1986). *Center for Reliable Computing, Technical Report,* 86-7, Stanford University, CA.

13. Friedman, A.D. and Menon, P.R. (1971). "Fault Detection in Digital Circuits ", Prentice-Hall, Engelwood-Cliffs, N.J.

14. Friedman, A.D. (1974). *IEEE Trans. on Computers.* C-23, 746-752.

15. Frohwerk, R.A. (1977). *Hewlett-Packard J.* 2-8.

16. Hayes, J.P. (1986). *Proc. IEEE ,* 70, 1140-1151.

17. Hayes, J.P. (1986). *IEEE Transactions on Computers,* 35, 602-612.

18. Hughes, J.L.A. and McCluskey, E.J. (1984). *Proc. Int'l Test Conf.* 52-58.

19. Hughes, J.L.A., Mourad, S. and McCluskey, E.J.(1985). *Proc., COMPCON Spring 85,* 384-387.

20. Jain, S.K., and Agrawal, V.D. (1984). *Proc., 21st Design Automation Conf.,* 21, 18-23.

21. Jha, N.K., and Abraham, J.A. (1985). *IEEE Trans. Computer-Aided Design,* CAD-4, 264-269.

22. Kampel, I. (1985). "A Practical Introduction to the New Logic Symbols", Butterworth, Stonam, MA.

23. Knaizuk, J., Jr., and Hartman, C.R.P. (1977). *IEEE Trans. Computers.,* C-26, 1141-1144.

24. Lenk, J.D. (1972)."Handbook of Logic Circuits," Reston Publishing Co., Reston, VA.

25. Liu, D., and E.J. McCluskey (1986). *Custom Integrated Circuits Conf.*

26. Malaiya, Y.K. (1984). *Proc. IEEE Int'l Test Conf.*, 237-245.

27. Maley, G.A.(1970). "Manual of Logic Circuits," Prentice-Hall, Englewood Cliffs, NJ.

28. Maly, W., Fergsun, F.G. and Shen, J.P. (1985). *Proc. IEEE Int'l Test Conf.*, 384-389.

29. McCluskey, E.J. and Bozorgui-Nesbat, S. (1981). *IEEE Trans. Comput.*, 866-875.

30. McCluskey, E.J. (1984). *IEEE Trans. Comput..*, C-33, 541-546.

31. McCluskey, E.J., "Logic Design Principles: With Emphasis on Testable Semicustom Circuits ", Prentice-Hall, Inc., Englewood Cliffs, NJ.

32. Mei, K.C.Y.(1974). *IEEE Trans. Comput.*, C-23, 720-727.

33. Moore, E.F. (1955). In "Automata Studies" (Eds. C.E.Shannon and E. J. McCarthy). 129-153.

34. Mourad, S., and D. Andrews, D. (1985). *Proc., Fifteenth Int'l Symp. on Fault-Tolerant Computing.* 15, 93-98.

35. Mourad, S., Hughes, J.A.L.and McCluskey, E.J. (1986). *Proc. of COMPCON Spring 86* 441-444.

36. Mourad, S. and McCluskey, E.J. (1987). In preparation.

37. Parker, K.P. (1976). *Dig., Sixth Ann. Symp. on Fault-Tolerant Computing*, 6, 93-98.

38. Patshnik, O. (1983). *Center for Reliable Computing, Technical Report,* 83-14, Stanford University, Stanford, CA.

39. Pfister, G.F. (1982). *Proc. 19th Design Automation Conf.,* 51-54.

40. Reddy, S.M.,. Reddy, M.K.and Kuhl, J.G. (1983). *Proc. nt'l Test Conf.* 435-445.

41. Roberts, M.W. and Lala, P.K. (1984). *IEE Proc.* 133, 1231-1240.

42. Sakov, J. and McCluskey, E.J. (1986). *Center for Reliable Computing, Technical Report,* in publication, Stanford University, Stanford, CA.

43. Seth, S.C. and Agrawal V.D. (1984). *IEEE Transactions on Computer Aided Design*, 3, 123-126.

44. Sperling, I. and McCluskey, E.J. (1986). *Center for Reliable Computing, Technical Report,* in publication, Stanford University, Stanford, CA.

45. Stover, A.C. (1984). "ATE: Automatic Test Equipment," Mc Graw-Hill Book Co, New York, NY.

46. Thatte, S.M. (1979). Report R-842, Coordinated Science Lab., Univ. of Illinois, Urbana, IL.

47. Turner, M.E., Leet, D.G., Prilik, R. J. and McLean, D.J. (1985). *Proc., Int'l Test Conf.* 322-328.

48. Udell, J. and McCluskey, E.J. (1986). *Proc. of the Int'l Conference on Computer Aided Design.* ICCAD 86.

49. Van Burnt, D. (1983). *Proc. of Int'l Conf. on Computer Design.* ICCD-83, 232-233.

50. Wagner, K.D., Chin, C. and McCluskey, E. J. (1986). *Center for Reliable Computing, Technical Report,* 86-7, Stanford University, Stanford, CA.

51. Williams, T.W. and Brown, N.C. (1981). *IEEE Trans. on Computers.* C-30, 987-988.

52. Williams, T.W. (1985). *IEEE Design and Test of Computers,* 2, 59-63.

53. Zasio, J.J. (1985). *Proc. COMPCON Spring 85,* 26-28, 388-391.

MATRIX METHODS IN THE DETECTION AND ELIMINATION OF REDUNDANCY IN COMBINATIONAL CIRCUITS

A.R. Fleming and G.E. Taylor

Department of Electronic Engineering
University of Hull
Hull, U.K.

A.C. Pugh

Department of Mathematics
Loughborough University of Technology
Loughborough, U.K.

I. INTRODUCTION

The increasing complexity of digital devices has brought about a corresponding concern over component reliability. Testing digital devices involves applying a sequence of input patterns, called a 'test set' and observing the corresponding sequence of output patterns.

Because of their logical/topological structure, some circuits are inherently easier to test, and it is this fact which has motivated research on the problem of deriving transformations which preserve the behaviour of logic circuits, but improve their 'testability'. There are obvious links here with much of the work in the theory of linear systems, where the use of equivalence transformations has greatly simplified the problems of analysis and design.

However, very few digital devices can be regarded as linear, so that existing results are not immediately of use.

II. REDUNDANCY AND TESTABILITY

The standard fault model for digital devices is the permanent stuck-at fault model, and this will be adopted here. This model assumes that logical faults in digital devices manifest themselves as signal lines which are either permanently stuck-at-0 (s-a-0) or stuck-at-1 (s-a-1). Another common assumption made with this model is that only one such fault will occur at any time, and this convention will also be adopted here, except where multiple faults are specifically referred to. Tests then, are derived to check whether a circuit has a s-a-0, or a s-a-1 fault on a particular signal line. A test set will contain tests for all faults on all the signal lines in the circuit (or on an 'acceptable proportion' of the signal lines).

Test set generation has its own particular problems, but these will not be discussed here. However, the effectiveness of any test set is reduced by a digital circuit design problem referred to generally as 'redundancy'.

Historically, a digital circuit has always been regarded as exhibiting redundancy if a signal line (or lines) can be deleted without affecting the logic function (or functions) realised by the circuit. (Deletion of a signal line may in turn involve the deletion of a gate (or gates), and other signal lines from the circuit.)   It can be shown that the presence of redundant signal lines in a circuit is equivalent to the presence of undetectable stuck-at faults. This has led to a more concise definition of redundancy in terms of testability (24):-

Definition 1: A signal line S is 'irredundant' if there exist input vectors $x(0)$, $x(1)$ such that $x(0)$ is a test for S s-a-0, and $x(1)$ is a test for S s-a-1.

Definition 2: A signal line S is 'partially redundant' if there exists a test for S s-a-0, but there exists no test for S s-a-1, or vice versa.

Definition 3: A signal line S is 'fully redundant' if there exist tests for neither S s-a-0, nor S s-a-1.

To aid future discussion, the following definitions are also included here:

Definition 4: A partially redundant signal line S for which there is no test for the fault s-a-0 [s-a-1] is 's-a-0 [s-a-1] redundant'.

Definition 5: A 'redundant' signal line S is one that is partially or fully redundant.

In order to extend the definition of redundancy to circuits in an unambiguous manner, the relationship between circuit outputs and signal lines must be clarified. In a multi-output circuit, each output has its own corresponding subcircuit formed from the original circuit by deleting each gate for which there exist no paths from the gate to the output in question. This leads to the following:

Definition 6: For a particular choice of output Y in a multi-output circuit, a signal line S of the circuit is 'trivially redundant with respect to Y' if it does not form part of the subcircuit corresponding to Y.

Definition 7: A signal line S of a multi-output circuit, is (non-trivially) 'redundant with respect to an output Y' if S is a signal line of the subcircuit corresponding to Y, and S is redundant in that subcircuit.

Definition 8: A combinational circuit is 'redundant' if and only if it contains a redundant signal line.

Note that, for a signal line in a multi-output circuit to be redundant, it must be 'consistently' redundant with respect to each of the circuit's outputs (i.e. s-a-0 redundant with respect to each output, or s-a-1 redundant with respect to each output - being trivially redundant with respect to an output includes both of these possibilities).

The equivalence of the two notions of redundancy is due to the fact that if a signal line is s-a-0 redundant [s-a-1 redundant], then setting that signal line to logic 0 [logic 1] permanently will not affect the output functions at all (otherwise the fault would be detectable), but does, in fact, remove the redundancy. It is this equivalence that provides the means of eliminating redundant signal lines from a circuit after they have been located. Hence, in order to correctly delete partial redundancy, it is necessary to determine whether an undetectable fault is a s-a-0 or a s-a-1 fault.

The effect on a circuit of setting various signal lines to permanent logic values depends upon the topology of the circuit and the logic gates concerned, and involves the concept of dominance and non-dominance of logic values with respect to logic gates:

Definition 9: A logic value is 'dominant with respect to a logic gate' if, when applied to an input of that gate, it forces a specific logic value onto the output of the gate, independent of all other input values. Otherwise a logic value is 'non-dominant with respect to a logic gate'; in this case, logic values can be propagated through the gate from the other inputs, as if the input with the non-dominant logic value did not exist.

For example, in fig. 1 a logic 0 on any input of the AND gate forces a logic 0 onto the output, regardless of the other inputs. On the other hand, if input a takes the logic value 1, the output of the gate would be bc (as if input a did not exist), permitting the signals on b and c to be propagated through the gate. Hence for an AND gate, 0 is the dominant value and 1 is the non-dominant value. Similarly, for an OR gate, the dominant and non-dominant values are 1 and 0 respectively.

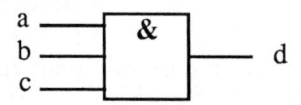

Fig. 1. Dominant (0) and non-dominant (1) values of an AND gate.

Setting an input line of a logic gate permanently to a non-dominant value makes that input line, and only that input line, redundant in the sense that it can be deleted from the circuit. However, setting an input line of a logic gate permanently to a dominant value, makes that input line, the logic gate and its output redundant too, and the redundancy may well be propagated through several other gates, before terminating. Conversely, if a signal line can be deleted from a circuit, the deletion can always be achieved by assuming an appropriate permanent logic value on the signal line concerned.

Superficially, redundancy seems to be a 'harmless' condition, since the presence of an undetectable fault means that the circuit functions correctly in spite of the fault. However, the presence of such redundancy not only makes the job of test generation much more difficult (23), but may invalidate some of the tests (12).

Redundancy then is clearly an undesirable attribute. However, redundancy is often included deliberately, in order to avoid static hazards for example, and attempts have been made to overcome the associated testability problem by including extra inputs and/or outputs or other circuitry for control purposes (15),(22).

Generally however, research in this area of testability has been directed at the problem of detecting redundant signal lines and deleting them. Alternative approaches to this problem have been put forward by Clegg (6), Dandapani and Reddy (7) and Lee and Davidson (20).

In earlier work on transforms for NAND network design, Lee and Davidson (19) recognised that particular logical and topological combinations created redundancy. Apart from this observation and the rather obvious fact that a fan-out free network is irredundant, little work appears to have been done on characterising redundancy in terms of logical and/or topological constraints. (The phrase 'fan-out free' is sometimes used to allow for fan-out on inputs; here it is used to mean completely fan-out free.)

III. CHARACTERISATION OF REDUNDANCY IN COMBINATIONAL CIRCUITS

Clearly then, redundancy only occurs in circuits with reconvergent fan-out. Conversely, there are plenty of irredundant circuits with reconvergent fan-out, as the example in fig. 2 illustrates.

In their paper on NAND network design, Lee and Davidson (19) indicated that redundancy occurs where reconvergent fan-out takes a particular form. Bossen and Hong (3) and To (23) also stressed the importance of fan-out nodes and branches in test-pattern generation,

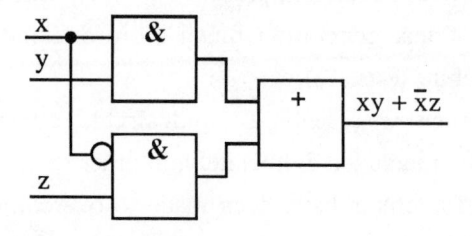

Fig. 2. Each literal is vital to the sum-of-products expression; hence each signal line is necessary and cannot be deleted.

especially To, who explicitly considered redundant circuits. This earlier work has provided the motivation for much of the theory contained in this chapter.

Clearly then, there is something particular about the logical/topological nature of some reconvergent paths, which causes redundancy. In fact, it will be shown that redundancy in combinational circuits is always due to redundant fan-out branches, so that in any search for redundancy, only the fan-out branches need to be considered.

Before qualifying the above statements, there are the following definitions:-

Definition 10: For each pair of paths originating from a fan-out node and reconverging at some logic gate in the circuit, each such path between the fan-out node and the gate of reconvergence will be referred to as a 'reconvergent path'.

Definition 11: A collection of fan-out nodes are '(logically) independent' if each can take the logic values 0 and 1 independently of any other.

Clearly, in a circuit where fan-out is restricted to input lines (as in the tree structures adopted by Dandapani and Reddy (7) and Lee and Davidson (20)) all fan-out nodes are independent. However, in general, a path from a primary input to an output may pass through several fan-out nodes, so that it is quite possible for some of these to be logically dependent upon each other, thereby complicating the problem of redundancy detection. For the moment, it will be assumed that all fan-out nodes are independent. The general case will be considered later.

In spite of the previous comments, the following results apply to any combinational circuit:-

Lemma 1: In a combinational circuit, full redundancy only occurs in the presence of partial redundancy.

Proof: Let signal line S be such that S s-a-0, S s-a-1 are undetectable. Setting S permanently to 0, or permanently to 1 will not affect the outputs of the circuit, since both faults are undetectable. However the two distinct logic assignments will affect the circuit in two distinct ways; the non-dominant assignment merely removes the signal line S (and any preceding circuitry), and the dominant assignment will remove the signal line S, its preceding circuitry and the logic gate of which it is an input, and hence the output of that gate (and possible further circuitry). However, taking the first option removes the redundant signal line S, but leaves a redundant logic gate which must therefore have another input which is also redundant. In considering the redundant output of this gate, if it is fully redundant then by the above argument it forms an input to another redundant gate whose output is also redundant. Hence if there is no partial redundancy in the circuit, then by a repeated use of the above argument, the signal lines corresponding to one or more of the outputs of the circuit are fully redundant, which is impossible, even for the trivial case of constant 0 or 1 outputs. Hence the result.

In the example shown in fig. 3, the input b is fully redundant, since signal line m has a dominant stuck-at fault (i.e. s-a-1) which also makes signal line n s-a-1 redundant.

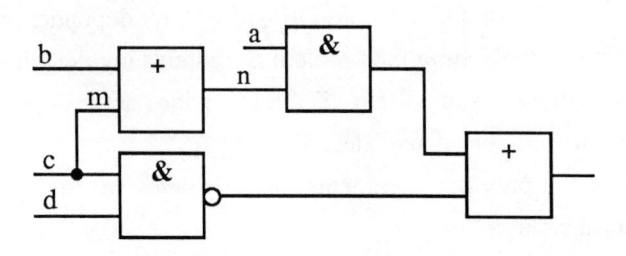

Fig. 3.

Lemma 2: In a combinational circuit, partial redundancy can only occur in a reconvergent path.

Proof: Let signal line S be such that S s-a-0 is undetectable, but S s-a-1 is detectable. Every combination of input variables making S equal to 1 (a necessary part of testing for S s-a-0) effectively makes S unobservable from any output. As S is not permanently unobservable from each output (S s-a-1 is detectable) this situation implies that in every path from S to an output there exists a logic gate, another input of which is logically dependent on one or more of the input variables of S. Such a gate is a gate of reconvergence and S lies on a reconvergent path. Interchanging 0 and 1 in the above does not invalidate the argument; hence the result.

(In the proof of this and the previous lemma, it is assumed that there are no cases of input variables being set permanently to logic 0 or 1 intentionally; each input variable must be able to take logic values independently of any other.)

Theorem 1: A combinational circuit is irredundant if and only if every signal line in every reconvergent path is irredundant.

Proof: A circuit is redundant if it contains a redundant signal line, trivially.

Conversely, if the circuit is redundant, then by Lemmas 1 and 2 there exists a partially redundant signal line in a reconvergent path. Hence the result.

Theorem 2: A combinational circuit is irredundant if and only if each fan-out branch is irredundant.

Proof: A redundant fan-out branch implies that the circuit is redundant, trivially.

Conversely, suppose a signal line S in the circuit is redundant. In view of Lemma 1 assume that S is partially redundant. Assume also that S is not a fan-out branch (otherwise there is nothing to prove). Just as redundancy can be propagated 'forwards' through a circuit, it can also be propagated 'backwards', since a redundant signal line which is not a fan-out branch, must be the output of some logic gate; this gate and its inputs must also be redundant, and each such input which is not a fan-out branch must also be the output of another redundant logic gate, and so on. Propagation of redundancy backwards stops at fan-out branches (unless all the branches from a fan-out node are redundant, and in a consistent way) or primary inputs. By Lemma 2, S must lie in a reconvergent path, so that propagating the redundancy backwards must therefore involve a redundant fan-out branch. Hence the result.

In a single-output circuit every fan-out node is a source of reconvergent fan-out and hence a candidate for redundancy. This is not the case for multi-output circuits which may have two kinds of fan-out nodes; those which are the source of reconvergent fan-out, and those which fan-out to different outputs (this classification does not exclude the possibility of a fan-out node exhibiting both characteristics). However, the above theorem is only valid if no distinction is made between the two types of fan-out nodes. As an illustration, consider example 3 in section VIII. The circuit contains both types of fan-out nodes; x4 is a source of reconvergent fan-out, and x7 is an output fan-out node. Considering subcircuits, the first subcircuit has no redundancy, but in the second, fan-out branch x5 is s-a-1 redundant making x9 s-a-1 redundant too. However, x5 is irredundant with respect to the first output and hence irredundant in the circuit as a whole. In contrast, x9 is (trivially) redundant with respect to the first output, and hence is redundant overall.

However, had the gate preceding x7 been an AND gate, the circuit as a whole would be irredundant even though it contained a redundant subcircuit. This situation will be considered later, in section VIII.

Returning to Theorem 2 and rephrasing it; a combinational circuit is redundant if and only if it contains a redundant fan-out branch. This result forms the basis of the rest of the chapter.

IV. SIMPLE AND COMPLEX REDUNDANCY

For the present, only single-output circuits with independent fan-out nodes will be considered; extensions to the general case will be considered later.

Theorem 3 (Simple redundancy): In a combinational circuit with a single case of reconvergent fan-out, where there is a fan-out of two, and where the gate of reconvergence is an OR gate, then a fan-out branch a is partially redundant if and only if the logic signal at the input to the gate of reconvergence, from the other fan-out branch b, is of the form $b+V$ or $\bar{b}+V$, where V is a logic term independent of b.

Proof: Let x be the signal line fanning out to lines a and b, and let a and b reconverge as signals A and B respectively on the inputs to the OR gate. Let the output of the OR gate be z. (See fig. 4.)

As there is only one case of reconvergent fan-out, and a fan-out restriction of two, it is necessary to consider only that part of the circuit between x and z. Assuming an even number of inversions between a and A, the logic at A

Fig. 4.

can be expressed in the form aS+T where S,T are logic terms independent of x (since x, a and b are logically equivalent). B can be expressed in the form B=b̊U+V where b̊ is the logic variable b or its complement $\bar{b}$, and U,V are logic terms independent of x, and independent of S and T. Then a will exhibit partial redundancy if and only if every input combination that makes S=1 and T=0 (so that a is propagated to the gate of reconvergence) also makes B take the 'dominant' value 1 (so that A is blocked at the OR gate).

Assume B=b+V where V is independent of b (and hence x). (The proof for B=$\bar{b}$+V is exactly analogous, and hence omitted.) When x=1 (so that b=1), the output z must equal 1 regardless of the condition of a, and thus a s-a-0 is undetectable. Also, when x=0, B can assume a non-dominant value so that a s-a-1 is detectable. Hence a is partially redundant.

Conversely, assume that a is partially redundant with a s-a-0 being undetectable. (The proof for a s-a-1 undetectable is exactly analogous, and hence omitted.) When x=1, every combination of the remaining inputs which make S=1 and T=0, must also ensure that B=b̊U+V equals 1. By the independence of S,T from U,V B=b̊U+V must be identically equal to 1, whenever x=1. As a s-a-1 is detectable, V can take arbitrary values, so that b̊U must be 1 whenever x is 1, implying that b̊=b=1 and U=1. Thus if a is s-a-0 redundant, then B=b+V as required.

An analogous argument can be given for A=$\bar{a}$S+T, hence completing the proof.

The restriction on the level of fan-out can be lifted with a little consideration. With three or more fan-out paths, reconvergence may or may not take place at the same gate. However, if there is a logic signal of the form g+T (or $\bar{g}$+T) on the input to any gate of reconvergence, where g is logically equivalent to the fan-out node under consideration, then all other fan-out branches originating from the same fan-out node and reconverging at that gate will be partially redundant, so each such gate of reconvergence must be considered separately.

Theorem 3 is really saying that partial redundancy in one fan-out branch is indicated by the logic of another branch from the same fan-out node. Such redundancy will be referred to as 'simple redundancy'. To be more specific, a signal g+T [$\bar{g}$+T] on the input to a gate of reconvergence, where g is logically equivalent to the corresponding fan-out node, indicates that all other fan-out branches from the same node have the property that a s-a-0 [s-a-1] fault would be undetectable. Hence the number of inversions on the 'indicator' branch (i.e. odd or even) determines the type of partial redundancy on the other branches.

To detect simple redundancy, a method is required that will 'track' logic signals from fan-out nodes to gates of reconvergence, and pick out any of the form g+T (or $\bar{g}$+T). Adopting an augmented-OR gate structure as in section VI reduces the problem to one of tracing reconvergent paths which, after the first gate, are inverter-free.

The requirement that the gate of reconvergence be an OR gate is no restriction, since this can always be achieved by an appropriate application of De Morgan's Laws. An equivalent theorem can alternatively be given to cover the case where the gate of reconvergence is an AND gate.

Allowing two or more cases of reconvergent fan-out creates the possibility of further types of partial redundancy which will be referred to generally as 'complex redundancy'. For the sake of clarity, the situation of two cases of reconvergent fan-out, each with a fan-out of two, will be considered initially.

Theorem 4 (Complex Redundancy): In a combinational circuit with two cases of reconvergent fan-out, each with a fan-out of two, and sharing the same OR gate of reconvergence, then if x fans out to a and b, and y fans out to g and h, and the circuit has no simple redundancy, then a is partially redundant if and only if for one choice of y (logic 0 or 1) all paths from a to the output are 'blocked', and for the other choice of y, the circuit 'appears' to have a simple redundancy at a (i.e. the circuit obtained by permanently setting y to the second choice, has a simple redundancy at a).

Proof: As there are, by assumption, only two fan-out points it is only necessary to consider that part of the circuit shown in fig. 5. Note that De

Morgan's Laws ensure that there is no loss of generality in the assumptions made in the choice of logic gates M and P.

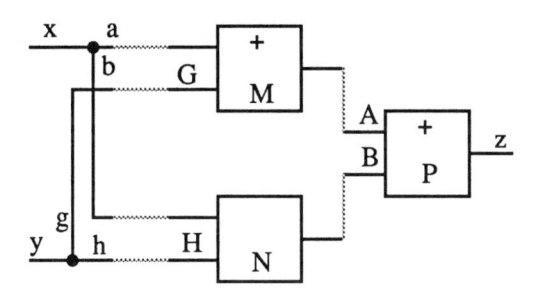

Fig. 5.

Using the notation in the above diagram, assume that when y=0 either G or B equals 1 and that when y=1 a is 'apparently' s-a-0 redundant, so that a will certainly be s-a-0 redundant for the complete circuit. (The proof for a s-a-1 is similar and hence omitted.) Repeating the above arguments with y instead of y completes the first part of the proof.

Conversely, assume that a is s-a-0 redundant, so that when x=1, z is independent of a, implying that either G or B takes the dominant value 1 whenever x=1. (The proof for a s-a-1 is similar and hence omitted.) Ignoring any dependence on x for the moment, for a given value of y, G=1 or assumes a sum of products expression independent of y. If G≠1 for either choice of y then a s-a-0 redundant implies that when x=1, B=1 independently of y, which implies a simple redundancy at a - a contradiction. Clearly G cannot equal 1 for both choices of y, so that a s-a-0 redundant implies that for one value of y, G=1 (blocking a from the output) and for the other value of y, G assumes a sum of products expression independent of y (so the circuit reduces to the single fan-out form as in fig. 4). Note that when G assumes a sum of products expression under the assumption that a is s-a-0 redundant, B cannot be forced into taking the dominant value 1 independently of x, as this would imply the existence of a simple redundancy in fan-out branch g. Hence, under the assumption that a is s-a-0 redundant, when G assumes a sum of products expression, so must H, so that B is still dependent on x. Hence the proof is completed.

If the two cases of reconvergent fan-out had separate gates of reconvergence, only simple redundancy would be possible.

The condition that the circuit has no simple redundancy is only included to show that in such circumstances, there is an alternative form of partial redundancy. Removing this condition from the theorem, merely complicates the proof by allowing other possibilities for signal line a being partially redundant.

Theorem 4 can be generalised by allowing three or more cases of reconvergent fan-out. To show that a particular fan-out branch is partially redundant then involves showing that for all combinations of logical values on the other fan-out nodes, the particular branch is either blocked from all outputs, or 'appears' to be a simple redundancy. The least number of other fan-out nodes which need to be assigned values in order to detect a specific redundancy will be referred to as the order of the complex redundancy. Hence if a circuit has four or more fan-out nodes, and all 2^3 combinations need to be considered on three of them in order to detect a redundant branch from a fourth, this redundancy will be called a third-order complex redundancy. (Hence a simple redundancy is really a zeroth-order complex redundancy.)

Generalising further by allowing three or more fan-out branches per fan-out node, to show that a particular fan-out branch is partially redundant, then involves showing that for all combinations of logical values on the other fan-out nodes, the particular branch is either blocked from all outputs, or 'appears' to be a simple redundancy in a consistent way (only s-a-0 redundant, or only s-a-1 redundant).

To detect such redundancy a method is required that will not only identify simple redundancy, but will also show the effects of setting various fan-out nodes to 0 or 1. The particular realisation adopted in section V and its corresponding matrix representation are particularly well suited to this particular requirement.

Although the possibility of having to consider all combinations of logical values on other fan-out branches sounds a daunting task for a large circuit, this is only necessary where several cases of reconvergent fan-out overlap. And even in such cases it will often only be necessary to consider a few combinations until a partial redundancy is found, and deleted, and the process begins again, often greatly simplified.

V. MATRICES AND AUGMENTED GATES

Matrix representations of digital circuits have been shown to be particularly useful in several areas.

The theory of Boolean matrices (5) has been developed for use in circuit analysis, design and testing (9), (10), (11), (21). One particularly fruitful offshoot of this line of research has been the theory of Rademacher-Walsh transforms and the Rademacher-Walsh spectra of digital circuits (18).

Another matrix representation which can be shown to be closely related to the Boolean matrix of Campeau is the 'Connection Matrix' (2), (17). This theory has had a more specific application in the analysis, design and testing of sequential circuits (2), (13), (17), (21), (25).

Both representations referred to above, are however, realisation independent; they represent the logical behaviour of circuits without any reference to circuit topology. A third contrasting matrix representation attributable to Acken (1), and originally derived as an aid to automatic test generation, has the advantage that it describes a circuit's topological, as well as its logical, behaviour, and it has been adopted here as particularly useful for redundancy detection and deletion.

Taking Hayes' definition of an extended gate (14), and restricting it further yields:-

Definition 11: An 'augmented gate' is a logic AND or OR gate, with inverters incorporated into one or more of its inputs. For examples, see fig. 6.

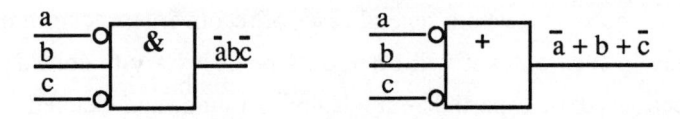

Fig. 6. Augmented logic gates

In fact, in what follows, only augmented OR gates will be used.

Any logic gate realisation can be transformed into an equivalent augmented-OR gate realisation, containing no independent NOT gates, except possibly on an output. By De Morgan's Laws

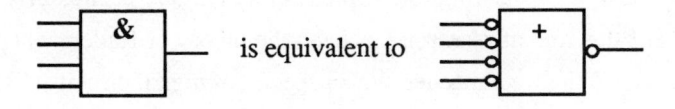

[& gate] is equivalent to [+ gate]

and the output inverter can be 'pushed' along to be incorporated into the input of the next gate, if there is one, or to take up an independent existence on one of the circuit's output lines. It is possible then to create a cascade of inverters, but it is assumed that in any such cascade of inverters, each adjacent pair will combine and cancel, to leave at most just one inverter on a gate input, or an output line.

This particular realisation effectively removes all cascades of inverters (making the circuit b-irredundant by Hayes' classification of circuit redundancy (16)). It also has the particularly useful property that the output of any gate in the circuit can be expressed as a (logical) sum of its inputs, using the four unary functions $(0,1,i,c):\{0,1\} \to \{0,1\}$ as defined by: $0(a)=0$; $1(a)=1$; $i(a)=\bar{a}$; $c(a)=a$ (where a is logic 0 or 1 in each case). An example of this is given in fig. 7.

$$z = a + \bar{b} + \bar{c} = i(a) + c(b) + c(c)$$

Fig. 7.

This feature makes it possible to represent a combinational circuit in the form of a matrix of unary functions, where each line in the matrix represents the output of a logic gate in the circuit in terms of its inputs, as in the following:

Example 1 (Simple Redundancy):

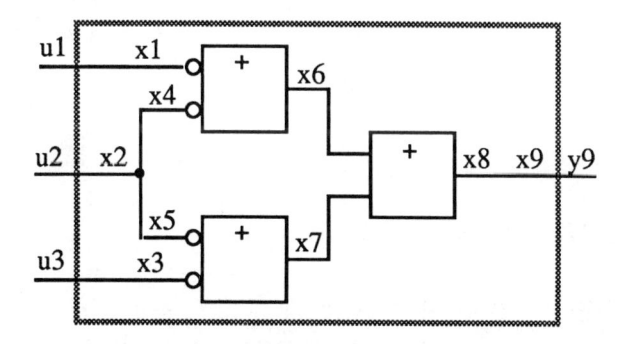

Fig. 8. Example due to Acken (1).

In fig. 9, zero entries have been indicated with periods for reasons of clarity. The first three rows of the matrix represent the relabelling of the inputs, and the last row corresponds to the relabelled output. Rows 4 to 8 are the logic rows, and represent the logic gates of the circuit. This basic matrix will be referred to as the 'Structural Matrix' of the circuit in all the following examples.

$$\begin{bmatrix} x1 \\ x2 \\ x3 \\ x4 \\ x5 \\ x6 \\ x7 \\ x8 \\ y9 \end{bmatrix} = \begin{bmatrix} i & . & . & . & . & . & . & . & . \\ . & i & . & . & . & . & . & . & . \\ . & . & i & . & . & . & . & . & . \\ . & i & . & . & . & . & . & . & . \\ . & i & . & . & . & . & . & . & . \\ c & . & . & c & . & . & . & . & . \\ . & . & c & . & c & . & . & . & . \\ . & . & . & . & . & i & i & . & . \\ . & . & . & . & . & . & . & i & . \end{bmatrix} \begin{bmatrix} u1 \\ u2 \\ u3 \\ x4 \\ x5 \\ x6 \\ x7 \\ x8 \\ x9 \end{bmatrix}$$

Fig. 9.

VI. MATRICES AND REDUNDANCY

In this section it will be shown how the particular matrix representation adopted here can be developed to produce a procedure for detecting and eliminating redundancy.

Changing every non-zero entry in the matrix into a 1, yields the 'Adjacency Matrix' (8), and successive powers of this matrix can be generated to show the number of paths of a particular length between signal lines. Hence an accumulative sum of successive powers can be used to show points of fan-out and reconvergence. An adaptation of this idea can be used to indicate the logic changes that a signal encounters as it proceeds through the circuit, and hence highlight not only the points of fan-out and reconvergence, but also the number of inversions involved and their pattern of occurrence.

The 'Diagnostic Matrix' is obtained from the 'Structural Matrix' by ignoring the input rows, changing all lower case entries to upper case to indicate a change of purpose, and by changing all I entries in fan-out columns to F (fan-out columns are immediately identified, after deleting the input rows, by having two or more I entries). Hence the diagnostic matrix for Example 1 is:-

$$D = \begin{bmatrix} . & . & . & . & . & . & . & . \\ . & . & . & . & . & . & . & . \\ . & . & . & . & . & . & . & . \\ . & F4 & . & . & . & . & . & . \\ . & F5 & . & . & . & . & . & . \\ C & . & . & C & . & . & . & . \\ . & . & C & . & C & . & . & . \\ . & . & . & . & I & I & . & . \\ . & . & . & . & . & . & I & . \end{bmatrix}$$

The fan-out entries are indexed according to the row in which they occur (and hence according to their original signal line label).

Successive powers of the diagnostic matrix can then be generated using functional composition as multiplication, and logical OR as addition, using:-

```
C * F = U   (One inversion)
I * F = Z   (Zero inversions)
C * U = N   (Null entries)
I * U = U   (One inversion)
C * Z = N   (Null entries)
I * Z = Z   (Zero inversions)
F * X = X   (For all entries X)
```

Only a subset of the possible compositions is used, since by section III, only the paths between fan-out nodes and gates of reconvergence are of interest, and of those paths, only those which apart perhaps from an initial inversion undergo no inversions at all will indicate circuit redundancy.

In order to keep track of all lengths of path simultaneously, an accumulative logic sum (denoted by $D^{(n)}$) of successive powers of the diagnostic matrix must be calculated. As the successive powers are only 'partially' obtained using the above subset of compositions, conventional matrix multiplication is rather inefficient in this context, so that an equivalent 'diagonal' method has been devised, which takes advantage of the lower triangular form of the matrix.

Taking each row in turn, the diagonal is used to identify the next dependent row, and the entry in the column corresponding to the original row is then used to 'bring down' (after having undergone functional composition as appropriate) entries from that row to the next dependent one.

(For clarity in the use of the diagonal method, diagonal entries in the matrix will be replaced by asterisks.)

As shown in fig. 10, the position of the new entry completes the rectangle, and its value is found by functional composition as C * F4 = U4.

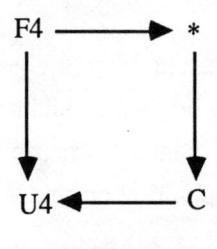

```
      F4 ────────▶ *
      │            │
      │            │
      ▼            ▼
      U4 ◀──────── C
```

Fig. 10.

By Theorem 3 partial redundancy will be indicated by U or Z in a position with multiple entries (multiple entries correspond to reconverged paths).

Hence, for the circuit given in example 1:-

$$D = \begin{bmatrix} * & . & . & . & . & . & . & . & . \\ . & * & . & . & . & . & . & . & . \\ . & . & * & . & . & . & . & . & . \\ . & F4 & . & * & . & . & . & . & . \\ . & F5 & . & . & * & . & . & . & . \\ C & . & . & C & . & * & . & . & . \\ . & . & C & . & C & . & * & . & . \\ . & . & . & . & . & I & I & * & . \\ . & . & . & . & . & . & . & I & * \end{bmatrix}$$

Taking the logic sum of successive powers of D yields:-

$$D = \begin{bmatrix} * & . & . & . & . & . & . & . & . \\ . & * & . & . & . & . & . & . & . \\ . & . & * & . & . & . & . & . & . \\ . & F4 \rightarrow & . & * & . & . & . & . & . \\ . & F5 & . & \downarrow & * & . & . & . & . \\ C & U4 \leftarrow & C & . & * & . & . & . \\ . & U5 & C & . & C & . & * & . & . \\ . & [+] & . & . & . & I & I & * & . \\ . & [+] & . & . & . & . & . & I & * \end{bmatrix}$$

[U4+U5]

The entry U4+U5 shows that fan-out branches x4,x5 reconverge at x8, in the form (x4+S)+(x5+T). So the fault x4 s-a-1 is undetectable because of x5 (or vice versa). Setting x4 equal to 1 will remove this redundancy:-

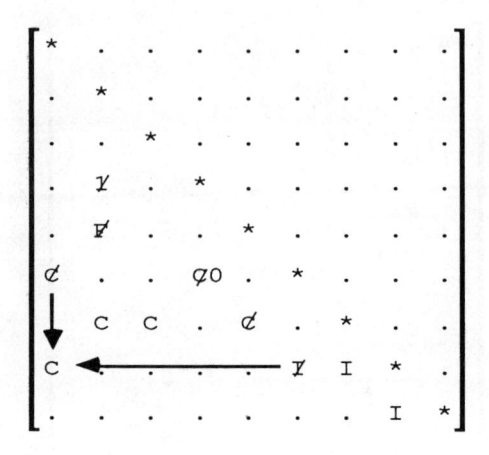

$$\begin{bmatrix} \star & . & . & . & . & . & . & . & . \\ . & \star & . & . & . & . & . & . & . \\ . & . & \star & . & . & . & . & . & . \\ . & \cancel{F} & . & \star & . & . & . & . & . \\ . & \cancel{F} & . & . & \star & . & . & . & . \\ \cancel{C} & . & . & \cancel{F0} & . & \star & . & . & . \\ \downarrow & C & C & . & \cancel{C} & . & \star & . & . \\ C & \leftarrow & . & . & . & \cancel{F} & I & \star & . \\ . & . & . & . & . & . & . & I & \star \end{bmatrix}$$

With F4 equal to 1, F5 is no longer a valid fan-out entry. Row 5 is equivalent to a single entry row with an I entry (rather than its redundant F status) and so must be accommodated on the input to the next gate. This is achieved by the diagonal method, except that in this case the original entries are deleted and only the composed one remains - in this example, the C entry in row 7, column 2. The 1 entry is propagated through the matrix in an obvious manner, all constant entries being deleted on a second pass of the matrix, and any resultant single entry rows are handled as for row 5.

Hence the matrix becomes:-

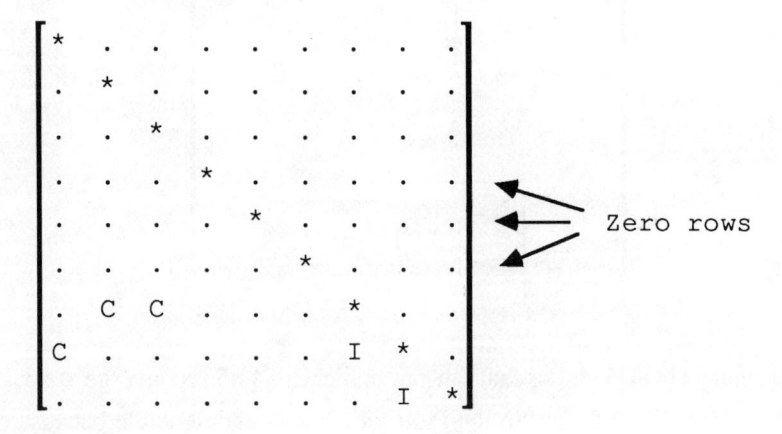

$$\begin{bmatrix} \star & . & . & . & . & . & . & . & . \\ . & \star & . & . & . & . & . & . & . \\ . & . & \star & . & . & . & . & . & . \\ . & . & . & \star & . & . & . & . & . \\ . & . & . & . & \star & . & . & . & . \\ . & . & . & . & . & \star & . & . & . \\ . & C & C & . & . & . & \star & . & . \\ C & . & . & . & . & . & I & \star & . \\ . & . & . & . & . & . & . & I & \star \end{bmatrix} \begin{matrix} \leftarrow \\ \leftarrow \\ \leftarrow \end{matrix} \text{Zero rows}$$

Removing zero rows/columns reduces the matrix to:-

$$\begin{bmatrix} * & . & . & . & . & . \\ . & * & . & . & . & . \\ . & . & * & . & . & . \\ . & c & c & * & . & . \\ c & . & . & I & * & . \\ . & . & . & . & I & * \end{bmatrix}$$

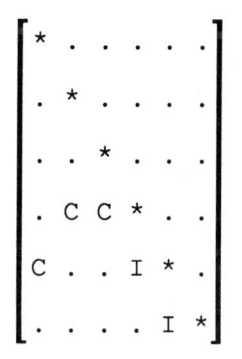

Clearly the circuit has no more redundancy, so the structural matrix and its equivalent circuit can be reformed:-

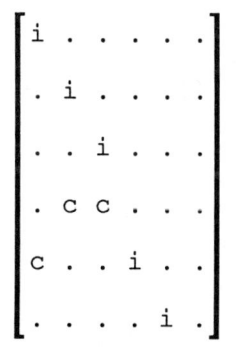

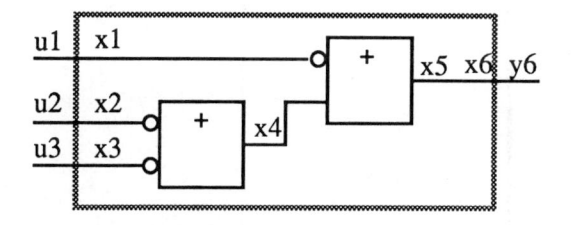

$$\begin{bmatrix} i & . & . & . & . & . \\ . & i & . & . & . & . \\ . & . & i & . & . & . \\ . & c & c & . & . & . \\ c & . & . & i & . & . \\ . & . & . & . & i & . \end{bmatrix}$$

Fig. 11. Equivalent irredundant circuit.

VII. FURTHER EXAMPLES

In view of theorems 3 and 4 and the subsequent comments included in section IV, the most efficient way to detect and eliminate redundancy is to search for simple redundancy initially, then first and subsequent orders of complex redundancy, since detecting an nth order complex redundancy involves computing and analysing 2^n matrices.

Example 2 (Complex redundancy):

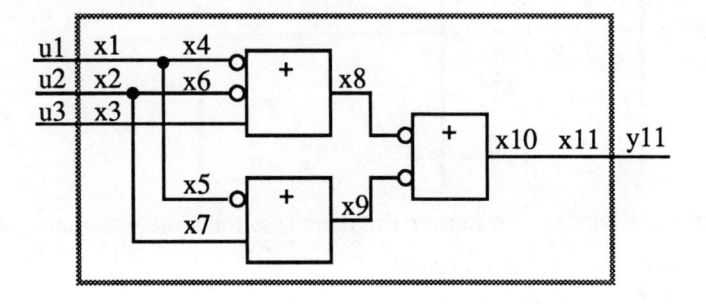

Fig. 12. Example derived from an example of Dandapani and Reddy (7).

This circuit has a structural matrix:-

$$M = \begin{bmatrix} i & . & . & . & . & . & . & . & . & . & . \\ . & i & . & . & . & . & . & . & . & . & . \\ . & . & i & . & . & . & . & . & . & . & . \\ i & . & . & . & . & . & . & . & . & . & . \\ i & . & . & . & . & . & . & . & . & . & . \\ . & i & . & . & . & . & . & . & . & . & . \\ . & i & . & . & . & . & . & . & . & . & . \\ . & . & i & c & . & c & . & . & . & . & . \\ . & . & . & . & c & . & i & . & . & . & . \\ . & . & . & . & . & . & . & c & c & . & . \\ . & . & . & . & . & . & . & . & . & i & . \end{bmatrix}$$

The corresponding diagnostic matrix is:-

$$D = \begin{bmatrix} \star & . & . & . & . & . & . & . & . & . \\ . & \star & . & . & . & . & . & . & . & . \\ . & . & \star & . & . & . & . & . & . & . \\ F4 & . & . & \star & . & . & . & . & . & . \\ F5 & . & . & . & \star & . & . & . & . & . \\ . & F6 & . & . & . & \star & . & . & . & . \\ . & F7 & . & . & . & . & \star & . & . & . \\ . & . & I & C & . & C & . & \star & . & . \\ . & . & . & . & C & . & I & . & \star & . \\ . & . & . & . & . & . & C & C & \star & . \\ . & . & . & . & . & . & . & . & I & \star \end{bmatrix}$$

Generating $D^{(n)}$ by the diagonal method:-

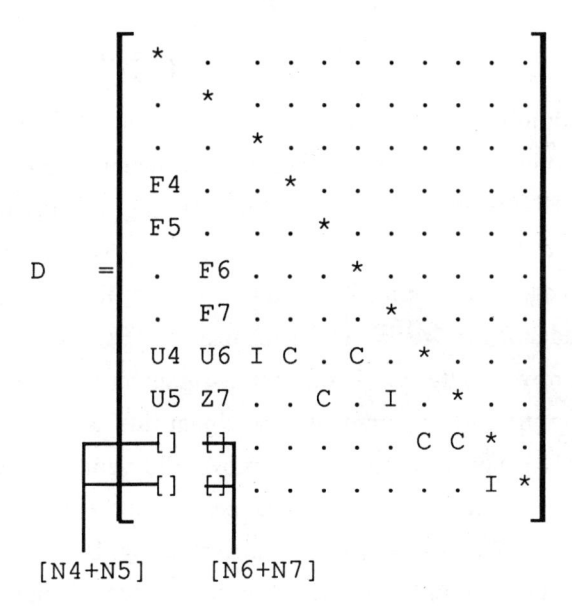

$$D = \begin{bmatrix} \star & . & . & . & . & . & . & . & . & . \\ . & \star & . & . & . & . & . & . & . & . \\ . & . & \star & . & . & . & . & . & . & . \\ F4 & . & . & \star & . & . & . & . & . & . \\ F5 & . & . & . & \star & . & . & . & . & . \\ . & F6 & . & . & . & \star & . & . & . & . \\ . & F7 & . & . & . & . & \star & . & . & . \\ U4 & U6 & I & C & . & C & . & \star & . & . \\ U5 & Z7 & . & . & C & . & I & . & \star & . \\ [] & [] & . & . & . & . & C & C & \star & . \\ [] & [] & . & . & . & . & . & . & I & \star \end{bmatrix}$$

[N4+N5]   [N6+N7]

Clearly, the circuit has no simple redundancy. To test for complex redundancy set:

(1) x4=x5=0 and then x4=x5=1;
(2) x6=x7=0 and then x6=x7=1.
With x4=x5=0:-

$$\begin{bmatrix}
\star & . & . & . & . & . & . & . & . & . & . \\
. & \star & . & . & . & . & . & . & . & . & . \\
. & . & \star & . & . & . & . & . & . & . & . \\
0 & . & . & \star & . & . & . & . & . & . & . \\
0 & . & . & . & \star & . & . & . & . & . & . \\
. & F6 & . & . & . & \star & . & . & . & . & . \\
. & F7 & . & . & . & . & \star & . & . & . & . \\
. & U6 & I & \cancel{1} & . & C & . & \star & . & . & . \\
. & Z7 & . & . & \cancel{1} & . & I & . & \star & . & . \\
. & . & . & . & . & . & . & \cancel{0} & \cancel{0} & \star & . \\
. & . & . & . & . & . & . & . & . & \cancel{0} & \star
\end{bmatrix}$$

Setting x4=x5=0 initially, and then generating $D^{(n)}$ again with this temporary assignment yields the previous matrix, from which it can be observed that F6 and F7 are absent from the output (in any form), and hence are possible candidates for redundancy. Setting x4=x5=1 for the second assignment and again generating $D^{(n)}$ will determine which fan-out branches (if any) are redundant. x4=x5=1 yields the matrix in fig. 13.

The entry N6+U7 in row 10 indicates that for this assignment x6 s-a-1 is undetectable. Hence by the earlier comments, the signal line x6 is s-a-1 redundant. Setting x6=1 permanently will remove the redundancy. Generating $D^{(n)}$ with x6=1 will then indicate whether or not there is any possibility of further redundancy. x6=1, yields the matrix in fig. 14.

In the remaining circuit there is only one case of reconvergent fan-out, but there is clearly no simple redundancy, and hence no further redundancy of any kind. Hence the matrix can be reduced by removing zero rows and columns, and it can then be transformed back into a structural matrix from

$$\begin{bmatrix}
* & . & . & . & . & . & . & . & . & . \\
. & * & . & . & . & . & . & . & . & . \\
. & . & * & . & . & . & . & . & . & . \\
1 & . & . & * & . & . & . & . & . & . \\
1 & . & . & . & * & . & . & . & . & . \\
. & F6 & . & . & . & * & . & . & . & . \\
. & F7 & . & . & . & . & * & . & . & . \\
. & U6 & I & \cancel{0} & . & C & . & * & . & . \\
. & . & . & . & \cancel{0} & . & \cancel{I} & . & * & . \\
. & [\!\!\dashv & . & . & . & . & C & C & \cancel{C} & * & . \\
. & [\!\!\dashv & . & . & . & . & . & . & . & I & * \\
\end{bmatrix}$$

[N6+U7]

Fig. 13.

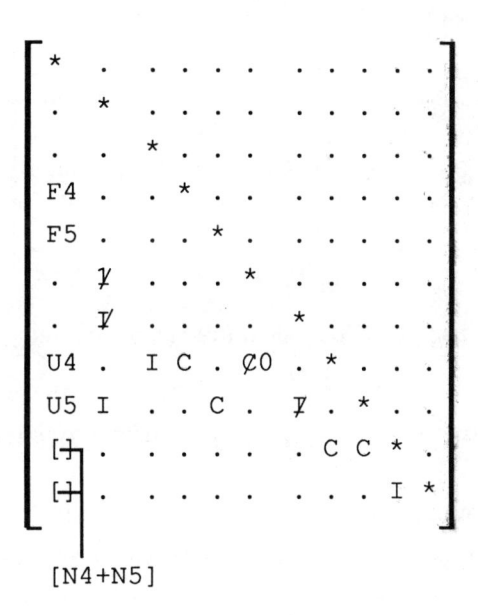

$$\begin{bmatrix}
* & . & . & . & . & . & . & . & . & . \\
. & * & . & . & . & . & . & . & . & . \\
. & . & * & . & . & . & . & . & . & . \\
F4 & . & . & * & . & . & . & . & . & . \\
F5 & . & . & . & * & . & . & . & . & . \\
. & \cancel{I} & . & . & . & * & . & . & . & . \\
. & \cancel{I} & . & . & . & . & * & . & . & . \\
U4 & . & I & C & . & \cancel{0} & . & * & . & . \\
U5 & I & . & . & C & . & \cancel{I} & . & * & . \\
[\!\!\dashv & . & . & . & . & . & C & C & * & . \\
[\!\!\dashv & . & . & . & . & . & . & . & I & * \\
\end{bmatrix}$$

[N4+N5]

Fig. 14.

which the equivalent irredundant circuit can be derived. Removing rows/columns 6 and 7 and transforming yields:-

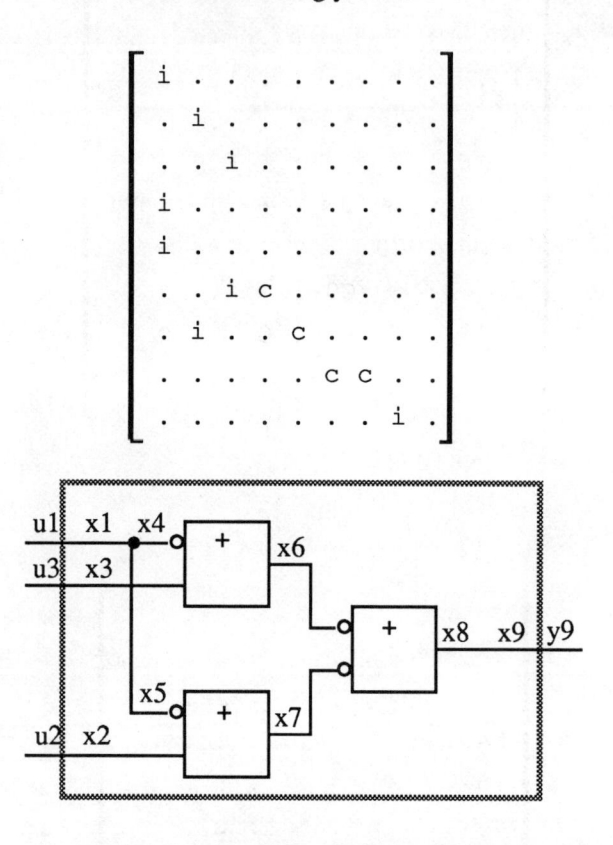

$$\begin{bmatrix} i & . & . & . & . & . & . & . & . \\ . & i & . & . & . & . & . & . & . \\ . & . & i & . & . & . & . & . & . \\ i & . & . & . & . & . & . & . & . \\ i & . & . & . & . & . & . & . & . \\ . & . & i & c & . & . & . & . & . \\ . & i & . & . & c & . & . & . & . \\ . & . & . & . & . & c & c & . & . \\ . & . & . & . & . & . & . & i & . \end{bmatrix}$$

Fig. 15. Equivalent irredundant circuit.

VIII. EXTENSION TO MULTI-OUTPUT CIRCUITS

Multi-output circuits pose particular problems. It has already been mentioned that it is possible for an irredundant multi-output circuit to have a redundant subcircuit; the presence of other outputs may well make an otherwise undetectable fault, detectable. Hence for a particular signal line to be redundant, it must be 'consistently' redundant with respect to every output. This condition can be accomodated by an adaption of the redundancy detection and deletion procedure detailed earlier.

Given that a circuit has n outputs, its corresponding matrix can be 'sliced' into n separate copies, one for each output, by deleting, in each case, all other outputs and all preceding signal lines which are independent of the remaining output. (Because of the structure of the matrix this can be achieved quite easily by a succession of related row and column operations.) Retaining the original indexing of the signal lines, each single output circuit can be reduced as indicated previously, and then the separately reduced submatrices can be reassembled into a composite matrix at the end, adjusting the structure as necessary. With this method, elimination of single entry rows poses a particular problem in the reassembly stage, so they have to be tackled in a different way. The real purpose of eliminating single entry rows, is to avoid cascades of single-entry gates which could include a number of inverters. The solution, is to retain the single entry rows, but 'push' the C entries along the cascades until they are incorporated into the input of some gate with one or more other inputs, or until they 'cancel out' another single C entry in the cascade. If, during the analysis stage, a redundancy is found, the original matrix is recalled and the redundancy removed from that, again retaining all single entry non-constant rows. In this way, only lines which are redundant with respect to every output are found to be deleted from the whole matrix after reassembly.

Example 3:

The circuit in fig. 16 has the diagnostic matrix shown in fig. 17. Splitting by output yields two matrices, fig. 18. Isolated F entries are regarded as just relabelled signal lines so that F*X=X for all entries X and for all such F. The matrix in fig. 18(1) has no fan-out and hence no redundancy.

The matrix in fig. 18(2) has fan-out in column 4, and hence may have redundancy. Detecting and eliminating redundancy as previously produces the matrix in fig. 19.

76 A. R. Fleming *et al.*

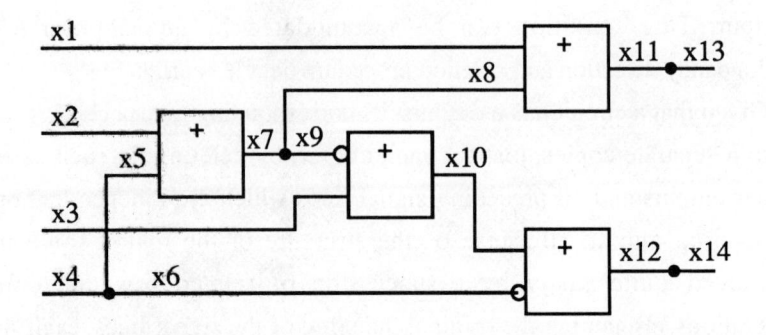

Fig. 16. Example of redundant multi-output circuit.

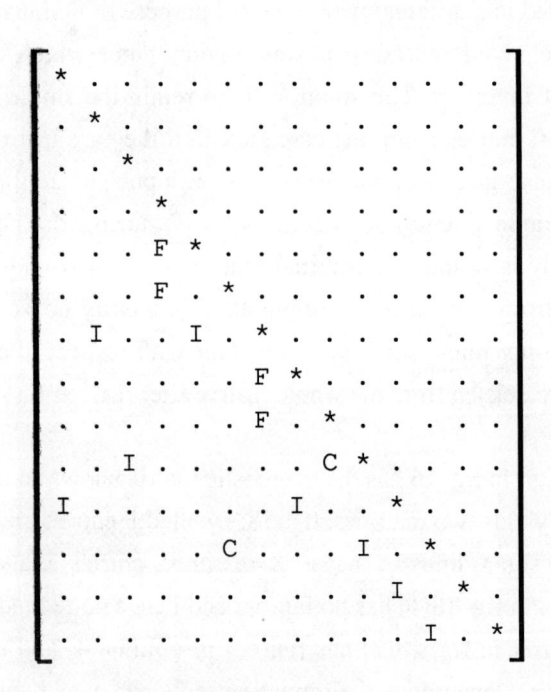

$$\begin{bmatrix}
* & . & . & . & . & . & . & . & . & . & . & . & . & . \\
. & * & . & . & . & . & . & . & . & . & . & . & . & . \\
. & . & * & . & . & . & . & . & . & . & . & . & . & . \\
. & . & . & * & . & . & . & . & . & . & . & . & . & . \\
. & . & . & F & * & . & . & . & . & . & . & . & . & . \\
. & . & . & F & . & * & . & . & . & . & . & . & . & . \\
. & I & . & I & . & * & . & . & . & . & . & . & . & . \\
. & . & . & . & . & F & * & . & . & . & . & . & . & . \\
. & . & . & . & . & F & . & * & . & . & . & . & . & . \\
. & . & I & . & . & . & . & C & * & . & . & . & . & . \\
I & . & . & . & . & . & . & I & . & * & . & . & . & . \\
. & . & . & . & . & C & . & . & I & . & * & . & . & . \\
. & . & . & . & . & . & . & . & . & I & . & * & . & . \\
. & . & . & . & . & . & . & . & . & . & I & . & * & . \\
\end{bmatrix}$$

Fig. 17.

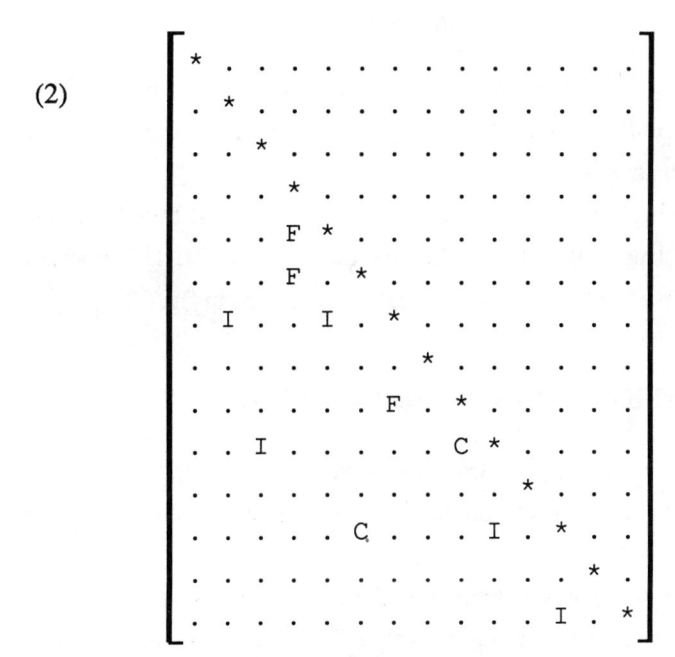

(1)
$$\begin{bmatrix}
\star & . & . & . & . & . & . & . & . & . & . & . & . & . & . \\
. & \star & . & . & . & . & . & . & . & . & . & . & . & . & . \\
. & . & \star & . & . & . & . & . & . & . & . & . & . & . & . \\
. & . & . & \star & . & . & . & . & . & . & . & . & . & . & . \\
. & . & . & F & \star & . & . & . & . & . & . & . & . & . & . \\
. & . & . & . & . & \star & . & . & . & . & . & . & . & . & . \\
. & I & . & . & I & . & \star & . & . & . & . & . & . & . & . \\
. & . & . & . & . & . & F & \star & . & . & . & . & . & . & . \\
. & . & . & . & . & . & . & . & \star & . & . & . & . & . & . \\
. & . & . & . & . & . & . & . & . & \star & . & . & . & . & . \\
I & . & . & . & . & . & I & . & . & \star & . & . & . & . & . \\
. & . & . & . & . & . & . & . & . & . & \star & . & . & . & . \\
. & . & . & . & . & . & . & . & . & . & I & . & \star & . & . \\
. & . & . & . & . & . & . & . & . & . & . & . & . & \star & .
\end{bmatrix}$$

(2)
$$\begin{bmatrix}
\star & . & . & . & . & . & . & . & . & . & . & . & . & . & . \\
. & \star & . & . & . & . & . & . & . & . & . & . & . & . & . \\
. & . & \star & . & . & . & . & . & . & . & . & . & . & . & . \\
. & . & . & \star & . & . & . & . & . & . & . & . & . & . & . \\
. & . & . & F & \star & . & . & . & . & . & . & . & . & . & . \\
. & . & . & F & . & \star & . & . & . & . & . & . & . & . & . \\
. & I & . & . & I & . & \star & . & . & . & . & . & . & . & . \\
. & . & . & . & . & . & . & \star & . & . & . & . & . & . & . \\
. & . & . & . & . & F & . & \star & . & . & . & . & . & . & . \\
. & . & I & . & . & . & . & C & \star & . & . & . & . & . & . \\
. & . & . & . & . & . & . & . & . & \star & . & . & . & . & . \\
. & . & . & . & . & C & . & . & I & . & \star & . & . & . & . \\
. & . & . & . & . & . & . & . & . & . & . & \star & . & . & . \\
. & . & . & . & . & . & . & . & . & . & I & . & \star & . & .
\end{bmatrix}$$

Fig. 18

$$\begin{bmatrix} * & . & . & . & . & . & . & . & . & . & . & . & . \\ . & * & . & . & . & . & . & . & . & . & . & . & . \\ . & . & * & . & . & . & . & . & . & . & . & . & . \\ . & . & . & * & . & . & . & . & . & . & . & . & . \\ . & . & . & . & * & . & . & . & . & . & . & . & . \\ . & . & . & F & . & * & . & . & . & . & . & . & . \\ . & . & . & . & . & . & * & . & . & . & . & . & . \\ . & . & . & . & . & . & . & * & . & . & . & . & . \\ . & . & . & . & . & . & . & . & * & . & . & . & . \\ . & . & I & . & . & . & . & . & * & . & . & . & . \\ . & . & . & . & . & . & . & . & . & . & * & . & . \\ . & . & . & . & . & C & . & . & . & I & . & * & . \\ . & . & . & . & . & . & . & . & . & . & . & * & . \\ . & . & . & . & . & . & . & . & . & . & I & . & * \end{bmatrix}$$

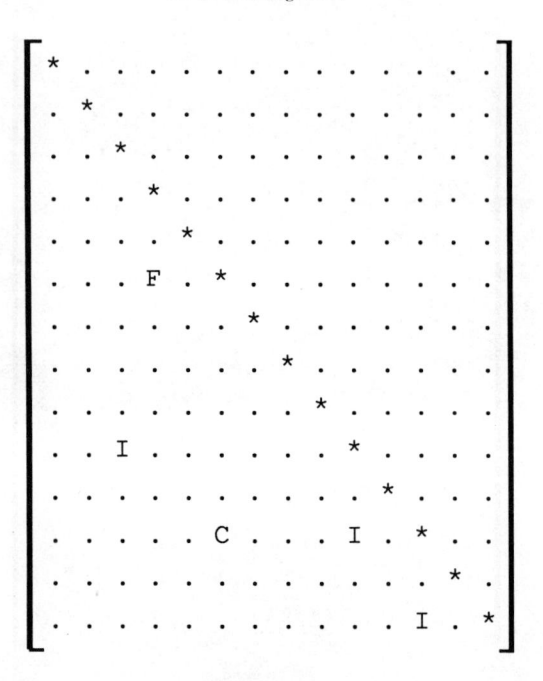

Fig. 19.

Reassembling the whole circuit by superimposing (a valid operation since no signal line indices have been changed, only redundant entries deleted) produces the matrix in fig. 20.

Removing single entries, zero rows and columns and transforming the matrix into a structural matrix yields the matrix in fig. 21. This matrix can be used to derive the equivalent irredundant circuit, as in fig. 22.

IX. EXTENSION TO DEPENDENT FANOUT

The previous theory applies only to circuits with independent fan-out. Extending the theory to more general circuits involves several complications.

When seeking complex redundancy, the fan-out nodes cannot in general take the values 0 and 1 independently of all other fan-out nodes. The solution is to take a maximal logically independent subset of the fan-out

$$\begin{bmatrix}
\star & . & . & . & . & . & . & . & . & . & . & . & . & . \\
. & \star & . & . & . & . & . & . & . & . & . & . & . & . \\
. & . & \star & . & . & . & . & . & . & . & . & . & . & . \\
. & . & . & \star & . & . & . & . & . & . & . & . & . & . \\
. & . & . & F & \star & . & . & . & . & . & . & . & . & . \\
. & . & . & F & . & \star & . & . & . & . & . & . & . & . \\
. & I & . & . & I & . & \star & . & . & . & . & . & . & . \\
. & . & . & . & . & . & F & \star & . & . & . & . & . & . \\
. & . & . & . & . & . & . & \star & . & . & . & . & . & . \\
. & . & I & . & . & . & . & . & \star & . & . & . & . & . \\
I & . & . & . & . & . & . & I & . & \star & . & . & . & . \\
. & . & . & . & . & C & . & . & I & . & \star & . & . & . \\
. & . & . & . & . & . & . & . & . & I & . & \star & . & . \\
. & . & . & . & . & . & . & . & . & . & I & . & \star & . \\
\end{bmatrix}$$

Fig. 20.

$$\begin{bmatrix}
i & . & . & . & . & . & . & . & . & . & . \\
. & i & . & . & . & . & . & . & . & . & . \\
. & . & i & . & . & . & . & . & . & . & . \\
. & . & . & i & . & . & . & . & . & . & . \\
. & . & . & i & . & . & . & . & . & . & . \\
. & . & . & i & . & . & . & . & . & . & . \\
. & i & . & . & i & . & . & . & . & . & . \\
i & . & . & . & . & . & i & . & . & . & . \\
. & . & i & . & . & c & . & . & . & . & . \\
. & . & . & . & . & . & . & i & . & . & . \\
. & . & . & . & . & . & . & . & i & . & . \\
\end{bmatrix}$$

Fig. 21.

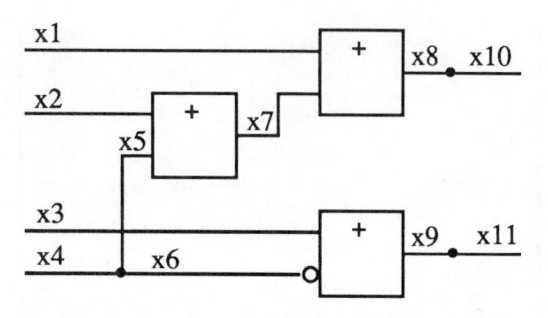

Fig. 22. Equivalent irredundant circuit.

nodes, as the only fan-out nodes of interest, at least initially. This solution in turn creates another problem with multiple paths, even in single-output circuits; fan-out nodes for which there exist multiple paths between them and an output, must be shown to be 'consistently' redundant with respect to each path. Research work in this area is continuing, but initial results using the matrix representation are encouraging.

X. CONCLUSIONS

A matrix method has been devised for the detection and elimination of redundancy in general combinational circuits. It has been shown that, in contrast to previous methods, only fan-out nodes need to be considered as candidates for redundancy.

An accurate assessment of the computational complexity is difficult, since the search part of the procedure stops immediately a redundancy is found, and a circuit may or may not have a great deal of complex redundancy. In the worst case, the number of matrices to be analysed could be of the order $3^N\text{-}2^N$ (where N is the number of independent fan-out nodes) before a redundancy is found, or before the circuit is declared irredundant. Multiplying this figure by the number of fan-out branches will give an extremely generous upper bound for the computational complexity of the method; it is hoped that further research will yield a more realistic and acceptable bound. Software, for the VAX 11/750, and based on the redundancy detection procedures contained in this chapter has been developed and is constantly being revised as research into this aspect of circuit design continues. The main problem is with the detection of high order complex redundancy; it is hoped that with further research such redundancy will be identified at an early stage without having to generate and analyse large numbers of matrices.

The way that the procedure tests for simple redundancy, then first-order redundancy, etc., can be modified to test for redundancy of any order, with a computational complexity of $Nx2^{N-1}$ for each redundancy found. However, such a procedure would be inappropriate for circuits with lots of simple, or low order, redundancy, but it could be useful as a confirmation test for circuits believed to be irredundant.

XI. REFERENCES

1. Acken, C.F. (1982). IEEE International Conference on Circuits and Computers, New York. 464-467.
2. Aufenkamp, D.D. and Hohn, F.E. (1957). Transactions IRE. EC-6, 299-306.
3. Bossen, D.C. and Hong, S.J. (1971). IEEE Transactions on Computers. C-20, 1252-1257.
4. Breuer, M.A. and Friedman, A.D. (1976). "Diagnosis and Reliable Design of Digital Systems". Computer Science Press, Woodland Hills.
5. Campeau, J.O. (1957). Transactions IRE. EC-6, 23-41.
6. Clegg, F.W. (1973). IEEE Transactions on Computers. C-22, 229-234.
7. Dandapani, R. and Reddy, S.M. (1974). IEEE Transactions on Computers. C-23, 1139-1149.
8. Deo, M. (1974). "Graph Theory with Applications to Engineering and Computer Science". Prentice Hall, Englewood Cliffs.

9. Edwards, C.R. (1972). Computer Journal (GB). V15, 247-253.
10. Edwards, C.R. (1972). Electronics Letters (GB). V8, 113-115.
11. Ektare, A.B. (1978). International Journal of Electronics. V44, 157-159.
12. Friedman, A.D. (1967). IEEE Transactions on Electronic Computers. EC-16, 99-100.
13. Gill, A. (1962). "Introduction to the theory of finite state machines". McGraw Hill, New York.
14. Hayes, J.P. (1971). IEEE Transactions on Computers. C-20, 1496-1506.
15. Hayes, J.P. (1974). IEEE Transactions on Computers. C-23, 56-62.
16. Hayes, J.P. (1976). IEEE Transactions on Computers. C-25, 884-892.
17. Hohn, F.E., Seshu, S. and Aufenkamp, D.D. (1957). Transactions IRE. EC-6, 154-161.
18. Hurst, S.L., Miller, D.M. and Muzio, J.C. (1985). "Spectral Techniques in Logic Design". Academic Press, London.
19. Lee, H-P. and Davidson, E.S. (1972). IEEE Transactions on Computers. C-21, 12-20.
20. Lee, H-P. and Davidson, E.S. (1974). IEEE Transactions on Computers. C-23, 1029-1047.
21. Page, E.W. (1976). Electronic Letters (GB). V12, 49-50.
22. Sutton, J.C. and Bredeson, J.G. (1980). IEEE Transactions on Computers. C-29, 648-656.
23. To, K. (1973). IEEE Transactions on Computers. C-22, 1008-1015.
24. Whitney, G.E. (1971). IEEE Transactions on Computers. C-20, 141-148.
25. Yau, S.S. (1965). IEEE Transactions on Electronic Computers. EC-14, 467-472.

THE INTERRELATIONSHIPS BETWEEN FAULT SIGNATURES BASED UPON COUNTING TECHNIQUES

S.L. Hurst

*The Open University
Faculty of Technology
Walton Hall
Milton Keynes
England
MK7 6AA*

ABSTRACT

Compressed fault signatures for combinatorial networks based upon a counter (accumulator) which records the number of occurances of a certain output characteristic of the network under test have been proposed by several authorities. Such proposals have included syndrome counting, arithmetic coefficients, spectral coefficients, and others. Here we will briefly review the bases of these several approaches, and develop the simple recursive matrix relationships which exist between them.

LIST OF SYMBOLS USED

x_i, i = 1 to n independent binary variables,

$x_i \in \{0,1\}$, x_1 least significant

$f(x_1, x_2, ..., x_n)$ or merely $f(X)$	binary function of the x_i input variables, $f(X) \in \{0,1\}$
X]	truthtable column vector for function $f(X)$, entries $\in \{0,1\}$, in mintern decimal order
Y]	truthtable column vector for $f(X)$, entries $\in \{+1, -1\}$, in mintern decimal order
$a_j, j = 0$ to $2^n - 1$	coefficients in canonic mintern expansion for $f(X)$
A]	coefficient column vector of a_j, in decimal order
$b_j, j = 0$ to $2^n - 1$	coefficients in arithmetic canonic expansion for $f(X)$
B]	coefficient column vector of b_j, in decimal order
$c_j, j = 0$ to $2^n - 1$	coefficients in positive Reed-Muller canonic expansion for $f(X)$
C]	coefficient column vector of c_j, in decimal order

r_j, j = 0 to 2^n -1   spectral coefficients of f(X), transformed from **X**]

R]   coefficient column vector of r_j, in decimal order

s_j, j = 0 to 2^n -1   spectral coefficients of f(X), transformed from **Y**]

S]   coefficient column vector of s_j, in decimal order

[**T^n**]   general 2^n x 2^n transform matrix

[**Hd**] 2^n x 2^n Hadamard matrix

[**RW**] 2^n x 2^n Rademacher-Walsh matrix

[**I**] 2^n x 2^n identity matrix

I. INTRODUCTION

Developments in testing for VLSI have increasingly emphasised (a) appropriate partitioning of complete networks, since global testing no longer remains a feasible method, and (b) the adoption of scan-in/scan-out design methodologies in order to allow realistic testing of the sequential elements. This is illustrated in Fig. 1. Ongoing research interests are therefore largely concerned with considerations of combinatorial functions and the methods of test for both random logic nets and regular array structures.

In considering testing of buried combinatorial networks, there is the practical problem of providing controllability and accessibility of all the network interconnections. Partitioning and scan techniques may be adopted for such purposes. We will not be concerned with these important practical design aspects herewith. However, assuming access is available, testing methods for combinatorial networks may be divided into two broad categories, namely (i) exhaustive testing, where the input test set consists of all possible (2^n) input vectors, and (ii) non-exhaustive testing, where the input test set consists of a reduced set of vectors ($<2^n$), chosen such that particular network failings may be detected.

This scenario is shown in Table 1. Non-exhaustive testing based upon fault models such as stuck-at-0 or stuck-at-1, (1) - (9), is becoming increasingly difficult and suspect as network size increases (10) - (11). The alternative strategy of exhaustively energising the network under test, appropriately partitioned, is now receiving increasing emphasis, since it is based upon functionality rather than upon possibly inadequate fault modelling. It also eliminates the complexity in formulating a minimum (minimal) test set and

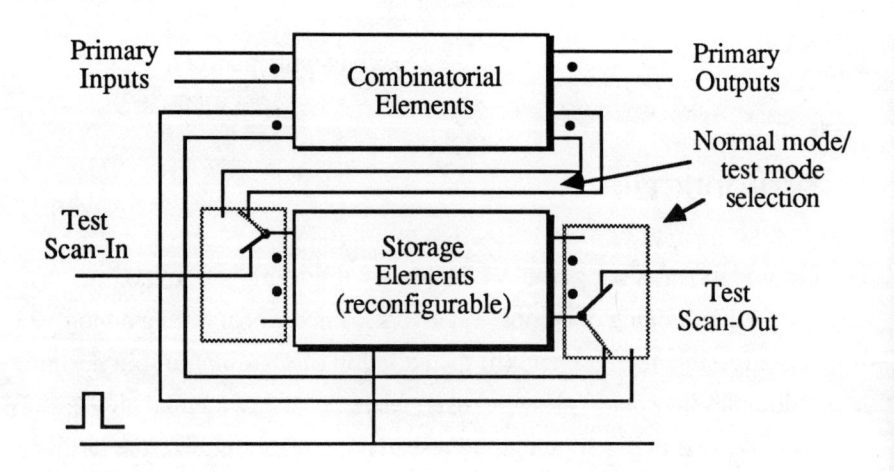

Fig. 1. The present favoured method of testing sequential networks, with the reconfiguration of the sequential elements into some form of

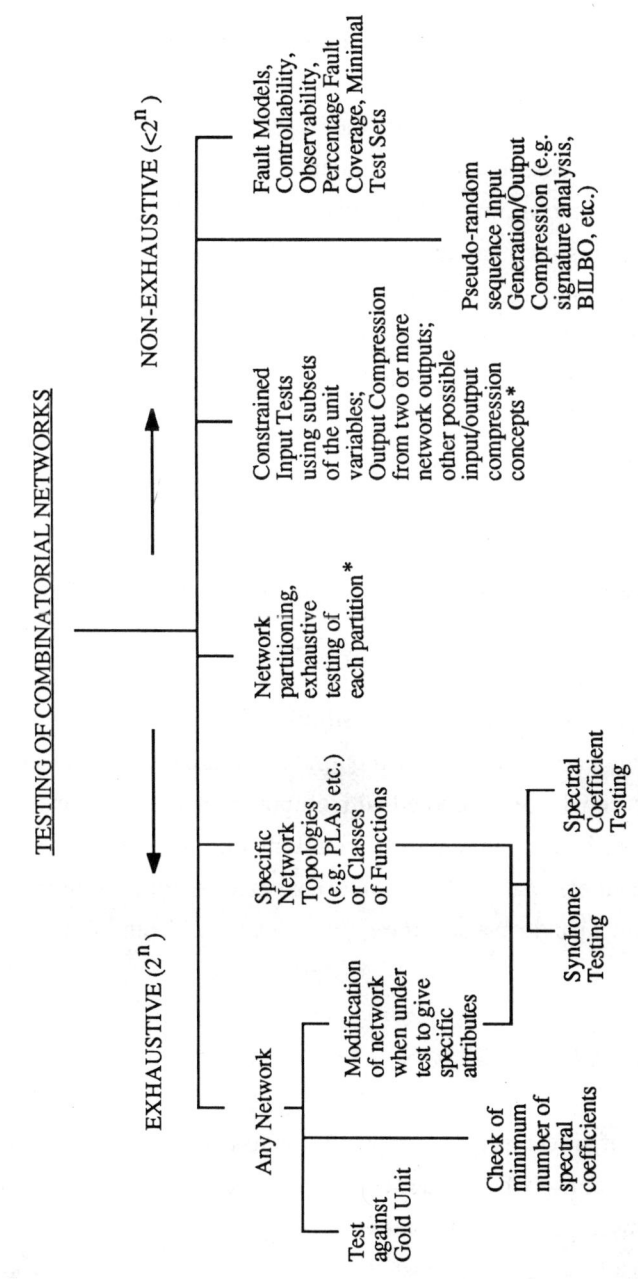

Table 1.

TESTING OF COMBINATORIAL NETWORKS

EXHAUSTIVE (2^n) ←→ **NON-EXHAUSTIVE ($<2^n$)**

- **Any Network**
  - Test against Gold Unit
  - Check of minimum number of spectral coefficients
  - Modification of network when under test to give specific attributes

- **Specific Network Topologies (e.g. PLAs etc.) or Classes of Functions**
  - Syndrome Testing
  - Spectral Coefficient Testing

- **Network partitioning, exhaustive testing of each partition***

- **Constrained Input Tests using subsets of the unit variables; Output Compression from two or more network outputs; other possible input/output compression concepts***

- Fault Models, Controllability, Observability, Percentage Fault Coverage, Minimal Test Sets

- Pseudo-random sequence Input Generation/Output Compression (e.g. signature analysis, BILBO, etc.)

The general scenario of testing of combinatorial networks. In the partitioned and constrained methods marked *, the total number of tests should be less than 2^n or else there may be little advantage over a single exhaustive global test.

the determination of the resulting fault coverage, and in most cases is independent of the realising topology for any given function. For further details see (10) - (23).

When exhaustive testing is adopted, the major interests are in compaction of the output response of the net under test. The simple method of comparing the output response of a function against a known good copy is not ideal, since this involves memory to hold the correct response, and a step-by-step comparison of this data against the function under test. Therefore some means of output compaction of network responses rather than step-by-step comparison is desirable, which should:-

(a) be simple to implement, using minimum silicon area when implemented on-chip,
and
(b) give an unambiguous signature to indicate function fault-free/function faulty.

A further desirability may be the possibility of combining two or more function outputs into one compressed output signature (10), (19), (35).

In this paper we will be concerned with output compaction proposals, and in particular with the mathematical relationships which exist between separate disclosures. We will not have occasion to refer to multiple-output compaction into a single test signature, since this is still in an initial R and D stage.

II. OUTPUT COMPACTION

The objective of single-output compaction is to avoid having to compare every output response when under test against the truthtable record of the healthy response. However the network response on every applied input test vector must be a part of the test, and hence it is clear that some form of accumulation at the network output is necessary. An appropriately driven counter forms an obvious practical mechanism for this purpose.

Figure 2 therefore defines the area of our present considerations. The combinatorial network under test has independent binary inputs x_i, $i = 1$ to n, and binary output $f(X)$. The input test set therefore consists of 2^n input vectors, which fully exercise the network. The output $f(X)$ controls the accumulation of the count in the output test counter. It will be appreciated that this constitutes the basic situation for all published developments; ongoing research which is investigating (a) reduced input test sets from the full 2^n set, and (b) the combining of more than one output function into one accumulator, will not be considered in the following relationship discussions. The accumulator (counter) must record the presence (absence) of a certain output characteristic on each input test vector, the final count being a summation of these characters. For the procedure to be a satisfactory test, the final count (signature) must be unique for a healthy network, any output fault resulting in a dissimilar final value.

We will first review the several methods by which a function $f(X)$ may be defined, other than by conventional Boolean algebraic expressions. These involve a vector of coefficient values, not necessarily two-valued, which in

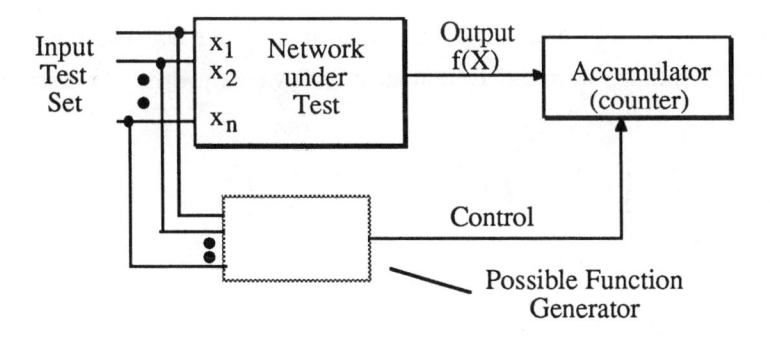

Fig. 2. The basic topology with output compaction into one numerical signature. The full 2^n input test set or a subset may be present; here we will only consider the exhaustive test set of 2^n vectors.

total fully define any given f(X). It is this, together with other counting methods, which leads to the consideration of their use as fault signatures.

A. Canonic Minterm Expansion of f(X)

The coefficients most directly related to Boolean algebra are those associated with the canonic minterm expansion of f(X). For any three-variable function we have:

$$f(X) = a_0\bar{x}_1\bar{x}_2\bar{x}_3 + a_1x_1\bar{x}_2\bar{x}_3 + a_2\bar{x}_1x_2\bar{x}_3 + a_3x_1x_2\bar{x}_3 + a_4\bar{x}_1\bar{x}_2x_3$$
$$+ a_5x_1\bar{x}_2x_3 + a_6\bar{x}_1x_2x_3 + a_7x_1x_2x_3$$

$$a_j, j = 0 \text{ to } 2^n - 1, \in \{0,1\} \tag{1}$$

The vector $A]$ of the a_j in correct canonic order fully defines f(X).

This is a trivial case, since the a_j are merely defining the zero-valued and one-valued minterms of f(X), with $A]$ being identical to the normal output truthtable of f(X) in minterm order. In matrix form we may write (See List of Symbols):

$$[I]X] = A], \text{ giving } A] = X] \tag{2}$$

B. Canonic Arithmetic Expansion of f(X)

An arithmetic canonic expansion for f(X) is also available (20), (24). For any 3-variable function we have:

$$f(X) = b_0 + b_1 x_1 + b_2 x_2 + b_3 x_1 x_2$$
$$+ b_4 x_3 + b_5 x_1 x_3 + b_6 x_2 x_3 + b_7 x_1 x_2 x_3,$$

$$b_j, j = 0 \text{ to } 2^n - 1, \tag{3}$$

where the addition is arithmetic, not Boolean disjunction. The general expression for any $f(X)$ is given by:

$$f(X) = \sum_{j=0}^{2^n - 1} b_j X_j, \tag{4}$$

$$X_j \in \{1, x_1, x_2, ..., x_n, x_1 x_2, ..., x_1 x_2 ... x_n\},$$

and where the binary subscript identifiers of the x_i product terms are read as decimal numbers j. As an example, the parity function $\bar{x}_1 \bar{x}_2 x_3 + \bar{x}_1 x_2 \bar{x}_3 + x_1 \bar{x}_2 \bar{x}_3 + x_1 x_2 x_3$ may be written as:

$$f(X) = x_1 + x_2 - 2x_1 x_2 + x_3 - 2x_1 x_3 - 2x_2 x_3 + 4x_1 x_2 x_3$$

Note that b_j is not confined to positive values. The vector $B]$ of the b_j coefficient is the arithmetic spectrum of $f(X)$. For any 3-variable function ($n = 3$) it may be determined by the transform:

$$\begin{bmatrix} 1 & 0 & 0 & 0 & 0 & 0 & 0 & 0 \\ -1 & 1 & 0 & 0 & 0 & 0 & 0 & 0 \\ -1 & 0 & 1 & 0 & 0 & 0 & 0 & 0 \\ 1 & -1 & -1 & 1 & 0 & 0 & 0 & 0 \\ -1 & 0 & 0 & 0 & 1 & 0 & 0 & 0 \\ 1 & -1 & 0 & 0 & -1 & 1 & 0 & 0 \\ 1 & 0 & -1 & 0 & -1 & 0 & 1 & 0 \\ -1 & 1 & 1 & -1 & 1 & -1 & -1 & 1 \end{bmatrix} X = B \begin{bmatrix} b_0 \\ \\ \\ \\ \\ \\ \\ b_7 \end{bmatrix} \quad (5)$$

The transform for any n is given by:

$$T^n = \begin{bmatrix} T^{n-1} & 0 \\ -T^{n-1} & T^{n-1} \end{bmatrix}$$

$$T^0 = +1. \quad (6)$$

C. Canonic Positive Reed-Muller Expansion of f(X)

The Reed-Muller canonic expansion for f(X) involves exclusive-OR (addition modulo 2) relationships rather than inclusive-OR or arithmetic. There are 2^n different possible Reed-Muller expansions available for any f(X), involving all possible permutations of true and complemented x_is; here we will confine ourselves to the "positive polarity" RM expansion involving all x_is true and none complemented (24), (25). For any 3-variable function we have:

$$f(X) = c_0 \oplus c_1x_1 \oplus c_2x_2 \oplus c_3x_1x_2 \oplus c_4x_3$$
$$\oplus c_5x_1x_3 \oplus c_6x_2x_3 \oplus c_7x_1x_2x_3 \qquad (7)$$

$c_j, j = 0$ to $2^n - 1, \in \{0,1\}$,

where the addition is mod_2. The general expression for an $f(X)$ is given by:

$$f(X) = \sum_{j=0}^{2^n-1} c_j X_j \qquad (8)$$

X_j as previously defined.

The vector C] of the c_j coefficients is the Reed-Muller positive canonic spectrum of $f(X)$. For any three-variable function we have:

$$\begin{bmatrix} 1 & 0 & 0 & 0 & 0 & 0 & 0 & 0 \\ 1 & 1 & 0 & 0 & 0 & 0 & 0 & 0 \\ 1 & 0 & 1 & 0 & 0 & 0 & 0 & 0 \\ 1 & 1 & 1 & 1 & 0 & 0 & 0 & 0 \\ 1 & 0 & 0 & 0 & 1 & 0 & 0 & 0 \\ 1 & 1 & 0 & 0 & 1 & 1 & 0 & 0 \\ 1 & 0 & 1 & 0 & 1 & 0 & 1 & 0 \\ 1 & 1 & 1 & 1 & 1 & 1 & 1 & 1 \end{bmatrix}_{\mathrm{mod}_2} \begin{bmatrix} X \end{bmatrix} = \begin{bmatrix} c_0 \\ \\ \\ \\ \\ \\ \\ c_7 \end{bmatrix} \qquad (9)$$

where the matrix additions are modulo 2, not arithmetic (24), (26). It may readily be shown by consideration of all the 2^n equations which have to be satisfied between the input vectors and the function output that no arithmetic matrix is possible for the transformation of **X]** to **C]**. A signal flow graph equivalent of the matrix of Eq. 9 may be found in Besslich (34).

The transform for any n is given by:

$$T^n = \begin{bmatrix} T^{n-1} & 0 \\ T^{n-1} & T^{n-1} \end{bmatrix}_{mod_2}$$

$$T^0 = +1. \tag{10}$$

The relationships between the c_j coefficients for the positive canonic case (above) and any other polarity expansion may be given by a further matrix operation; this will not concern us here.

D. Hadamard spectrum of f(X)

It is possible to consider the transformation of the function output vector into some alternative (spectral) domain by means of $2^n \times 2^n$ complete orthogonal matrix transforms, giving unique spectral coefficients for any given f(X). The simplest well-structured transform is the Hadamard transform, defined as:

$$Hd^n = \begin{bmatrix} Hd^{n-1} & Hd^{n-1} \\ Hd^{n-1} & -Hd^{n-1} \end{bmatrix}$$

$$Hd^0 = +1 \tag{11}$$

Counting Techniques

For n = 3, this gives the following, shown here for the example function
$$f(X) = \bar{x}_1\bar{x}_2x_3 + \bar{x}_1x_2\bar{x}_3 + x_1\bar{x}_2\bar{x}_3 + x_1x_2x_3$$

$$\begin{bmatrix} 1 & 1 & 1 & 1 & 1 & 1 & 1 & 1 \\ 1 & -1 & 1 & -1 & 1 & -1 & 1 & -1 \\ 1 & 1 & -1 & -1 & 1 & 1 & -1 & -1 \\ 1 & -1 & -1 & 1 & 1 & -1 & -1 & 1 \\ 1 & 1 & 1 & 1 & -1 & -1 & -1 & -1 \\ 1 & -1 & 1 & -1 & -1 & 1 & -1 & 1 \\ 1 & 1 & -1 & -1 & -1 & -1 & 1 & 1 \\ 1 & -1 & -1 & 1 & -1 & 1 & 1 & -1 \end{bmatrix} \begin{bmatrix} 0 \\ 1 \\ 1 \\ 0 \\ 1 \\ 0 \\ 0 \\ 1 \end{bmatrix} \begin{bmatrix} m_0 \\ m_1 \\ m_2 \\ m_3 \\ m_4 \\ m_5 \\ m_6 \\ m_7 \end{bmatrix} = \begin{bmatrix} 4 \\ 0 \\ 0 \\ 0 \\ 0 \\ 0 \\ 0 \\ -4 \end{bmatrix} \begin{bmatrix} r_0 \\ r_1 \\ r_2 \\ r_3 \\ r_4 \\ r_5 \\ r_6 \\ r_7 \end{bmatrix} \quad (12)$$

[Hd]   **X]**   **R]**

For completeness, we identify above the function minterms m_0 to m_7 and the resulting spectral coefficients r_0 to r_7. The spectral coefficients may be re-designated and expressed in rearranged order as follows:

$$r_0 \quad r_1 \quad r_2 \quad r_{12} \quad r_3 \quad r_{13} \quad r_{23} \quad r_{123}$$
$$(r_0) \quad (r_1) \quad (r_2) \quad (r_3) \quad (r_4) \quad (r_5) \quad (r_6) \quad (r_7)$$

Spectrum of f(X):   4   0   0   0   0   0   0   -4

It is equally relevant to transform the output truth-table vector **Y]** representing f(X), where **Y]** is a recording of **X]**, recoded logic 0→ +1, logic 1 → -1. **[Hd]Y]** then gives the spectral coefficient vector **S]** for f(X), which for the above function would yield the spectrum in functional order:

s_0	s_1	s_2	s_{12}	s_3	s_{13}	s_{23}	s_{123}
0	0	0	0	0	0	0	+8

Both [**Hd**]**Y** and [**Hd**]**X** will be found extensively used in the literature (22). The relationships between the r_j and s_j spectral coefficients is linear, being

$$r_0 = \frac{1}{2}(2^n - s_0), \quad r_j, j \neq 0, = -\frac{1}{2} s_j \qquad (13)$$

There are a number of alternative complete orthogonal transforms to the Hadamard which are row re-orderings (21), (22). The most prominent is the Rademacher-Walsh variant, [**RW**], which directly generates the spectral coefficients in the re-arranged logical order shown above rather than in the Hadamard ordering. It will be appreciated that with such variants the recursive structure of the Hadamard transform is lost, but there is no difference in the information content of the resulting r_j (s_j) spectral coefficients.

III. PRESENTLY DISCLOSED COUNTING PROCEDURES

There have been a number of techniques which involve some form of count at the function output, which gives a faulty/fault-free signature for the network under test. Here we will consider those which have as their basis the several canonic expansions and transforms considered above, and which in their original concept involve an exhaustive input test set requiring no knowledge of the network topology or derivation and application of a minimum test set.

We will briefly survey these presently disclosed methods, in order to compare them in the developments given in Section 4.

A. Syndrome Testing

The conceptually most simple test signature is to count the number of logic 1's which appear at a function output when the full 2^n input test set is applied (27). This is (effectively) summing the a_j coefficients in the canonic minterm expansion of Eq. 2. However it may readily be shown that for many networks this simple test is inadequate; for example any Exclusive-OR function $x_1 \oplus x_2 \oplus ...$ will have the same minterm count with any input stuck at 0 or stuck at 1. Developments from this approach may be found, including transition count testing (15), (28), which can provide compressed test signatures for any given network.

Savir's work on syndrome testing (12) - (14), (22), (23), is directly equivalent to a count of the number of logic 1s in the output of the function under test. As originally defined by Savir (12), the syndrome of any combinatorial function f(X) is given by

$$S = \frac{1}{2^n} K \qquad (14)$$

where K is the number of 1-valued minterms in f(X). The value of S thus defined therefore lies between $0 \leq S \leq 1$. However for most practical purposes the normalising factor of $1/2^n$ is omitted (10), (19), (22), (23).

Savir's work includes consideration of networks which are syndrome-testable and syndrome-untestable. In the former the value of S is unique for a fault-free network, and therefore its value may be taken as the test signature for the network. In the latter S is not unique, and circuit additions are suggested to make the extended circuit syndrome-testable. In general this involves the addition of gating and control inputs which effectively add minterms to f(X) when under test.

B. Arithmetic Coefficient Testing

The arithmetic canonic expansion of Eq. 4 has also been suggested for fault signature purposes (20). Since the complete vector **B**] uniquely defines any given f(X), any functional fault in f(X) will modify one or more of the 2^n b_j values. However, since it is obviously uneconomical to consider the determination and inspection of all 2^n values, some reduced subset of b_j which will act as a fault signature is sought. Work using the b_j coefficients for fault indication, which has a certain mathematical commonality with the so-called probabilistic treatment of general combinatorial networks (24), (29), (30), has not been pursued in such detail as work using spectral coefficients, see below; however the purpose of this paper is to show later the numerical relationships between different counting approaches, rather than expand upon their detailed development.

C. Reed-Muller Coefficient Testing

As far as is known, the coefficient vector **C**] of the Reed-Muller canonic expansion, Eq. 8, does not appear to have been widely considered for fault diagnostics. As seen later, the c_j values are distinct from other coefficient values, not being related to others by normal arithmetic relationships. There is, however, no fundamental reason why they may not be considered for test signatures, although it is not obvious that advantages over other coefficients may accrue. Some work and further references may be found in Upadhyaya and Saluja (10).

D. Spectral Coefficient Testing.

As previously noted, the Hadamard, Rademacher-Walsh and other row variants of these transforms produce the spectral domain coefficients **R**] or **S**] which fully define any f(X). The values of the r_j (s_j) are the same for each transform variant, merely being generated in a different order. As with the previous case of the arithmetic coefficient vector **B**], it is uneconomical to consider the determination and inspection of the complete spectral coefficient vector **R**] (**S**]) for diagnostic purposes. Rather a subset, ideally one coefficient only, is sought.

Considerable work using individual r_j (s_j) coefficients and subsets of coefficients, p = integer, $1 \leq p \leq n$, has been reported (16) - (18), (22), (23), (31), (32). As may be appreciated and will be specifically noted later, Savir's syndrome S without the $1/2^n$ normalisation is identical to the first spectral coefficient r_0 ; the higher order spectral coefficients have no parallel with Savir's work. It may also be appreciated that with a range of 2^n spectral coefficients available, each of which is a many-valued parameter representing some global picture of a given function f(X), there is a rich source of information from which fault signatures for f(X) may be possible. Research developments are continuing in this area (22), (23).

Finally we should note that we will not here consider the important and increasingly used method of numerical fault analysis, usually termed Signature Analysis (3), (7), (8), (33). Our omission is firstly because it does not conform to the same mathematical bases as the above, and secondly because there is at present no theoretical justification for its success as a fault signature. The logical relationships in generating the final count signature make it difficult to envisage the generation of a fault-free count by a faulty network, but firm mathematical qualification is not available.

IV. THE RELATIONSHIPS

In the following developments we will seek to show the mathematical relationships which exist between the several alternative coefficient vectors which may be used to represent any given function f(X), from which their commonality for test signature purposes will subsequently be shown.

Since the spectral coefficients have received considerable attention for network analysis and synthesis, and more recently test purposes, we will use the Hadamard spectral coefficients as the basis to which other coefficients will be related. We will use the r_j coefficients obtained from [Hd]X] rather than the s_j coefficients obtained from [Hd]Y], since the {0,1} valued vector X] is common to all conventional digital logic synthesis. The linear relationships between r_j and s_j have been given in Eq. 13.

A. The r_j to a_j Coefficient Relationships

The value of the individual a_js, being identical to the {0,1} values in vector X], may be obtained from the spectral coefficient vector R] by the inverse of Eqs. 11 and 12, that is:

$$A] = X]$$
$$= [Hd]^{-1} R]$$

$$\text{where} \quad [Hd]^{-1} = \frac{1}{2^n}[Hd] \tag{15}$$

In particular

$$a_0 = \frac{1}{2^n}\left[\sum_{j=0}^{2^n-1} r_j\right] \tag{16}$$

Savir's syndrome parameter S, See Eq. 14, without the normalising factor $1/2^n$ is merely:

$$S = \sum_{j=0}^{2^n-1} a_j = r_0. \qquad (17)$$

B. The r_j to b_j Coefficient Relationships

From Eqs. 5, 6 and 15, the b_j arithmetic coefficients are related to the r_j spectral coefficients by:

$$\begin{aligned} B] &= [T^n] \, X] \\ &= [T^n] \left[\frac{1}{2^n} [Hd] \, R] \right] \\ &= \frac{1}{2^n} \left[[T^n] \, [Hd] \, R] \right] \end{aligned} \qquad (18)$$

where $[T^n]$ is the transform given in Eq. 6. Evaluation of $[T^n][Hd]$, see Eqs. 6 and 11, will give the resulting conversion matrix $[Trb]$ from $R]$ to $B]$, giving:

$$\mathbf{Trb}^n = \begin{bmatrix} \mathbf{Trb}^{n-1} & \mathbf{Trb}^{n-1} \\ 0 & -2\,\mathbf{Trb}^{n-1} \end{bmatrix} \qquad (19)$$

$$\mathbf{Trb}^0 = +1.$$

For n = 3 this gives:

$$\mathbf{Trb} = \begin{bmatrix} 1 & 1 & 1 & 1 & 1 & 1 & 1 & 1 \\ 0 & -2 & 0 & -2 & 0 & -2 & 0 & -2 \\ 0 & 0 & -2 & -2 & 0 & 0 & -2 & -2 \\ 0 & 0 & 0 & 4 & 0 & 0 & 0 & 4 \\ 0 & 0 & 0 & 0 & -2 & -2 & -2 & -2 \\ 0 & 0 & 0 & 0 & 0 & 4 & 0 & 4 \\ 0 & 0 & 0 & 0 & 0 & 0 & 4 & 4 \\ 0 & 0 & 0 & 0 & 0 & 0 & 0 & -8 \end{bmatrix}$$

Eq. 18 may therefore be finally written as:

$$\mathbf{B}] = \frac{1}{2^n} [\mathbf{Trb}]\, \mathbf{R}] \qquad (20)$$

Again, the particular (simple) case of

$$b_0 = \frac{1}{2^n} \left[\sum_{j=0}^{2^n - 1} r_j \right] \qquad (21)$$

may be observed.

The inverse of [**Trb**] will provide the conversion matrix from **B**] to **R**], ie

$$\mathbf{R}] = [\mathbf{Trb}]^{-1}\, \mathbf{B}],$$
$$= [\mathbf{Tbr}]\, \mathbf{B}]$$

where it readily follows from Eq (19) that:

$$Tbr^n = \begin{bmatrix} 2Tbr^{n-1} & Tbr^{n-1} \\ 0 & -Tbr^{n-1} \end{bmatrix}$$

$$Tbr^0 = +1 \tag{22}$$

For n = 3 this gives:

$$Tbr = \begin{bmatrix} 8 & 4 & 4 & 2 & 4 & 2 & 2 & 1 \\ 0 & -4 & 0 & -2 & 0 & -2 & 0 & -1 \\ 0 & 0 & -4 & -2 & 0 & 0 & -2 & -1 \\ 0 & 0 & 0 & 2 & 0 & 0 & 0 & 1 \\ 0 & 0 & 0 & 0 & -4 & -2 & -2 & -1 \\ 0 & 0 & 0 & 0 & 0 & 2 & 0 & 1 \\ 0 & 0 & 0 & 0 & 0 & 0 & 2 & 1 \\ 0 & 0 & 0 & 0 & 0 & 0 & 0 & -1 \end{bmatrix}$$

C. The r_j to c_j Coefficient Relationships.

From Eqs. 9, 10 and 15, the Reed-Muller positive canonic c_j coefficients are related to the r_j spectral coefficients by:

$$\begin{aligned} C] &= [T^n]_{\mod 2} X] \\ &= [T^n]_{\mod 2} \left[\frac{1}{2^n} [Hd] R \right] \end{aligned} \tag{23}$$

where $[T^n]$ is the modulo 2 transform given in Eq. 10.

The modulo 2 (Exclusive-OR) relationships imposed by $[\mathbf{T^n}]$ imply that all odd numbers are expressed as +1 and all even numbers expressed as 0. Hence it is permissible to combine $[\mathbf{T^n}]$ $[\mathbf{Hd}]$ into a single conversion matrix $[\mathbf{Trc}]$ giving:

$$\mathbf{C]} = \frac{1}{2^n} [\mathbf{Trc}] \mathbf{R]} \tag{24}$$

where the final vector product summations are written as 1 or 0 depending upon odd or even summation values, respectively. From Eqs. 10 and 11 the conversion matrix $[\mathbf{Trc}]$ is given by:

$$\mathbf{Trc}^n = \begin{bmatrix} \mathbf{Trc}^{n-1} & \mathbf{Trc}^{n-1} \\ 2\mathbf{Trc}^{n-1} & 0 \end{bmatrix}_{\mathrm{mod}2}$$

$$\mathbf{Trc}^0 = +1. \tag{25}$$

For n = 3 this gives:

$$\mathbf{Trc} = \begin{bmatrix} 1 & 1 & 1 & 1 & 1 & 1 & 1 & 1 \\ 2 & 0 & 2 & 0 & 2 & 0 & 2 & 0 \\ 2 & 2 & 0 & 0 & 2 & 2 & 0 & 0 \\ 4 & 0 & 0 & 0 & 4 & 0 & 0 & 0 \\ 2 & 2 & 2 & 2 & 0 & 0 & 0 & 0 \\ 4 & 0 & 4 & 0 & 0 & 0 & 0 & 0 \\ 4 & 4 & 0 & 0 & 0 & 0 & 0 & 0 \\ 8 & 0 & 0 & 0 & 0 & 0 & 0 & 0 \end{bmatrix}_{\mathrm{mod}2}$$

The inverse of [**Trc**] will provide the conversion matrix [**Tcr**] from C] to R], ie

$$R] = [Trc]^{-1} C]$$
$$= [Tcr] C]$$

where it readily follows from Eq. 25 that:

$$Trc^n = \begin{bmatrix} 0 & Tcr^{n-1} \\ 2Tcr^{n-1} & -Tcr^{n-1} \end{bmatrix}$$

$$Tcr^0 = +1. \tag{26}$$

For n = 3 this gives:

$$Tcr = \begin{bmatrix} 0 & 0 & 0 & 0 & 0 & 0 & 0 & 1 \\ 0 & 0 & 0 & 0 & 0 & 0 & 2 & -1 \\ 0 & 0 & 0 & 0 & 0 & 2 & 0 & -1 \\ 0 & 0 & 0 & 0 & 4 & -2 & -2 & 1 \\ 0 & 0 & 0 & 2 & 0 & 0 & 0 & -1 \\ 0 & 0 & 4 & -2 & 0 & 0 & -2 & 1 \\ 0 & 4 & 0 & -2 & 0 & -2 & 0 & 1 \\ 8 & -4 & -4 & 2 & -4 & 2 & 2 & -1 \end{bmatrix}$$

D. Relationships Between the Hadamard and other Spectral Coefficients.

It has previously been noted that the spectral coefficients produced by the Hadamard transform and those produced by the Rademacher-Walsh and other row variants are identical, but generated in a dissimilar order (21), (22).

Other orthogonal transforms which are not row variants of the Hadamard may be found, the most prominent being the Haar transform. The relationship between the Hadamard and the Haar coefficients may be found in the literature (21).

However the majority of these alternative transform matrices contain zeros in certain of their row functions, and hence the resulting spectral coefficients are not defined by the whole function vector X]. Since they do not possess the global characteristics of all the Hadamard (and variant) spectral coefficients, they have not to date been actively considered for test signature purposes. We shall, therefore, have no cause to refer to them herewith. They may find specialist uses, however, should a particular "window" on the function output be required rather than a global response (22).

V. TEST SIGNATURES

The interpretation of the meaning of the spectral coefficients as correlation coefficients has been widely published. This interpretation is more readily appreciated when the $\{0,1\}$ function vector X] is recoded into the alternative $\{+1, -1\}$ vector Y], see Table 2. It is then apparent that each resulting coefficient value is numerically equal to the number of agreements between the given function and a matrix row function minus the number of disagreements. However the linear relationship between the two encodings (see Eq. 13) means that either X] or Y] produces a set of unique correlation coefficients between the given function vector and the function which the $\{+1, -1\}$ coding of each row of the transform represents.

The same considerations can be applied to all other coefficient vectors which represent a given function vector X]. All may therefore be regarded as correlation values between the network output f(X) and some further function of the primary inputs, the mathematical relationships between the actual values being as developed in Section 4 above. Hence if the numerical

Counting Techniques 107

values of coefficients are used as test signatures, all are mathematical variants on the basic test topology shown earlier in Fig. 2.

The commonality between many previously-disclosed test signature proposals may, therefore, be regarded as a search for the simplest test vector generator which provides a unique test signature for a given f(X) under all fault conditions, see Fig. 3. At one extreme we have the situation where f(X) is correlated with itself, see Fig. 3(a); clearly this is not a practical proposition except in exceptional circumstances, since the test vector generator is then a duplicate of, and therefore the same complexity as, the network under test. At the other extreme we have the situation where f(X) is correlated with logic 1 (or 0), see Fig. 3(b), which is the syndrome count. The remaining two examples shown in Fig. 3 represent the use of higher-order spectral coefficients as test signatures.

Table 2.

Illustration of the correlation interpretation of resulting coefficient values, sixth row of n = 3 [**Hd**] matrix shown operating on function $f(X) = x_1x_2 + x_1\bar{x}_3 + \bar{x}_1x_3$. Note the this row of [**Hd**] logically represents $x_1\bar{x}_3 + \bar{x}_1x_3$.

$$\begin{bmatrix} 1 -1\ 1 -1 -1\ 1 -1\ 1 \end{bmatrix} \begin{bmatrix} 0 \\ 1 \\ 0 \\ 1 \\ 1 \\ 0 \\ 1 \\ 1 \end{bmatrix} = -3 \qquad \begin{bmatrix} 1 -1\ 1 -1 -1\ 1 -1\ 1 \end{bmatrix} \begin{bmatrix} 1 \\ -1 \\ 1 \\ -1 \\ -1 \\ 1 \\ -1 \\ -1 \end{bmatrix} = -6$$

(a) [**Hd**] X   (b) [**Hd**] Y

VI. CONCLUSIONS

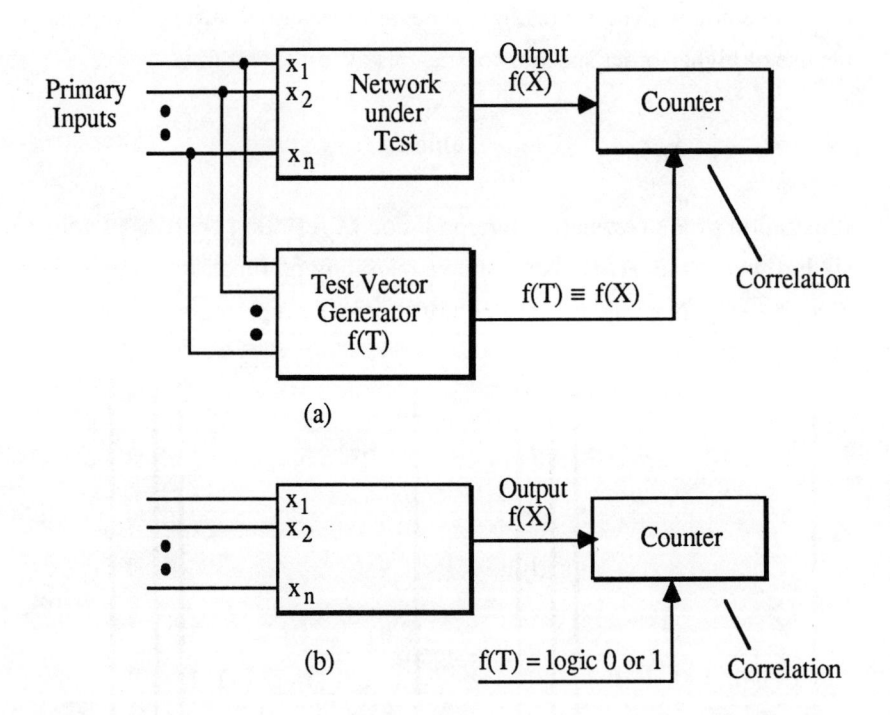

The commonality between many previously-disclosed procedures thus may be regarded as a search for the simplest test vector generator(s) which provide unique signatures for f(X) under all fault conditions. At present no one procedure has clear-cut advantages over others, and further research is necessary in order to determine the true merits and advantages of each. As with many engineering problems, there may not be one clear winner, and a repertoire of methods may evolve with guidance to show when each is most applicable. It is, however, hoped that ongoing and future research in this area will appreciate the commonality across all variants of these developments.

As a concluding thought for possible further consideration and evaluation, let us make the following observations:

(i) the simplest possible variant of Fig. 3 is the syndrome signature, which merely counts the number of logic 1s (or 0s) in the output f(X) when under test

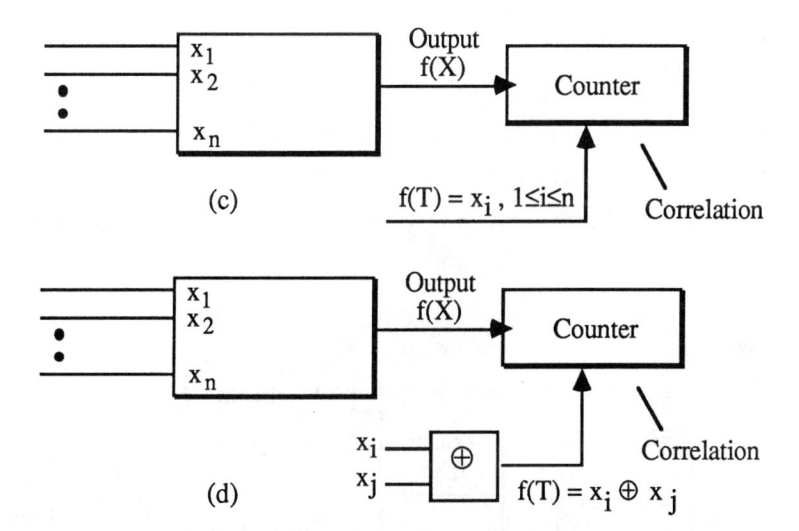

Fig. 3 Test signatures based upon the correlation between the output f(X) and a test vector generator f(T) over the exhaustive test cycle of 2^n.
  (a) The ultimate situation of f(X) compared against a copy of itself.
  (b) f(X) compared against logic 0 or logic 1, = syndrome testing.
  (c) f(X) compared against a single primary input x_i, = spectral coefficient r_i (s_i).
  (d) f(X) compared against the Exclusive-OR of two primary inputs, = spectral coefficient r_{ij}.

(ii) none of the counting signatures described above requires the exhaustive input test set to be applied in any pre-determined sequence,

and

(iii) no count is made of the number of transitions from 0 to 1 (or vice versa) during the exhaustive input test set.

Thus, as the simple syndrome signature is very desirable, but is not unique for many functions under fault conditions, possibly a combination of syndrome and transition counting may be a test method for any f(X) under all fault conditions. This is suggested in Fig. 4. The hardware circuitry for the two "count" boxes may possibly be merged to provide a compact silicon area on-chip circuit. Such a concept was originally suggested by Fujiwara and Kinoshita (28) but little further work seems to have been reported in published literature. Further research in this area would be profitable.

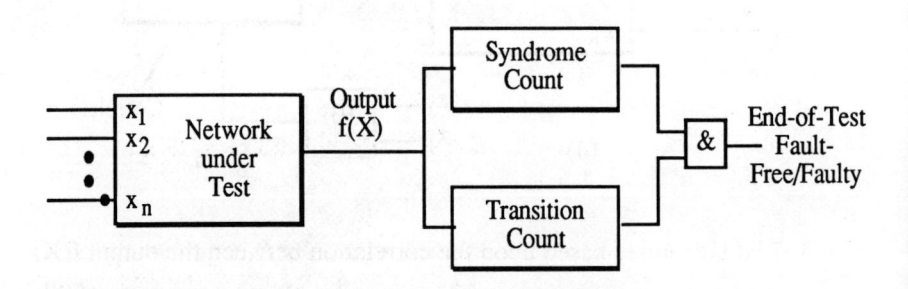

Fig. 4. A concept of combined syndrome and transition counting test signature.

References

1. Wadsack, R.L., "Fault modelling and logic simulation of CMOS and MOS integrated circuits", BSTJ, 57, p. 1449 - 1474, (1978).

2. Bryant, R.E., "A switch-level model and simulator for MOS digital systems", IEEE Trans. Comput, C.33, pp. 160 - 177, (1984).

3. Williams, T.W., and Parker, K.P., "Design for testability - a survey", PROC. IEEE, 71, pp. 98 - 112, (1983).

4. Breuer, M.A., and Friedman, A.D., Diagnosis and Reliable Design of Digital Systems, Computer Science Press, CA, (1976).

5. Siewiorek, D.P., and Swarz, R.S., The Theory and Practice of Reliable System Design, Digital Press, MA, (1982).

6. Muehldorf, E.I., and Savkar, A.D., "LSI logic testing - an overview", IEEE Trans. Comput., C.30, pp. 1 - 17, (1981).

7. Bennetts, R.G., Design of Testable Logic Circuits, Addison-Wesley, London, (1984).

8. Lala, P.K., Fault Tolerant and Fault Testable Hardware Design, Prentice-Hall, London, (1984).

9. Hayes, J.P., "Fault modelling", IEEE Design and Test, 2, pp. 88 - 95, (1985).

10. Upadhyaya, S.J., and Saluja, K.K., "Signature Techniques in Fault Detection", in Spectral Techniques and Fault Detection, Ed. M.G. Karpovsky, Academic Press, NY, (1985).

11. Akers, S.B., "On the use of linear sums in exhuastive testing", Proc. IEEE Design Automation Conf., pp. 148 - 153, (1985).

12. Savir, J., "Syndrome-testable design of combinational circuits", IEEE Trans. Comput., C.29, pp. 442 - 451, (1980).

13. Markowsky, G. "Syndrome-testability can be achieved by circuit modification", IEEE Trans. Comput., C.30, pp. 604 - 606, (1981).

14. Savir, J., "Syndrome testing of "Syndrome-untestable" combinational circuits", IEEE Trans. Comput., C.30, pp. 606 - 609, (1981).

15. Hayes, J.P., "Transition count testing of combinatorial logic networks", IEEE Trans. Comput., C.25, pp. 613 - 620, (1976).

16. Susskind, A.K., "Testing by verifying Walsh coefficients", IEEE Trans. Comput., C.32, pp. 198 - 201, (1983).

17. Miller, D.M., and Muzio, J.C., "Spectral fault signature for internally unate combinational networks", IEEE Trans. Comput., C.32, pp. 1058 - 1062, (1983).

18. Miller, D.M., and Muzio, J.C., "Spectral fault signatures for single stuck-at faults in combinational networks", IEEE Trans. Comput., C.33, pp. 745 - 769, (1984).

19. Serra, M., and Muzio, J.C., "Testing programmable logic arrays by sum of syndromes", (submitted to IEEE Trans. Comput.).

20. Heidtmann, K.D., "Arithmetic spectrum applied to stuck-at fault detection for combinational networks", (submitted to IEEE Trans. Comput.).

21. Beauchamp, K.G., Application of Walsh and Related Functions, Academic Press, London. (1984).

22. Hurst, S.L., Miller, D.M., and Muzio, J.C., Spectral Techniques in Digital Logic, Academic Press, London, (1985).

23. Miller, D.M., and Muzio, J.C., "Spectral techniques for fault detection in combinational logic", in Spectral Techniques and Fault Detection, Ed. M.G. Karpovsky, Academic Press, NY, (1985).

24. Davio, M., Deschamps, J.P., and Thayse, A., Discrete and Switching Functions, McGraw-Hill, London and New York, (1978).

25. Wu, X., Chen, X., and Hurst, S.L., "Mapping of Reed-Muller coefficients and the minimisation of Exclusive-OR switching functions; Proc. IEE, E.129, pp. 15 - 20, (1982).

26. Calingaert, P., "Switching functions: canonical forms based upon commutative and associative binary operations", AIEE Trans., 80, pp. 808 - 814, (1961).

27. Hayes, J.P., "Check sum methods for test data compression", J Design Automation and Fault Tolerant Computing, 1, pp. 3 - 17, (1976).

28. Fujiwara, H. and Kinoshita, K., "Testing logic circuits with compressed data", Proc. 8th IEEE Int. Symp. Fault-Tolerant Computing, pp.108 - 113, (1978).

29. Parker, K.P., and McClusky, E.J., "Probabilistic treatment of general combination networks", IEEE Trans. Comput., C.24, pp. 668 - 670, (1975).

30. Kumar, S.K., and Breuer, M.A., "Probabilistic aspects of Boolean switching functions via a new transform", J Association Comput. Machinery, 28, pp. 502 - 520, (1981).

31. Lui, P.K., and Muzio, J.C., "Spectral signature testing of multiple stuck-at-faults in irredundant combinational networks" (submitted to IEEE Trans. Comput.).

32. Eris, E. and Muzio, J.C., "Spectral testing of circuit realisations based upon linearization" (submitted to IEEE Trans. Comput.).

33. Smith, J.E., "Measures of the effectiveness of fault signature analysis", IEEE Trans, Comput., C.29, pp. 510 - 514, (1980).

34. Beeslich, P.W., "Spectral processing of switching functions", in Spectral Techniques and Fault Detection, Ed. M.G. Karpovsky, Academic Press, NY., (1985).

35. Hurst, S.L., "The use of linearisation in the input and output compaction testing of digital networks", Proc. 2nd International Workshop on Spectral Techniques, Montreal, Canada, (1986).

THE SPECTRAL TESTING OF MULTIPLE-OUTPUT CIRCUITS

M. Serra and J.C. Muzio

*Department of Computer Science
University of Victoria
Victoria, B.C., Canada*

ABSTRACT

In a multiple-output circuit a testing method is normally applied separately to each individual output. Here we present results for testing multiple-output circuits using a single test based on a syndrome-sum or a spectral sum. Syndrome counting is a simple and effective data compression technique and a weighted-syndrome-sum coding over all outputs has been proposed, although it can lead to cancellation problems.

We show that criteria exist for detecting untestable lines, making a single compressed signature a feasible testing scheme, which leads to an effective design for testability. Of special interest are highly structured multiple-output networks like PLA's, where it is shown that the weighted sum of syndromes of all the outputs very effectively covers all single stuck-at-faults, bridging faults, and cross-point faults.

I. INTRODUCTION.

Classical testing of a circuit consists of generating an appropriate set of input vectors, called the test set, such that for every covered fault in the circuit at least one of those input vectors provokes an incorrect output. Much computational work is involved in constructing the test set for each circuit, there are large storage requirements both for the input stimulations and for the output responses, and the test set is not applicable to any built-in test.

Data compression has been the other viable path for testing, where the output vector for a circuit is coded in the form of a "signature". The signature is an attribute or a set of attributes such that at least one of them changes in the presence of a fault. If the signature for the fault free circuit is the same as the signature for the faulty circuit then "aliasing" is said to occur. The aim is for the signature generator to be as small as possible, with minimal aliasing giving high fault coverage being proved by either a deterministic or probabilistic method, and simple supporting hardware which can be built on-chip or on-board. Examples include linear feedback shift register testing (LFSR) (17), transition count testing (5) and syndrome or spectral testing (8, 11). We are only concerned here with the latter approach. It has been shown recently (2, 12) that, for the worst case, syndrome testing has statistically the same probability of fault coverage as the LFSR.

The advantages of data compression for testing are clear in speed, efficiency and ease of use, compared to the more cumbersome test set. The disadvantages of data compression are primarily the lack of criteria as to the testability of particular faults using a given scheme. When syndrome testing was first introduced (11), theoretical conditions were given to verify when the syndrome test was effective in covering a particular fault, besides relying on simulation. The evaluation of the syndrome testing conditions was done in the boolean domain which resulted in complex manipulations. For LFSR testing, criteria do not yet exist.

The full conditions for syndrome testability for a single output were developed and proved in the spectral domain in (8). They result in straightforward calculations of a fast Rademacher-Walsh transform to determine the syndrome testability of all input lines, while the conditions for internal lines are somewhat more complex. These criteria have to be evaluated only once at design time, and their computational complexity remains far below that required for simulations for test sets. The testing itself is very rapid and the hardware requirement is minimal, being directly amenable to built-in testing.

General circuits usually have many outputs, and the testing scheme of choice is normally applied to each individual output, resulting in many signatures to be verified. Our approach is to introduce further space compression among the outputs in order to generate a single signature for the entire network. A similar approach has been proposed using parallel LFSR's with some increase in masking (4).

A weighted syndrome sum (WSS) was introduced in (3) but its applicability remained uncertain since no conditions were given to detect the occurrence of masking. Its feasibility has been proven by the authors in (13, 15) for particular circuits like PLA's. In this paper we present general results for the conditions for testability for any network using the Weighted Spectral Sum (WSPS), of which the WSS is one case. The conditions are developed in the spectral domain similarly to (8). However the actual testing is simply done by counters as for syndrome tests, involving no spectral transforms at testing time. The space compression so introduced gives more latitude for design for testability changes in the network, since different permutations of weights can be applied to enhance fault coverage.

In the next section we introduce some background material needed for the conditions for testability to be found in sections III and IV, where some examples are also described. In section V we discuss the special case of Programmable Logic Arrays, where the effectiveness of the method is optimal. In section VI the approach is given to design for testability, while in section VII the implementation of the tester is discussed.

II. DEFINITIONS AND BACKGROUND.

Let f(X) be a Boolean function with $X=\{x_1,x_2,...,x_n\}$ a set of Boolean variables. Let $\mathbf{Z}$ be the column vector (truth table vector) for f(X), ordered such that x_n is the high order variable.

Definition 1: The Rademacher-Walsh spectrum of f(X) is given by $\mathbf{R} = \mathbf{T}^n \cdot \mathbf{Z}$ where $\mathbf{T}^n$ is the nth order transform of size $2^n \times 2^n$. The simplest form is the recursive definition of $\mathbf{T}$ in Hadamard ordering as:

$$\mathbf{T}^0 = [1] \quad \mathbf{T}^1 = \begin{bmatrix} 1 & 1 \\ 1 & -1 \end{bmatrix} \quad \text{and} \quad \mathbf{T}^n = \begin{bmatrix} \mathbf{T}^{n-1} & \mathbf{T}^{n-1} \\ \mathbf{T}^{n-1} & -\mathbf{T}^{n-1} \end{bmatrix}$$

The spectrum of a Boolean function is unique. The column vector $\mathbf{R}$ consists of the spectral coefficients. We denote by $\mathbf{R}_f$ the spectrum of f. Details about spectra and their evaluation can be found in (6). We illustrate here only an example for the case of n=3.

Consider the example function: $f(x_1,x_2,x_3) = x_3 x_2 + \bar{x}_3 x_1$ whose truth vector is represented by $[\,0\,1\,0\,1\,0\,0\,1\,1\,]^t$. By premultiplying the truth vector by the appropriate size transform $\mathbf{T}^3$, the spectrum of f is obtained. It is written here in its transpose form with the proper labelling of the coefficients.

As shown below, the spectral coefficients are distinguished using subscripts identifying the variables directly correlated to them. The coefficients $r_1, r_2, ..., r_n$ are called first order coefficients and they measure the correlation of the function f(X) with the variables $x_1, x_2, ..., x_n$

respectively. All the other coefficients measure the correlation with particular Exclusive-OR functions. For example, r_{12} compares $f(X)$ with $x_1 \oplus x_2$. A general coefficient is denoted by r_α.

$$\begin{bmatrix} 1 & 1 & 1 & 1 & 1 & 1 & 1 & 1 \\ 1 & -1 & 1 & -1 & 1 & -1 & 1 & -1 \\ 1 & 1 & -1 & -1 & 1 & 1 & -1 & -1 \\ 1 & -1 & -1 & 1 & 1 & -1 & -1 & 1 \\ 1 & 1 & 1 & 1 & -1 & -1 & -1 & -1 \\ 1 & -1 & 1 & -1 & -1 & 1 & -1 & 1 \\ 1 & 1 & -1 & -1 & -1 & -1 & 1 & 1 \\ 1 & -1 & -1 & 1 & -1 & 1 & 1 & -1 \end{bmatrix} \cdot \begin{bmatrix} 0 \\ 1 \\ 0 \\ 1 \\ 0 \\ 0 \\ 1 \\ 1 \end{bmatrix} =$$

$$= \begin{bmatrix} r_0 & r_1 & r_2 & r_{12} & r_3 & r_{13} & r_{23} & r_{123} \\ 4 & -2 & -2 & 0 & 0 & -2 & 2 & 0 \end{bmatrix}^t$$

Definition 2: The weight of a function, $W(f)$, is defined as the number of ones in the corresponding Z. Since the first row of T^n is all ones, $r_0 = W(f)$. The weight itself is easily evaluated at test time and compared through the use of a register.

Definition 3: The syndrome of a function, $S(f)$, is defined as the normalized weight $S(f) = \dfrac{W(f)}{2^n}$. We follow other authors and use the term "syndrome testable" although the practical evaluation used is just the weight, both at the design stage and at testing time.

The spectral coefficients are used as an analytical tool for the syndrome testability analysis and have to be computed only at the design stage. The complexity of calculating a fast transform in Hadamard order is $n \cdot 2^n$ (6).

In order to explore the application of spectral testing some results about decompositions of spectra are necessary from (6). The subfunctions of $f(x_1,...,x_n)$ are defined by $f_u(x_1,...,x_m) = f(x_1,...,x_m,u_1,...,u_{n-m})$

where $(u_1, u_2,...,u_{n-m})$ is the binary expansion of u, with u_1 the least significant bit. For example, for $f(x_1, x_2)$ where n=2, and for m=1, $f_0(x_1) = f(x_1, 0)$ and $f_1(x_1) = f(x_1, 1)$. The spectra of the subfunctions are defined analogously: $\mathbf{R}_u = \mathbf{T}^m \cdot \mathbf{Z}_u$, where $\mathbf{Z}_u$ is the column vector for f_u. Initially we give the result for the relationship between the spectra of the subfunctions when m=n-1, that is for the decomposition for the nth input variable.

Consider halving the vector $\mathbf{Z}$ of f(X) as $\mathbf{Z} = \begin{bmatrix} \mathbf{Z}^0 \\ \mathbf{Z}^1 \end{bmatrix}$ where $\mathbf{Z}^0$ and $\mathbf{Z}^1$ are the vectors for $f_0(x_1,...,x_{n-1}) = f(x_1,...,x_{n-1}, 0)$ and $f_1(x_1,...,x_{n-1}) = f(x_1,...,x_{n-1}, 1)$ respectively, so that

$$\mathbf{R}_0 = \mathbf{T}^{n-1} \cdot \mathbf{Z}^0 \text{ and } \mathbf{R}_1 = \mathbf{T}^{n-1} \cdot \mathbf{Z}^1. \tag{1}$$

The spectrum $\mathbf{R}_f$ can be partitioned as in:

$$\mathbf{R}_f = \begin{bmatrix} \mathbf{R}^0 \\ \mathbf{R}^1 \end{bmatrix} = \begin{bmatrix} \mathbf{T}^{n-1} & \mathbf{T}^{n-1} \\ \mathbf{T}^{n-1} & -\mathbf{T}^{n-1} \end{bmatrix} \begin{bmatrix} \mathbf{Z}^0 \\ \mathbf{Z}^1 \end{bmatrix}$$

that is:

$$\mathbf{R}^0 = \mathbf{T}^{n-1}(\mathbf{Z}^0 + \mathbf{Z}^1) = \mathbf{R}_0 + \mathbf{R}_1 \tag{2a}$$

$$\mathbf{R}^1 = \mathbf{T}^{n-1}(\mathbf{Z}^0 - \mathbf{Z}^1) = \mathbf{R}_0 - \mathbf{R}_1 \tag{2b}$$

which are equivalent to:

$$\mathbf{R}_0 = \frac{1}{2}(\mathbf{R}^0 + \mathbf{R}^1) \tag{2c}$$

$$\mathbf{R}_1 = \frac{1}{2}(\mathbf{R}^0 - \mathbf{R}^1). \tag{2d}$$

This process can be done in general with m < n-1, such that $\mathbf{R}$ and $\mathbf{Z}$ are partitioned into 2^{n-m} subvectors, as in:

$$R = \begin{bmatrix} R^0 \\ R^1 \\ . \\ . \\ R^\beta \end{bmatrix} \quad Z = \begin{bmatrix} Z^0 \\ Z^1 \\ . \\ . \\ Z^\beta \end{bmatrix} \quad \text{where } \beta = 2^{n-m}-1$$

It can be shown that (6)

$$[R^0 R^1 \cdots R^\beta] = [R_0 R_1 \cdots R_\beta] T^{n-m} \tag{3a}$$

and $\quad [R_0 R_1 \cdots R_\beta] = \dfrac{1}{2^{n-m}} [R^0 R^1 \cdots R^\beta] T^{n-m}. \tag{3b}$

Since we are dealing with networks of many outputs, we must identify the vectors and spectra for the individual outputs. We choose square brackets, so that $[Z]_j$ and $[R]_j$ indicate respectively the truth vector and the spectrum for the jth output in the network.

Definition 4: Define the Weighted Spectral Sum (WSPS) for a network of m outputs as:

$$K = \sum_{j=1}^{m} w_j T^n [Z]_j = \sum_{j=1}^{m} w_j [R]_j$$

with individual entries in K denoted by k_α. The weights w_j are integers, and so are the coefficients in the sum of the spectra of the m functions in the network.

Lemma 5: The Rademacher-Walsh transform is distributive for the WSPS:

$$K = \sum_{j=1}^{m} w_j T^n [Z]_j = w_1 T^n [Z]_1 + w_2 T^n [Z]_2 + \cdots + w_m T^n [Z]_m =$$

$$= T^n (w_1 [Z]_1 + w_2 [Z]_2 + \cdots + w_m [Z]_m) = T^n \cdot Y$$

Proof: From (7), the Rademacher-Walsh transform is an instance of a discrete Fourier transform defined in general over a field. The WSPS is a linear combination of matrix multiplications, where the operations upon the boolean vectors $[Z]_j$ yield integer results. The weights act as scalar multipliers. Δ

This lemma means that in order to find the WSPS for a network of m functions, we do not need to calculate m spectra and add them up. Instead we calculate a superfunction $Y = w_1[Z]_1 + \cdots + w_m[Z]_m$ over the relevant finite field and apply the transform only once to Y.

Note that the first coefficient of the weighted spectrum, k_0, is the weighted sum of the first coefficients of the individual spectra for each function $[f]_j$, and it is thus the weighted sum of the weights of all functions in the network. The weighted syndrome sum is formally the normalized value of this coefficient:

$$WSS = \frac{1}{2^n}k_0 = \frac{1}{2^n}\sum_{j=1}^{m} w_j [W(F)]_j$$

For all practical purposes we use the unnormalized sum of the syndromes. By formally considering all outputs as functions of n inputs, all syndrome properties apply to the calculations without normalization. Thus we consider all $[Z]_1, \ldots, [Z]_m$ to be vectors of 2^n entries, even if, in fact, not all corresponding functions may depend on all input variables.

When a fault occurs on a line in the circuit, we denote by $[f]_j^*$ the function produced by the faulty circuit at output j. The syndromes or some spectral coefficients at the outputs which depend on the faulty line change if the fault is testable at those outputs. When using some coefficients from the WSPS as a signature for a circuit there may be faults which are considered untestable for two reasons: (a) they do not change the signature at any of the outputs, (b) there is cancellation induced by the summation process inherent in the WSPS. In the former case the problem is the same as that for separately testing all the single outputs. If the signatures alias at all outputs for a particular fault, then the design must be changed. One straightforward solution is to introduce an extra output in the network. This adds minimal or no complexity to a signature from a space compressed WSPS. More discussion on this appears in a later section. For case (b) the aliasing is a new problem introduced by the space compression itself: one solution is to select different weights to avoid the cancellation effect. In the implementation this can be achieved by changing the layout with no extra hardware necessary. These aspects are discussed in a later section.

Definition 6: Let K^* be the weighted spectral sum for a network in the presence of a fault that causes the circuit outputs to be $[f]_1^* \cdots [f]_m^*$ respectively. Let k_α be a coefficient from K chosen as a signature for the good circuit. Then the faulty circuit is testable by this signature if $k_\alpha^* \neq k_\alpha$.

Example 1: Consider the circuit of Figure 1, with 4 inputs and 2 outputs. This circuit is the example used in (3): there exhaustive simulation had to be performed in order to show it to be WSS untestable for their chosen weights. We show below a set of weights to make it testable. The functions are $[f]_1 = (x_1x_4+x_3)x_2$ and $[f]_2 = \bar{x}_1x_2+x_1\bar{x}_2\bar{x}_3$. The weights chosen in (3) are $w_1=2$ and $w_2=3$ giving $K = T^4 (2[Z]_1 + 3[Z]_2)$. Consider at first only the syndromes, represented by k_0 in the WSPS. The value for the fault-free circuit is $k_0 = 2 \cdot 5 + 3 \cdot 6 = 28$. Let input line x_3 be stuck-at-0. Then $[f]_1^* = x_1x_2x_4$ and $[f]_2^* = \bar{x}_1x_2+x_1\bar{x}_2$, and $k_0^* = 2 \cdot 2 + 3 \cdot 8 = 28$. Since $k_0 = k_0^*$ the fault is k_0 (WSS) untestable. However, if the weights for the WSPS are changed to $w_1=1$ and $w_2=2$, then $k_0=17$ while $k_0^*=18$ and the fault becomes k_0 (WSS) testable. In general a signature from the WSPS can contain any arbitrary coefficient from K. In this example the first order coefficient k_1 from the WSPS covers the fault $x_3/0$ under both sets of weights. For the first set ($w_1=2$, $w_2=3$), then $(k_1=4) \neq (k_1^*=-4)$; and for the second set ($w_1=1$, $w_2=2$), then $(k_1=3) \neq (k_1^*=-2)$.

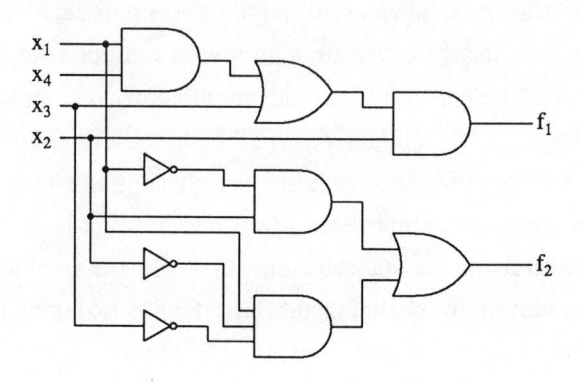

Fig. 1: Example Circuit.

In the following sections we present conditions to verify at design time whether a particular fault is testable by some coefficient from the WSPS for a given set of weights, thereby avoiding simulating all the faults as was proposed in (3). It is useful to consider separately input lines and internal lines.

III. INPUT LINES TESTABILITY CONDITIONS.

We consider initially the case for a single stuck-at-fault on an input line for a network N with n inputs and m outputs.

Theorem 7: Let $k_{\alpha i}$ be a coefficient from a WSPS with some set of weights, where $k_{\alpha i} = \sum_{j=1}^{m} w_j [r_{\alpha i}]_j$. The input stuck-at-faults $x_i/0$ and $x_i/1$ are k_α and $k_{\alpha i}$ testable iff $k_{\alpha i} \neq 0$.

Proof: Without loss of generality we consider the case of x_n, the high order input. For each output j we have $[f]_j = \bar{x}_n[f_0]_j + x_n[f_1]_j$. Let $[R_0]_j$ and $[R_1]_j$ be the spectra for $[f_0]_j$ and $[f_1]_j$ respectively. Let $Y = \sum_{j=1}^{m} w_j[f]_j$, and K be its spectrum. Also let $Y_0 = \sum_{j=1}^{m} w_j[f_0]_j$ and $Y_1 = \sum_{j=1}^{m} w_j[f_1]_j$, with corresponding spectra K_0 and K_1.

Consider $x_n/0$: then $[f]_j^* = 1 \cdot [f_0]_j + 0 \cdot [f_1]_j$, i.e. $[f]_j^* = [f_0]_j$. But $[f_0]_j$ is a function of n-1 inputs: for all n input combinations of $[f]_j$, $[f]_j^*$ repeats the output vector $[f_0]_j$ for both $x_n=0$ and $x_n=1$. Thus $[R_0]_j^* = [R_1]_j^* = [R_0]_j$, and $K_0^* = K_1^* = K_0$. In the previous section the properties of the decomposition of spectra were stated; in particular (6) $K^0 = K_0 + K_1$. It follows that $K^{0^*} = K_0^* + K_1^* = 2K_0$. Another decomposition property is that $2K_0 = K^0 + K^1$: by substituting above we have $K^{0^*} = K^0 + K^1$. Looking at the coefficients $k_\alpha \in K^0$, $k_{\alpha n} \in K^1$, $n \notin \alpha$, from the appropriate rows, we have $k_\alpha^* = k_\alpha + k_{\alpha n}$. For k_α testability, we require that $k_\alpha^* \neq k_\alpha$, and the result follows that $x_n/0$ is testable by k_α iff $k_{\alpha n} \neq 0$.

Note that $k_{\alpha n}^*$ is evaluated for a set of functions all of which are vacuous in x_n, since $x_n/0$. Thus $k_{\alpha n}^* = 0$, and the fault is $k_{\alpha n}$ testable iff $k_{\alpha n} \neq k_{\alpha n}^*$, i.e. $k_{\alpha n} \neq 0$.

Similarly for $x_n/1$. Δ

It is useful to state explicitly the conditions for k_0 testability.

Corollary 8: Input stuck-at-faults $x_i/0$ and $x_i/1$ are k_0 testable iff $k_i \neq 0$.

Proof: k_0 testability is a special case of Theorem 7 when $\alpha = \phi$. Δ

Example 2: Consider again the circuit of Fig. 1.

$[f]_1 = x_2(x_1x_4+x_3)$ and $[f]_2 = \bar{x}_1x_2+x_1\bar{x}_2\bar{x}_3$

so that $[\mathbf{Z}]_1 = [0\ 0\ 0\ 0\ 0\ 0\ 1\ 1\ 0\ 0\ 0\ 1\ 0\ 0\ 1\ 1]^t$

and $[\mathbf{Z}]_2 = [0\ 1\ 1\ 0\ 0\ 0\ 1\ 0\ 0\ 1\ 1\ 0\ 0\ 0\ 1\ 0]^t$.

Using weights $w_1=2$ and $w_2=3$ we have $\mathbf{K} = \mathbf{T}^n(2[\mathbf{Z}]_1 + 3[\mathbf{Z}]_2)$ as follows:

k_0 k_1 k_2 k_{12} k_3 k_{13} k_{23} k_{123}
28 4 −16 −16 0 −8 12 −4

k_4 k_{14} k_{24} k_{124} k_{34} k_{134} k_{234} k_{1234}
−2 2 2 −2 −2 −2 2 −2

Indeed $k_3=0$ and $x_3/0$ is WSS untestable (k_0 untestable) as seen in example 1. However $k_{13}\neq 0$ and $x_3/0$ is k_1 testable. If we change the weights to $w_1=1$ and $w_2=2$ then we have

k_0 k_1 k_2 k_{12} k_3 k_{13} k_{23} k_{123}
17 3 −9 −11 1 −5 1 −3

k_4 k_{14} k_{24} k_{124} k_{34} k_{134} k_{234} k_{1234}
−1 1 1 −1 −1 1 1 −1

Note that $k_3=1$ and $x_3/0$ is k_0 (WSS) testable; moreover since all $k_{\alpha i}\neq 0$ all input stuck-at-faults are both k_0 (WSS) testable and k_α testable for the appropriate α.

The choice of weights is thus fundamental to the WSPS testability of a fault and to the corresponding design for testability. A discussion about this is presented in section VI.

IV. INTERNAL LINES TESTABILITY CONDITIONS.

As expected, the testability of internal lines is much more complex than for input lines. Consider an internal line g: the appropriate functional model is given by $[f(X)]_j = h_j(X, x_{n+1})$ where $x_{n+1} = g(X)$, for each j, $1 \le j \le m$. This is illustrated as a diagram in Figure 2. Each h_j is formally considered as a function of n+1 inputs and measures the boolean dependence of an output $[f]_j$ on the line g. Let $[\hat{R}]_j$ be the spectrum of h_j, $[V]_j$ its vector representation, and let $\hat{K} = \sum_{j=1}^{m} w_j [\hat{R}]_j$. As before, we do not need to calculate the $[\hat{R}]_j$ separately: we construct the superfunction $H = w_1[V]_1 + \cdots + w_m[V]_m$ where "+" is over the integer field, and only apply the transform once to obtain $\hat{K} = T^{n+1} H$. The check for the WSPS testability of an internal line requires the construction of the boolean h_j functions from the circuit description. In some cases only a subset of the

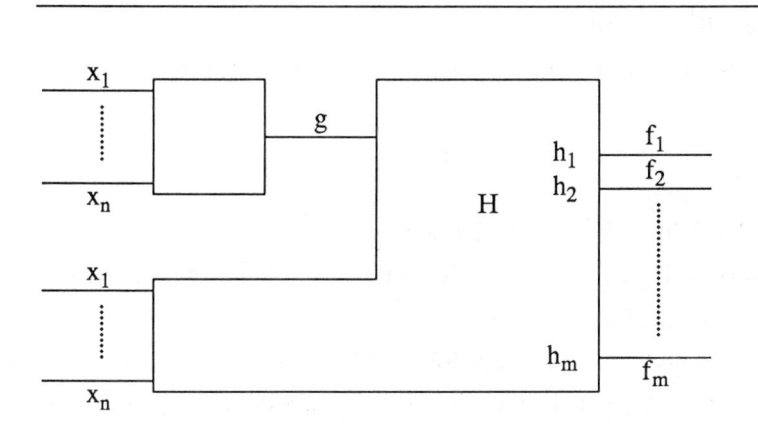

Fig. 2: Internal Line Model.

m outputs may depend on a line g and only that subset of corresponding h_j functions need be explicitly evaluated. The remaining h_j functions are equal to the corresponding $[f]_j$, although they must formally be considered as functions of n+1 variables.

Theorem 9: The fault g/0 is k_α testable iff

$$k_\alpha \neq \frac{1}{2}(\hat{k}_\alpha + \hat{k}_{\alpha n+1})$$

The fault g/1 is k_α testable iff

$$k_\alpha \neq \frac{1}{2}(\hat{k}_\alpha - \hat{k}_{\alpha n+1})$$

Proof: Let $[V_0]_j$ be the vector for $h_j(X,0)$ and $[V_1]_j$ for $h_j(X,1)$.
Let $H_0 = \sum_{j=1}^{m} w_j[V_0]_j$ and $H_1 = \sum_{j=1}^{m} w_j[V_1]_j$, with the corresponding spectra being $\hat{K}_0$ and $\hat{K}_1$.

We know from the properties of decomposition of spectra that $\hat{R}_0 = \frac{1}{2}(\hat{R}^0 + \hat{R}^1)$ for each j (see (6)), so that $\hat{K}_0 = \frac{1}{2}(\hat{K}^0 + \hat{K}^1)$ and conversely $\hat{K}^0 = \hat{K}_0 + \hat{K}_1$.

Note that $\hat{K}^0$ and $\hat{K}^1$ are the partitions of the spectrum $\hat{K}$, i.e

$$\hat{K} = \begin{bmatrix} \hat{K}^0 \\ \hat{K}^1 \end{bmatrix}$$

When g/0, then $[f]_j^* = h_j(X,0)$ and thus:
$H_0^* = H_1^* = H_0$ which implies $\hat{K}_0^* = \hat{K}_1^* = \hat{K}_0$. (The "*" always indicate the faulty functions or spectra).

Since the high order input, g, for each h_j is stuck-at-0, each faulty spectrum for h_j is equal to the spectrum of $h_j(X,0)$ repeated twice, that is $[R^*]_j = [\hat{R}_0]_j$. Similarly for the sum of spectra, $K^* = \hat{K}_0$.

We can apply the rules for decomposition of spectra and by substitution we obtain: $K^* = \hat{K}_0 = \frac{1}{2}(\hat{K}^0 + \hat{K}^1)$.

Looking at the appropriate row we have

$$k_\alpha^* = \frac{1}{2}(\hat{k}_\alpha + \hat{k}_{\alpha n+1})$$

and we need $k_\alpha^* \neq k_\alpha$ for k_α testability, i.e. $k_\alpha \neq \frac{1}{2}(\hat{k}_\alpha + \hat{k}_{\alpha n+1})$. Similarly for g/1. Δ

Again it is useful to state explicitly the condition for k_0 (WSS) testability.

Corollary 10: The fault g/0 is k_0 (WSS) testable iff

$$k_0 \neq \frac{1}{2}(\hat{k}_0 + \hat{k}_{n+1})$$

The fault g/1 is k_0 (WSS) testable iff

$$k_0 \neq \frac{1}{2}(\hat{k}_0 - \hat{k}_{n+1})$$

Proof: k_0 testability is a special case of Theorem 9 when $\alpha=\phi$. Δ

Not all internal lines need to be explicitly checked: more discussion about this aspect can be found in section VI. In general, if all weights chosen for the WSPS are positive, then all internally unate lines are automatically WSPS testable. This is particularly relevant for all two-level circuits and for special layouts like PLA's.

Example 3: An interesting example comes from examining a binary to two's complement conversion circuit as presented in (3). In that paper the analysis was done only for WSS testability. To do so, a set of arbitrary weights was chosen, and exhaustive simulation was applied to a subset of 44 of the total 152 stuck-at-faults. The simulation results were used to verify whether a fault was testable or not, since the appropriate theoretical analysis was not available. Here we can constructively check a choice for the weights without simulation. The circuit in Figure 3 has 8 inputs and 6 outputs.

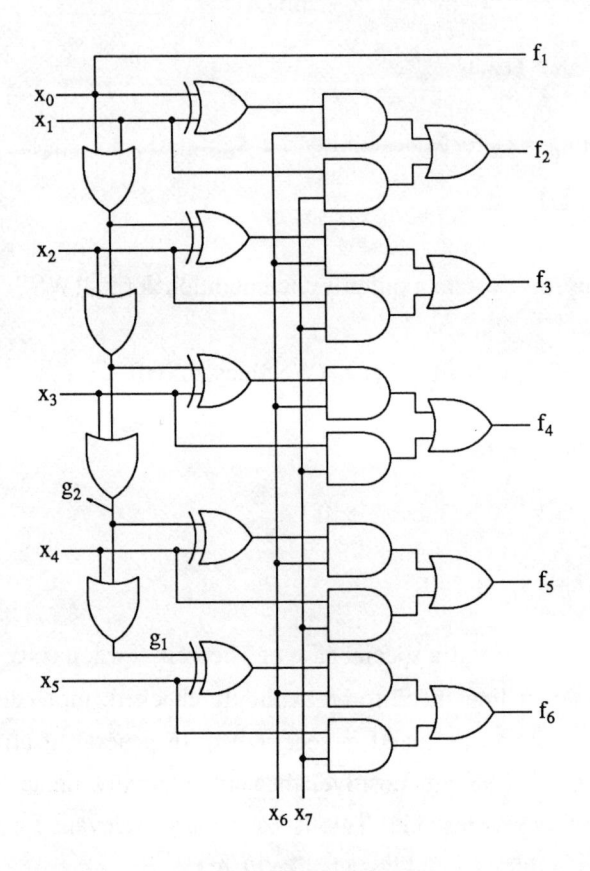

Fig. 3: Example Circuit.

$f_1 = x_0$

$f_2 = x_0\bar{x}_1 x_6 + \bar{x}_0 x_1 x_6 + x_1 x_7$

$f_3 = x_6(x_2 \oplus (x_0+x_1))+x_2x_7$

$f_4 = x_6(x_3 \oplus (x_0+x_1+x_2))+x_3x_7$

$f_5 = x_6(x_4 \oplus (x_0+x_1+x_2+x_3))+x_4x_7$

$f_6 = x_6(x_5 \oplus (x_0+x_1+x_2+x_3+x_4))+x_5x_7$

Assume all weights $w_j=1$, $1 \le j \le 6$. Here $k_0 = 585$ and the first order coefficients of **K**, the sum of the spectra, are:

k_1	k_2	k_3	k_4	k_5	k_6	k_7	k_8
-153	-393	29	25	-23	41	-71	-167

Since all $k_i \ne 0$, $1 \le i \le 8$, all input stuck-at-faults are k_0 testable.

Consider the internal line g_1 which affects only output f_6. Then $H = h_1+h_2+h_3+h_4+h_5+h_6$ where $h_6 = x_6(g_1 \oplus x_5) + x_5x_7$, and h_j, $1 \le j \le 5$, are equal to $[f]_j$. All of them must be considered as functions of 9 variables. Note however the relationships for the weights which are: $W([h]_j)=2W([f]_j)$, $1 \le j \le 5$.

After evaluating $\hat{K}$, the spectrum for H, we find that $\hat{k}_0 = 1204$ and $\hat{k}_9 = 0$.

For the k_0 testability of g_1 we need to apply the condition of Corollary 10: $k_0 \ne \frac{1}{2}(\hat{k}_0+\hat{k}_9)$. Here $k_0 = 585 \ne \frac{1}{2}(1204 \pm 0)$, and $g_1/0$ and $g_1/1$ are both k_0 (WSS) testable.

Looking at another coefficient, we can check if faults on g_1 are testable by k_1. Here $k_1 = -153$, $\hat{k}_1 = -244$, and $\hat{k}_{19} = 0$. Since we need $k_1 \ne \frac{1}{2}(\hat{k}_1+\hat{k}_{19})$ and $k_1 \ne \frac{1}{2}(\hat{k}_1-\hat{k}_{19})$, both stuck faults on g_1 are also k_1 testable.

Consider another line segment g_2; for stuck-at-faults on g_2 we only need to evaluate explicitly two functions, namely h_5 and h_6.

$$h_5 = (g_2 \oplus x_4)x_6 + x_4x_7$$

$$h_6 = ((g_2+x_4) \oplus x_5)x_6 + x_5x_7$$

The two truth vectors are added up together with the vectors for $[f]_1,\ldots,[f]_4$, which, considered as 9 variable functions, are used for $h_1, \ldots, h_4$. Then the spectral transform is applied to the result, obtaining the values

$$\hat{k}_0 = 1256 \text{ and } \hat{k}_{n+1} = \hat{k}_9 = -48$$

so that g_2 is verified to be k_0 testable after computing the condition of Corollary 10. We can also check if it is testable by some other k_α, say k_6. Here $k_6 = 41$, $\hat{k}_6 = 40$, and $\hat{k}_{69} = -48$. Hence, since

$$41 \neq \frac{1}{2}(40 \pm -48)$$

both stuck faults on g_2 are k_6 testable.

V. PROGRAMMABLE LOGIC ARRAYS AND SPECIAL CIRCUITS.

The design of circuits based on Programmable Logic Arrays (PLA's) has become of increasing interest and their testing, especially built-in, an important aspect. A PLA with n inputs and m outputs can realize m functions each of n variables as two level irredundant circuits in sum-of-products form. The internal organization of a PLA is divided into two parts and it is shown in Figure 4: an AND array which realizes selected product terms, and the OR array which connects together the product terms needed to form the output function. The WSPS as defined earlier proves to be very powerful in this case, and we show that the k_0 coefficient alone is sufficient to cover all single faults. A summary of the results is presented here as in (13): more details and proofs are in

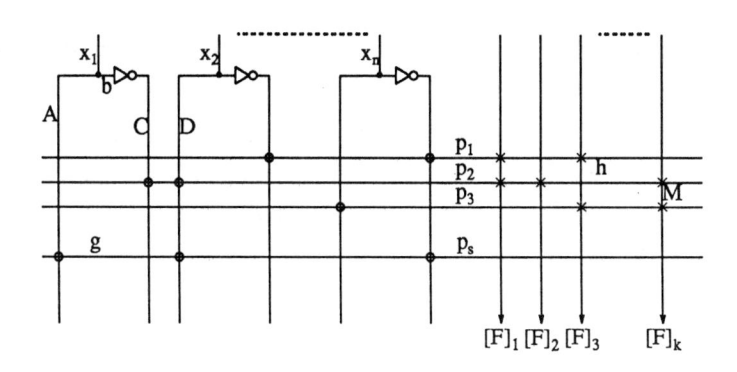

Fig. 4: AND-OR PLA.

(15).

There are three types of single faults to be considered for a PLA (see Figure 4):

(1) Stuck-at-faults: one line segment stuck at either logical 1 or 0.

(2) Bridging faults: two adjacent lines are shorted together assuming the value of the logical AND or OR of their individual signals (determined by the hardware implementation).

(3) Cross-point faults: the loss of contact of a connection or its spurious presence.

In a few situations the presence of a fault may have no functional effect on any of the outputs of the PLA. These undetectable trivial faults are ignored in the discussion below.

Any change in the syndrome of a function implies that the total number of minterms realized has increased or decreased. Clearly an increase or decrease in the syndrome of one output causes a corresponding increase or decrease in the weighted sum of the syndromes, if all weights are positive. We initially exclude from our analysis input faults which are covered separately.

We observe that a PLA implementation is based on sum-of-products expressions: all terms in such expressions are products of single literals and thus all are homogeneous in their nature, that is they contain no factors or subexpressions. The literals for each product are implemented as separate bit lines in their true or complemented form, which are then fed into the AND array. This means that any single fault on an internal line affects one or more product terms, but all are affected in a unidirectional manner. By examining the fault effect of each possible fault we can demonstrate that all the affected syndromes either increase or decrease, there being no possibility of cancellation.

Suppose a literal is stuck at constant 1 or is disconnected from the AND line. This is a growth fault (following the terminology of Smith (16)), and all affected product terms loose a literal leading to an expansion in the number of minterms they cover. If a literal is incorrectly connected to an AND line, all affected product terms gain that literal and there is a reduction in the number of covered minterms (shrinkage). If a segment in the AND array is stuck at constant 0, an entire product term disappears (disappearance): the syndromes of all affected outputs decrease.

Considering the different fault types separately, Table 1 gives a summary of the possible stuck-at-fault classes with the corresponding increase/decrease on the size of the product terms marked beside them. Each fault type has an example from Fig. 4 beside it for clarity, excluding input faults.

The examination of each stuck-fault type demonstrates that a single fault cannot cause a literal to be both removed and added in the sum-of-products expressions, due to the structure of the PLA itself. The

TABLE 1

Fault Type	Increase	Decrease	e.g.
1) s-a-0 between literals		X	g/0
2) s-a-1 between literals	X		g/1
3) s-a-0 sec. input		X	A/0
4) s-a-1 sec. input	X		A/1
5) s-a-0 product line		X	$p_1/0$
6) s-a-1 product line	X		$p_1/1$
7) s-a-0 between products		X	M/0
8) s-a-1 between products	X		M/1

corresponding fault effect on the syndromes is unidirectional, and the WSS, the k_0 coefficient, increases or decreases accordingly, assuming all weights have the same sign.

Cross-point faults, that is, erroneous connections between two crossing lines of the arrays, can be seen to be of four types, and for each we list its effect on the syndrome. For more details see (15).

(1) A spurious AND contact between a product line p_s and a secondary input line x_i where one should not exist causes either a *shrinkage* or a *disappearance* fault (e.g. AND at p_1 and C). A shrinkage fault occurs when the number of minterms covered by p_s is halved: e.g. from $p_s = x_1\bar{x}_2x_4$ to $p_s^* = x_1\bar{x}_2x_3x_4$. If p_s is already connected to $\bar{x}_i$ and the cross-point fault also connects it to x_i, then $p_s = 0$. In both cases, the weights of all functions involving p_s decrease. The same applies for a spurious contact between p_s and $\bar{x}_i$.

(2) The loss of an AND contact, where one should exist, causes the p_s involved to become independent of the variable x_i (e.g. no contact at p_2 and C). The product term doubles in size along x_i: e.g. from $p_s = x_1\bar{x}_2 x_4$ to $p_s^* = \bar{x}_2 x_4$. The weights for all functions sharing p_s increase.

(3) If a spurious OR contact appears between a product term p_s and an output line segment for $[F]_s$, an extra product term is realized for that function (e.g. contact at p_2 and $[F]_3$). This causes an increase in the number of minterms covered. If the new product term is redundant, the fault is undetectable as no change appears in the output truth vector.

(4) A loss of OR contact where one should exist causes the loss of a product term p_s for that function $[F]_s$, thus reducing the number of minterms that are covered (e.g. no contact at p_1 and $[F]_1$).

Finally, bridging faults are discussed using the same approach. We consider in detail the four classes for AND bridging, leaving the bridge between two input lines to be considered separately below. Similar results for OR bridging are easily deduced.

(1) Consider a bridge between x_i and $\bar{x}_i$. Some output $[F]_s$ that depends on x_i can be written in the form: $[F]_s = Ax_i + B\bar{x}_i + C$ for some A,B,C independent of x_i. The effect of the AND bridging fault substitutes $(x_i\bar{x}_i)$ for every x_i and $\bar{x}_i$. The expression becomes: $[F]_s^* = A(x_i\bar{x}_i) + B(\bar{x}_i x_i) + C = C$. Clearly the weight of $[F]_s$ decreases, as it will for any other output dependent on x_i. k_0, the WSS, decreases accordingly.

(2) Consider a bridging fault between $\bar{x}_i$ and x_j. This fault causes both $\bar{x}_i$ and x_j to be replaced by $(\bar{x}_i x_j)$. An output function dependent on $\bar{x}_i$ or x_j or both can be written in the form: $[F]_s = Ax_i x_j + Bx_i \bar{x}_j + C\bar{x}_i x_j + D\bar{x}_i \bar{x}_j + E$ for some A,B,C,D,E

independent of x_i, x_j. This changes to $[F]_s^* = Bx_i\bar{x}_j + C\bar{x}_i x_j + E$, with a corresponding decrease in the weight of $[F]_s$. Clearly the same is true for all outputs dependent on $\bar{x}_i$, or x_j, and k_0 decreases.

(3) Consider a bridging fault between two product term lines p_i and p_j. It causes both of them to be replaced by $(p_i p_j)$. An output $[F]_s = p_1 + ... + p_i + ...$ becomes $[F]_s^* = p_1 + ... + p_i p_j + ...$ with a resulting decrease in its weight.

(4) A bridging fault between two output lines $[F]_s$ and $[F]_t$ in the OR array causes the two outputs to become $[F]_s^* = [F]_t^* = [F]_s[F]_t$ with a decrease in the weights of both functions.

We are left with the two special cases concerning input faults, namely stuck-at input faults and input bridging faults. If an input line is stuck-at-0 or stuck-at-1, both secondary input lines receive the effect of the fault, where on the negative line the fault is also inverted. This is analogous to two separate faults: for stuck-at, for example, one of class 3 and one of class 4, as in Table 1, where we note that one fault increases the individual syndromes while the other decreases them: a cancellation effect is possible.

It is important to determine when and if this cancellation may occur in order to maintain the WSS testability for the PLA. This is quite simply solved by applying the conditions for input testability of Corollary 8. For each primary input x_i, the corresponding coefficient k_i from the WSPS must be nonzero for the fault to be testable by k_0 (the WSS). Thus only the n first-order coefficients need be examined. If any input fault is untestable, the addition of one extra product line in the PLA solves the problem: we discuss this in detail in section VI.

In the more problematic case of an AND bridging fault between two primary inputs, x_i, x_j, the effect is carried to four secondary input lines with the following multiple substitutions taking place: each of x_i, x_j is replaced by $x_i x_j$ and each of $\bar{x}_i, \bar{x}_j$ by $\bar{x}_i + \bar{x}_j$. If we write $[f]_s$ as

$[f]_s = Ax_ix_j+Bx_i\bar{x}_j+C\bar{x}_ix_j+D\bar{x}_i\bar{x}_j+E$, then under the fault it becomes $[F]_s^* = Ax_ix_j+D\bar{x}_i+D\bar{x}_j+E$, and a cancellation potential exists, since there is no guarantee that the number of minterms uniformly decreases or increases. To isolate this possibility a condition exists analogous to Theorem 7 which can be checked at the design stage; here we only state the special case for WSS testability. More detail about bridging faults and conditions for their testability can be found in (9, 10, 15).

Theorem 11: An AND bridging fault between two primary inputs x_i, x_j is k_0-testable if and only if $k_i + k_j + 2k_{ij} \neq 0$.

Proof: We apply the Shannon decomposition in order to write each $[F]_s$ in terms of the subfunctions $[f_{00}]_s$, $[f_{01}]_s$, $[f_{10}]_s$ and $[f_{11}]_s$ independent of x_i, x_j.

We have $[F]_s = \bar{x}_i\bar{x}_j[f_{00}]_s + \bar{x}_ix_j[f_{01}]_s + x_i\bar{x}_j[f_{10}]_s + x_ix_j[f_{11}]_s$. This means that the output vector for $[F]_s$ is partitioned into the four quarters according to the two-variable Shannon decomposition.

Let

$m_{0s}=W([f_{00}]_s)$, $m_{1s}=W([f_{01}]_s)$, $m_{2s}=W([f_{10}]_s)$, $m_{3s}=W([f_{11}]_s)$, $(1 \leq s \leq m)$, m = number of outputs and W(f) indicates the weight of a function f. Thus the m_{is}, $0 \leq i \leq 3$, $1 \leq s \leq m$ are the weights of the subfunctions. By definition

$$k_0 = \sum_{s=1}^{m} w_s[r_0]_s = \sum_{s=1}^{m} w_s(m_{0s}+m_{1s}+m_{2s}+m_{3s}),$$

$$k_i = \sum_{s=1}^{m} w_s[r_i]_s = \sum_{s=1}^{m} w_s(m_{0s}+m_{1s}-m_{2s}-m_{3s}),$$

$$k_j = \sum_{s=1}^{m} w_s[r_j]_s = \sum_{s=1}^{m} w_s(m_{0s}-m_{1s}+m_{2s}-m_{3s}), \text{ and}$$

$$k_{ij} = \sum_{s=1}^{m} w_s[r_{ij}]_s = \sum_{s=1}^{m} w_s(m_{0s}-m_{1s}-m_{2s}+m_{3s}).$$

When an AND bridging fault occurs between two primary inputs, the subfunctions $[f_{01}]_s$ and $[f_{10}]_s$, assume the same values as $[f_{00}]_s$, for $1 \leq s \leq m$. The respective weights become

$W([f_{01}]_s^*) = W([f_{10}]_s^*) = W([f_{00}]_s) = m_{0s}$.

Thus $k_0^* = \sum_{s=1}^{k} w_s(3m_{0s} + m_{3s})$.

It follows that $k_0^* \neq k_0$ if and only if $k_i + k_j + 2k_{ij} \neq 0$ by considering the appropriate linear combination of the above expressions for k_i, k_j and k_{ij}.
Δ

The result for OR bridging follows similarly.

Theorem 11A: An OR bridging fault between two primary inputs x_i, x_j is k_0-testable if and only if $k_i + k_j - 2k_{ij} \neq 0$. Δ

We have shown that k_0 (the WSS) is a complete test for all single faults in a PLA. The conditions for testability of input faults lead to a direct and simple hardware modification, discussed below. A similar analysis can be applied to other special circuits, namely Read Only Memories (ROM's) and Decoders, but they are not discussed here. For a complete set of results see (14).

VI. DESIGN FOR TESTABILITY.

The conditions presented so far are very useful to decide, in a deterministic fashion and without simulation, whether the syndrome or some other signature covers a particular fault or a set of faults. At that point in the design, it might be felt that some other testing method is more appropriate if and when the circuit presents too many problem lines. On the other hand, hardware changes may be very simple when the number of lines involved is small.

The WSPS adds one tool to the design: a line may be made testable by a simple change/permutation of the weights, with no extra hardware. One of the properties of the R-W spectrum of a particular function is that coefficients are either all odd or all even. This property carries through to the WSPS. Hence if at least one function in the network has odd weight, we can assign arbitrarily a weight of 1 to that output and an even weight to all other outputs for the WSPS calculations. Consequently all coefficients in **K** become odd and are thus non-zero. This not only ensures that all stuck-at-faults on input lines x_i are k_0 (WSS) testable (by Theorem 7), but also means that they are covered by any k_α for which $i \in \alpha$. For a set of sample weights like the powers of 2 ($w_j = 2^{j-1}$, $1 \le j \le m$), it is clear that $k_\alpha = [r_\alpha]_1 + 2A$ for some A. That is, k_α is even or odd according as $[r_\alpha]_1$ is even or odd, this being determined by whether $[f]_1$ is of even or odd weight. However we are free to choose the ordering of the output functions and consequently we can ensure that k_α is odd (and so nonzero) by selecting any of the functions of odd weight and routing it to be assigned $w_1=1$ in the WSPS.

Moreover, after an initial choice of weights for the implementation of the counter is made, and without any knowledge of the functions, the resulting WSPS may be odd. Note that this can be ascertained by checking whether only one coefficient is odd or even, as all the others follow. In this case no further effort is required. The main problem occurs when a fault on x_i is not testable at any output: for example an input fault is never syndrome testable if all functions in the network have even weight, and moreover all $[r_i]_j = 0$, $1 \le j \le m$,

In the case of PLA's, the solution is very simple: after calculating k_1 and possibly finding it equal to 0, the simplest hardware change is the addition of one extra product line and output line. The extra function realized in the PLA will have one product term covering only one minterm, $y_1 y_2 \cdots y_n$ where each $y_i \in \{x_i, \bar{x}_i\}$, $1 \le i \le n$. This extra output is assigned $w_1=1$ in the WSS and $[f]_1,...,[f]_m$ are assigned even weights. In this way the value of k_1 becomes +1 or -1 and (since we now have one of the k+1 functions of odd weight) all other k_i's are nonzero. In some cases it may be more appropriate to use an extra output line connected to some already existing product lines, to achieve a similar effect.

This solution also solves the input bridging fault testability if Theorem 11 indicates that a cancellation exists for a bridge between x_i, x_j (15). For this latter case the extra minterm is best implemented as $\bar{x}_1\bar{x}_2\cdots\bar{x}_n$. In this way the product term will change the weight of the output truth vectors by +1 only for the portion when both x_i and x_j are zero, making the WSPS odd. This hardware addition for the design of testable PLA's appears to be the simplest proposed so far in the literature. In other methods the tester itself may consume less area than the counter required for k_0 testing, but the changes in the design to achieve full testability may be more cumbersome (1) or difficult to determine (4).

A similar solution can be applied to general circuits: by embedding an extra output of odd weight, the input faults testability is guaranteed for the space compressed WSPS. Considering general circuits and internal lines, the design for testability can be seen from two angles: what to do once a line is considered untestable, and for which lines should the process of checking the testability conditions be carried out.

In the former case many paths are open to the designer: at times the introduction of very few additional gates solves the problem completely. Alternatively, the addition of an extra output may be very attractive: it would serve as an observability point for the untestable lines. The testing is still done by a single space compressed signature which includes the extra output. This whole subject is under ongoing research and will be dealt with separately in the future.

However, the other view to design for testability is to detect *a priori* a minimum set of internal lines for which doubts exist as to their testability. It is easily seen that all internally unate lines are WSPS testable. An extension of the marking algorithm (8) can be applied and it is currently under revision. This means that in time proportionally linear to the number of lines in the circuit, a labelling is given, denoting whether a line segment should be considered a "candidate" for untestability. Only lines marked in this fashion need to be checked by the criteria given earlier, after fault equivalence is done. This considerably reduces the amount of overhead work at the design for testability stage: for other methods such a tool is not always available. In fact for both LFSR and test sets, only fault collapsing can be used to reduce the number of faults to be examined.

VII. IMPLEMENTATION AND COMPLEXITY.

Some of the attractions of data compression testing are the small hardware requirement and the possibility of built-in testing. WSS testing always needs registers with fewer bit elements than testing individual syndromes or applying LFSR's to each output. Even if we build one parallel LFSR to compress all outputs together in one signature, the WSS still gives better storage for smaller circuits (as for the circuit of Fig. 3), since it needs as few as m+n bits. The actual number of elements needed for the accumulating register and the comparator depend on the weights chosen, so a direct comparison with LFSR's is not discussed here.

The complexity of pre-calculating the criteria for WSPS testability has to be taken into account. Simulation is needed for the fault-free circuit and more work is also necessary to obtain the truth vector of the h_j functions relative to the particular internal lines which need checking. Not all internal lines need the full evaluation of the conditions and not all outputs depend on all internal lines thus reducing the work. The amount of computation is much lower than for other schemes: only in the worst possible case, when all lines have to be checked, does it approach that for LFSR simulations or for test sets generation.

The choice of weights is crucial. The problem is to find a) a set of weights that is easily implemented (for example successive powers of 2 obtained directly by hardwiring the outputs into the successive bit positions in the register); b) to keep the value of the weights small so that the final result for the WSPS coefficient to be stored is as small as possible; c) to choose a set of weights such that all faults are detectable. The last item is clearly the most complex. While it can be proven that such a set exists if a fault is testable by at least one output (3), such a construction is quite impractical. For a given set of distinct weights, there are m! permutations that can be tried, keeping in mind that if at least one weight is equal to 1 for an output of odd syndrome, then all input faults are testable.

A design procedure should start by choosing an appropriate set of weights according to the hardware and storage requirements and such that the WSPS is odd and all input faults are covered by k_0. A selection is made of which internal lines need checking: for those the criteria of section IV are evaluated. If for some line the conditions reveal the untestability by k_0, other paths are open: either permute two weights for two outputs connected to that line, or embed that line through extra hardware towards an extra output, or select a testable k_α. If the line is not syndrome testable at any output, then similar hardware modifications applicable to single output circuits can be used.

The whole subject of design and modifications deserves a complete treatment by itself, and more research is continuing. The additional tool of weight permutations by rerouting looks very promising.

VIII. CONCLUSIONS.

In this paper it is shown that the weighted spectral sum is valuable for the testing of multiple-output circuits. General criteria are given for the testability of both input and internal stuck faults by any coefficient from the weighted spectral sum. It is noted that all input faults are easily covered by any of the coefficients (and in particular the weighted syndrome sum) when the set of output functions includes at least one of odd weight. The more difficult problem arises when a fault is untestable by a particular coefficient at every one of the individual outputs. In this case the simplest solution may be the addition of an extra output (of odd weight) to the set of functions.

For special applications including PLA's and ROM's, it is shown that the fault effect of any single fault on an internal line, which is either a stuck fault, a bridging fault or a crosspoint fault is unidirectional. The unidirectional fault effect guarantees that the fault is testable by the weighted syndrome sum alone. In the two instances where possible cancellation can occur (input stuck faults and input bridging faults), special cases of the general criteria are applicable. When such faults are indeed not testable a simple hardware modification to the PLA is proposed.

The testing approach presented here appears to be both practical and useful. It gives an attractive way to design a built-in test, especially for PLA's since they can be difficult to test effectively with some other data compression techniques. Ongoing research is looking at keeping the complexity of the procedures as low as possible and deciding the best choice of weights for the weighted spectral sum. It is not yet clear whether the approach proposed here is more appropriate to certain circuit types than others.

REFERENCES

1. Agrawal, V.K. (June 1980). "Multiple fault detection in programmable logic arrays". *IEEE Transac. Computers.* **C-29, 6,** 518-522.
2. Aitken, R., Muzio, J.C., Ruskey, F. and Serra, M. (1986). "Aliasing probabilities of data compression techniques". *Proc. Tech. Workshop: New Directions for IC Testing.* pp. 3.1-3.17. Univ. Victoria, Victoria, B.C.

3. Barzilai, Z., Savir, J., Markowsky, G. and Smith, M. (December 1981). "The weighted syndrome sum approach to VLSI testing". *IEEE Transac. Computers.* **C-30, 12,** 996-1000.

4. Hassan, S.Z. and McCluskey, E.J. (June 1983). "Testing PLA's using multiple parallel signature analyzers". *Proc. Symposium on Fault Tolerant Computer Systems.* **13,** 422-425.

5. Hayes, J.P. (June 1976). "Transition count testing of combinational logic circuits". *IEEE Transac. Computers.* **C-25, 6,** 613-620.

6. Hurst, S.L., Miller, D.M. and Muzio, J.C. (1985). "Spectral Techniques in Digital Logic". Academic Press, London and New York.

7. Lechner, R.J. (1971). *In* "Recent Developments in Switching Theory" (Ed. A. Mukhopadhyay). Academic Press, London and New York.

8. Miller, D.M. and Muzio, J.C. (August 1984). "Spectral fault signatures for single stuck-at-faults in combinational networks". *IEEE Transac. Computers.* **C-33, 8,** 765-769.

9. Muzio, J.C. and Serra, M. (1985). "Built-in-testing and design for testability of stuck-at-neighbour faults". *Canadian Conference on VLSI.* pp. 229-233. Toronto, Ont.

10. Muzio, J.C. and Serra, M. (October 1986). "Spectral criteria for the detection of bridging faults". *2nd International Workshop on Spectral Techniques.* Montreal, P.Q.

11. Savir, J. (June 1980). "Syndrome testable design of combinational networks". *IEEE Transac. Computers.* **C-29, 6,** 442-451.

12. Savir, J. and McAnney, W.H. (November 1985). "On the masking probability with one's count and transition count". *IEEE Int. Conference on Computer-Aided Design.* pp. 111-113, Santa Clara, CA.

13. Serra, M. and Muzio, J.C. (April 1985). "Testing programmable logic arrays by sum of syndromes." *8th IEEE Workshop on Design for Testability.* Beaver Creek, CO.

14. Serra, M. (1987). "New methods for the compaction testing of multiple-output digital circuits." Ph.D. Dissertation, University of Victoria, Dept. of Computer Science.

15. Serra, M. and Muzio, J.C. (1987) "Testing programmable logic arrays by sum of syndromes." *IEEE Transac. Computers.* forthcoming.

16. Smith, J.E. (November 1979). "Detection of faults in programmable logic arrays." *IEEE Transac. Computers.* **C-28, 11,** 845-853.

17. Smith, J.E. (June 1980). "Measures of effectiveness of fault signature analysis." *IEEE Transac. Computers.* **C-29, 6,** 510-514.

SIGNATURES AND SYNDROMES ARE ALMOST ORTHOGONAL

John P. Robinson

Department of Electrical and Computer Engineering
University of Iowa
Iowa City, Iowa 52242

Nirmal R. Saxena

Hewlett Packard
MS44UX
19111 Prune Ridge Avenue
Cupertino, California 95014

I. INTRODUCTION

In order to answer the question, "Is the gizmo working correctly?" one usually checks a listing of possible environments paired with correct gizmo behavior. For the first machine constructed a complete audit of all cases might have to be conducted. Once the design has been validated, the testing of additional copies is usually easier. Only manufacturing defects and component failures need to be considered.

For logic circuits, there is often a wide range of defective behavior even in this later situation. An appropriate list of test cases paired with correct behavior might have hundreds or thousands of entries. Each test input needs to be applied, the circuit behavior observed, then the behavior compared to that expected. When the test instrumentation is external to the circuit under test, there may be a substantial data flow between the testor and testee. This situation is costly in both test equipment and time. Various techniques to ameliorate the test problem for VLSI have been proposed (1-4).

One overall stratregy is to locate on the VLSI chip a test pattern generator and to reduce the amount of behavior data prior to comparison. For example a feedback shift register could be cycled to generate all possible input combinations to a logic block. A simple data reduction on the output would be to count the number of cases where a logic one is observed. The reference data needed is then just the count total for a correctly operating circuit. A test of length 1024 (all combinations of 10 variables) would only need 10 bits to specify the correct count total.

Counting as a test data compression technique will obviously detect any malfunction where the logic function has either expanded or contracted. If both new ones are created and old ones are lost in a balanced manner, counting will give the same total as correct operation. Such a loss of information is possible when data reduction is used. Whether this error masking is important depends on how the attribute retained (e.g. count total) depends on the test set and possible failures.

In this chapter two test data reduction methods are considered, polynomial division (signature) and ones counting (syndrome). The simultaneous use of both of these approaches in parallel as shown in Figure 1 is the organization under investigation.

Signature compression is a linear operation and thus partitions the space of binary m-tuples into regions all having 2^{m-L} members, where L is the length of the signature register. We assume m is larger than L. This

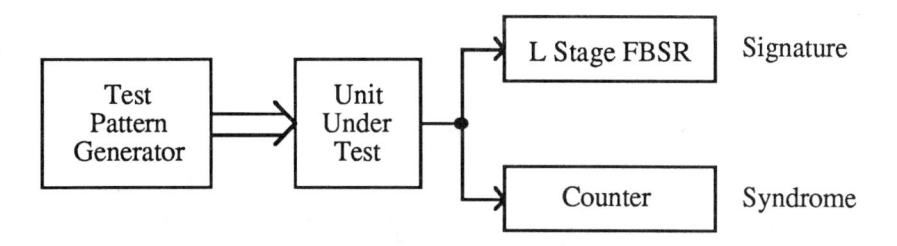

```
   ┌──────────┐     ┌────────┐  ┌─────────────┐
   │   Test   │     │  Unit  │─▶│ L Stage FBSR│  Signature
   │ Pattern  │────▶│ Under  │  └─────────────┘
   │ Generator│     │  Test  │  ┌─────────────┐
   └──────────┘     └────────┘─▶│   Counter   │  Syndrome
                                └─────────────┘
```

Fig. 1. Test Structure Under Investigation

signature partition is represented by the equal width bands in Figure 2a. All sequences in the same region have the same signature. For example if the compression polynomial is a primitive polynomial, then all those sequences in the zero signature zone are code words of a Hamming single error correcting code. All sequences with the same signature lie in a coset of the Hamming code.

Syndrome compression is a nonlinear operation and partitions the space of m-tuples into unequally sized regions. The number of sequences having the same syndrome value w is just the number of ways w thing can be chosen out of m or the binomial coefficient m choose w. This is represented in Figure 2b by the bands of varied width corresponding to different values for w. Each band represents those sequences which have the same syndrome.

When both signature and syndrome compression are used in parallel, certain misleading output data sequences can be confused with the correct output sequence. Such erroneous sequences or aliases would have the same number of ones, say w, as the valid sequence. A missed defective sequence would thus lie in the crosshatched area in Figure 2c.

What we will show in this chapter is that the syndrome region is almost equally divided up by the 2^L possible signatures. Hence, signature analysis will always reduce the error ambiguity of a counting based compression method. We first consider a special case for which an analytic solution is possible; then an empirical investigation is described.

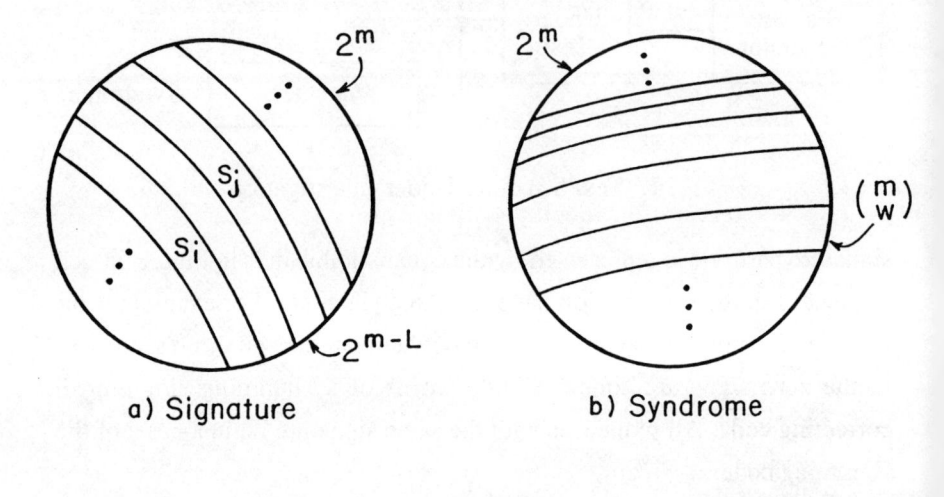

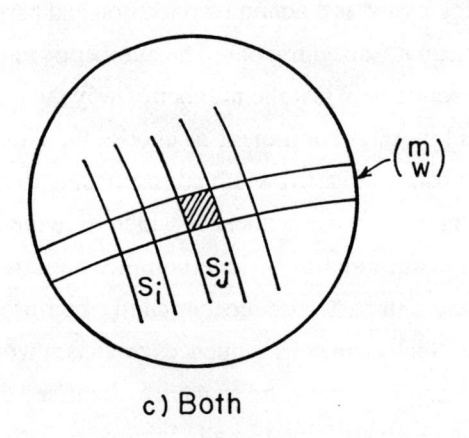

Fig. 2. Error Region Representation

II. MAXIMUM LENGTH CASE

Suppose that the length of the signature feedback shift-register is L and that the feedback polynomial is primitive. If we assume that the test length is $m = 2^L - 1$, then there is enough structure to compute the exact overlap between a signature and a syndrome. Specifically we will determine the number of binary sequences of length $2^L - 1$ which have the same syndrome count and the same signature when the L stage signature unit uses a primitive polynomial for feedback.

A sequence with a syndrome count of w has w ones in a span of m; thus there are exactly $m! / (m-w)! w!$ such sequences (6). These sequences can be divided into rotational equivalence classes. Two sequences A and B are equivalent if A can be obtained from B by rotating B one or more steps.

Definition: $A(x)$ is the polynomial corresponding to a sequence $A = a_0, a_1, \cdots, a_{m-1}$. The coefficient of x^i is a_i. Two sequences A and B are <u>rotationally equivalent</u> if $A(x) = x^t B(x)$ modulo x^m for some $t < m$.

Example 1: Let $A(x) = 1 + x^2 + x^5$ and $m = 7$, i.e. $A = 1010010$. If we rotate A two positions to the right we get $B = 1010100$. In polynomial form $B(x) = 1 + x^2 + x^4$. $B(x)$ and $A(x)$ are rotationally equivalent since $B(x) = x^2 A(x) \bmod x^7$ or $A(x) = x^5 B(x) \bmod x^7$.

Note that all the members of a rotational equivalence class have the same number of ones and will thus have the same syndrome. Let w denote the number of ones in a sequence. We next show that each member of an equivalence class has a distinct signature or all the signatures for the class are zero.

Example 2: The primitive polynomial $1 + x + x^3$ specifies a length 3 signature compressor. The $m = 2^L - 1 = 7$ bit sequence corresponding to p(x) and its 6 rotations are as follows:

$$1101000$$
$$0110100$$
$$0011010$$
$$0001101$$
$$1000110$$
$$0100011$$
$$1010001.$$

All of these weight 3 sequences compress to the 000 signature.

There are a total of 35 sequences of length 7 with 3 ones. These partition into 5 rotational equivalence classes each with 7 members. The following is a list of one representative form for each of these 5 classes.

$$1110000$$
$$1101000$$
$$1011000$$
$$1100100$$
$$1010100$$

Examining the first class above it is straightforward to establish the following correspondence between a member of the class and its signature.

sequence	signature
1110000	111
0111000	101
0011100	100
0001110	010
0000111	001
1000011	110
1100001	011

Notice that all 7 non-zero signatures appear on the above list once and only once.

With some additional computation it can be demonstrated that each of the last three equivalence classes also have distinct signatures within the class. Thus there are 7 sequences with 3 ones which have a zero signature and 4 sequences with 3 ones which have any other signature. For instance the 4 sequences

$$1110000$$
$$0101100$$
$$1001001$$
$$0010101$$

all have the signature 111.

For $(1 + x) p(x)$ the corresponding sequence is 1011100. This weight 4 sequence as well as all of its rotations compress to the 000 signature. The remaining 28 sequences with 4 ones also partition into groups of 4 sequences each with the same signature.

Example 2 can be extended to in the following theorem.

Theorem 1: Consider a feedback shift register where the feedback corresponds to a primitive polynomial of degree L. If this register is used to compute the syndrome of a sequence of length $2^L - 1$, then either

i) all the members of a rotational equivalence class have distinct nonzero signatures or,

ii) all the members of a rotational equivalence class have a zero signature.

Proof: Taking the last claim first, if a sequence has a zero signature then the sequence as a polynomial is a multiple of the feedback polynomial (9). Since the feedback polynomial is a primitive polynomial of degree L and m = $2^L - 1$, all such zero signature sequences are code words from a Hamming code in cyclic form. All cyclic shifts of a code word yield a code word; thus if any sequence has a zero signature, then all members of the equivalence class have a zero signature.

Note (as was demonstrated in Example 2) that each equivalence class is composed of members of the same weight but that several equivalence classes can have a zero signature. Next consider the case where some sequence has a nonzero signature. We assume that claim i) is false and show that a contradiction results. Suppose that some sequence A has a nonzero signature and that a rotation of t steps has the same signature and that the rotation is not A. Let S denote the nonzero signature. By definition the signature $S(x)$ is the remainder found when $A(x)$ is divided by $P(x)$ or

$$A(x) = P(x)Q_1(x) + S(x)$$

where $P(x)$ is the feedback polynomial, the degree of $S(x)$ is less than L, and $S(x)$ is not zero. The rotation of A by t steps has the same signature, $S(x)$. Then using the definition of a signature, we expand the rotation

$$x^t A(x) = P(x)Q_2(x) + S(x) \mod x^m.$$

Multiplying the first expression for $A(x)$ by x^t we get

$$x^t A(x) = x^t P(x)Q_1(x) + x^t S(x) \mod x^m.$$

Equating these last two expressions and then moving the terms involving $S(x)$ to the left hand side we get

$$(1+x^t)S(x) = P(x)[Q_2(x) + x^t Q_1(x)] \mod x^m.$$

Finally dividing by p(x) we obtain

$$\frac{(1+x^t)S(x)}{P(x)} = [Q_2(x) + x^t Q_1(x)] \quad \mod x^m$$

S(x) is not zero by hypothesis and has degree less than L. Therefore P(x) must divide $1 + x^t$ for the last equation to hold, where t is less than $2^L - 1$. By hypothesis P(x) is primitive hence the smallest n where P(x) divides $1+x^n$ is $n = 2^L - 1$, which leads to a contradiction.

Q.E.D.

The number of zero signature sequences can be found from the weight generating function for a Hamming code (9). The coefficient g_i of x^i is the number of code words of weight i. For binary codes

$$G(x) = \frac{1}{n+1} [(1+x)^n + n(1+x)^a (1-x)^{a+1}]$$

where $n = 2^L - 1$, $a = (n-1)/2$. Evaluating the first few terms we obtain

$g_0 = 1,$

$g_1 = g_2 = 0,$

$g_3 = \frac{n(n-1)}{3!},$

$g_4 = \frac{n(n-1)(n-3)}{4!},$

$g_5 = \frac{n(n-1)(n-3)(n-7)}{5!},$

$g_6 = \frac{n(n-1)(n-3)(n-5)(n-7)}{6!}.$

For w ≥ 9 the number of words of weight w is closely approximated by
(10)

$$(n-1)!/(n-w)!w!$$

When $2^L - 1$ is a prime (e.g., L = 3,5,7,13,17,18,31) then the Hamming weight coefficients and Theorem 1 can be combined to count exactly the number of sequences with the same signature and the same syndrome. For prime sequence length m all the rotational equivalence classes have m elements when w ≠ 0 or w ≠ m.

Theorem 2: Consider parallel signature and syndrome compression where:
  i) The feedback corresponds to a primitive polynomial P(x) of degree L.
  ii) The sequence length $m = 2^L - 1$ is a prime.

Then:
1. The number of sequences of weight w and zero signature is g_w, the number of code words of weight w in the Hamming code generated by the primitive polynomial P(x).
2. The number of weight w sequences with the same nonzero signature is

$$(m-1)!/(m-w)!w! \; - \; g_w/m,$$

for w ≠ 0 or w ≠ m.

Proof: There are binomial coefficient m choose w sequences of length m having weight w. If $m = 2^L - 1$ is a prime then each rotational equivalence class has exactly m members when w ≠ 0 or w ≠ m. From Theorem 1, each member has a distinct signature value if any member has a nonzero signature. Thus each equivalence class either contributes one sequence to each nonzero signature or all the class members have zero signatures. A sequence has a zero signature if and only if it is a code word of a Hamming code with n = m. There are exactly g_w such words. Subtracting away the g_w sequences with zero signature leaves those sequences with nonzero signature. All these sequences divide evenly across the $2^L - 1$ or m nonzero syndromes.

Q.E.D.

Example 3: Let L = 5 and m = $2^L - 1$ = 31. Consider w = 4. There are 1085 weight 4 code words in the Hamming code, i.e., g_4 = 1085. From Theorem 2, there are 980 sequences with any particular nonzero signature. When w = 6 the corresponding numbers are g_6 = 22,568 and for a nonzero signature 23,023. If all the weight 6 sequences of length 31 were to be partitioned as equally as possible into 32 classes, some classes would have 23,008 members and some 23,009. It can be seen that the signature partition is very close to the most uniform possible. As a measure of the improvement resulting from the addition of signature compression to syndrome compression we define a reduction factor R.

Definition: The reduction factor R of a signature unit with respect to a sequence length m and weight w is

$$R = \frac{\binom{m}{w}}{\text{Maximum Signature Volume}}$$

where the signature volume is the number of weight w sequences of length m which have a particular signature.

For example 3 where m = 31, suppose that some test data sequence had 6 ones, i.e. w = 6. If syndrome compression (ones counting) is used then a correct count total of 6 could result from the correct data sequence or from the 736,280 other length 31 sequences with 6 ones. If parallel signature compression with L = 5 is used then the ambiguity is reduced. When the correct sequence has a nonzero signature then 23,023 sequences could have the same syndrome total of 6 and the correct signature value; for a zero correct signature 22,568 sequences of weight 6 would be possible. The ratio of 736,281 to 23,023 or 31.980 corresponds to the reduction in the ambiguity when a signature compressor is added with nonzero signature value. The ratio of 736,281 to 22,568 or 32.625 is the reduction factor for a zero signature reference. All things being equal, the former value of 31.980 would occur 31 times out of 32 while the factor of 32.625 would occur once in 32 times.

An average reduction factor could be defined, but we will use the worst case situation or smallest value in defining the reduction factor R. This guarantees that the actual reduction for any specific case will be at least R. For the case of $m = 31$, $w = 6$ and signature compression with a degree 5 primitive polynomial, $R = 31.980$. Note that the very best that one could hope for would be an equal distribution accross all the syndromes or $R = 2^L = 32$. Note that $1 \leq R \leq 2^L$.

In the prime length case, using the expressions for the first few weight coefficients in Theorem 2, Table 1 results. It is conjectured that R approaches 2^L as w increases.

TABLE 1.

Reduction Factor R for Prime Length $m = 2^L - 1$

Sequence Weight w	Reduction Factor R
1	$2^L - 2$
2	
3	$2^L - 3$
4	
5	$2^L - \dfrac{15}{m^2 - 7m + 15}$
6	

III. ENUMERATIVE INVESTIGATION

The cases examined in the prior section include only a small portion of interest. To obtain information for a wider range of parameter values, a program was written to simulate the action of a signature compressor. A constant weight sequence was compressed, the signature noted, and the process repeated. For a range of values of weight w and sequence length m, all possible sequences were simulated. The general conclusion was that the signature unit partitioned the constant weight sequences into nearly equal sized regions as was found analytically for the prime length case. We next describe these experiments in detail.

The primitive feedback polynomial $P(x) = 1 + x + x^3 + x^4 + x^6$ was used in the cases summarized in Figure 3. The ordinate is the sequence length

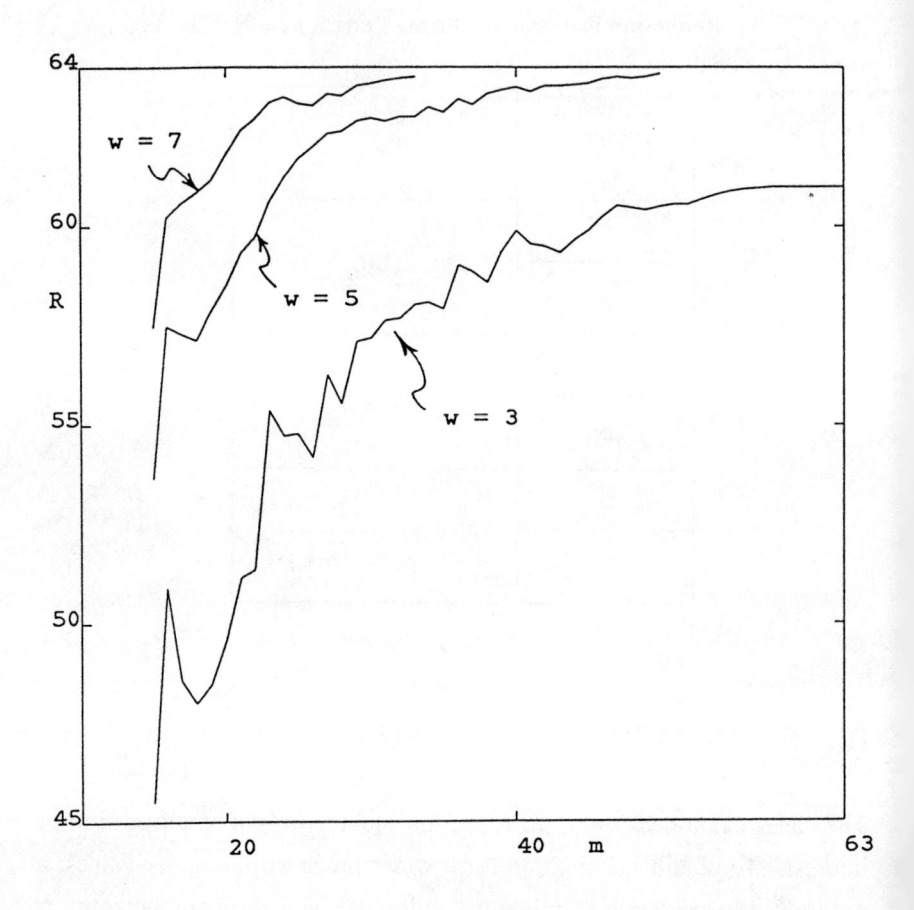

Fig. 3. Exact Reduction Factor R for $1 + x + x^3 + x^4 + x^6$

m. The reduction factor R would be at most $2^L = 64$. The vertical scale starts at 45 in Figure 3 and the sequence length m ranges from 15 to 63 for three different weights w = 3,5,7. Note that R is not monotonic in m but is generally increasing. For w = 3, R aproaches $2^L - 3$ which is the same value as that obtained analytically for prime lengths.

Next the characteristics for various polynomials were studied. All feedback polynomials of the form

$$1 + a_1 x + a_2 x^2 + \cdots a_{L-1} x^{L-1} + x^L$$

were considered. The packed representation was the number

$$PN = 2^L + \sum_{i=1}^{L-1} a_i 2^{i-1}.$$

For the polynomial used in Figure 3, PN = 45. Figure 4 is a plot of R for test length m = 63, signature compressor length L = 6, w = 3,4 and 5 as a function of PN, the packed polynomial coefficient. A total of 32 different polynomials are displayed. For each polynomial three values of R are plotted. The lowest set of connected points corresponds to w = 3, The highest to w = 5. The encircled points are the six primitive polynomials with packed values 33,45,48,51,54,57. The poorest performance, as noted by the entriangled point, is for packed 42 which is $(1+x)^6$.

All the polynomials which have a factor of 1 + x exhibit an R value less than or equal to 32. This behavior was observed for polynomials examined having 1 + x as a factor up through degree 12. It is conjectured that R ≤ 2^{L-1} whenever the feedback polynomial has 1 + x as a factor.

It appears that any polynomial without 1 as a root offers comparable performance for w large and that a primitive polynomial performs well for w small.

The cases reported so far have considered test lengths m less than 2^L. It is

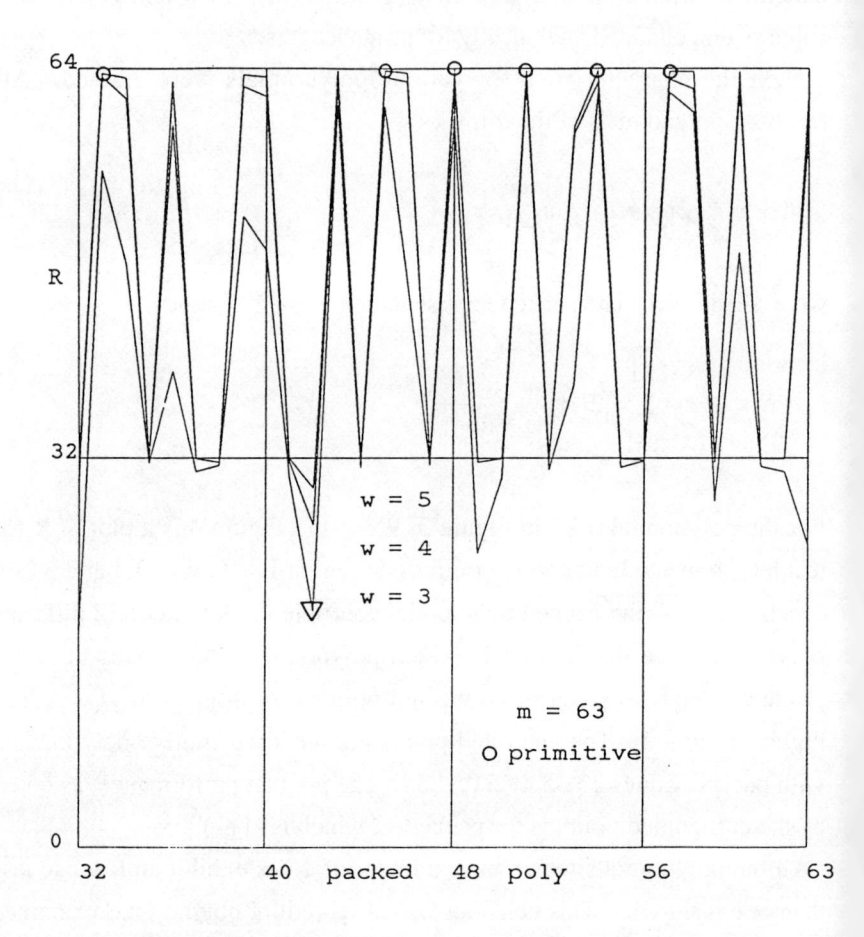

Fig. 4. R for Feedback Polynomials of Length L = 6

known (5) that when m is 2^L or larger there will be some double error patterns that will be missed by signature compression. However there are few of these patterns and from a parallel compression view they do not significantly change the error behavior.

The overall characteristics found in this enumerative study are illustrated in Figure 5. In this presentation each set of connected points is for a length L primitive polynomial used for signature compression. All sequences studied are of length 50 and the ordinate gives a range of weight values w from 3 to 7. Note that the vertical scale is logarithmic, and that for sequence weight w = 6 or more, the reduction Factor R is essentially 2^L.

IV. MONTE CARLO SIMULATION

Above about 10^9 vectors, complete enumeration consumes a substantial amount of simulation time. A Monte Carlo approach was programmed to estimate the reduction factor for some cases with larger values of m and w. A random weight w sequence of length m was generated then compressed. It is to be expected that the Monte Carlo estimates would be lower than the exact values since the largest sample bin count is used to estimate the reduction factor. Of course, it is not guaranteed that the Monte Carlo estimate will always be less than the exact value of R. It is conceivable that the random vectors might distribute evenly among the 2^L bins, when R is less than 2^L. However, this event is extremely unlikely.

The case of m = 50 and w = 6 was examined as a validation check on this approach. Values of L from 6 to 10 were studied for various numbers of random samples. For the average bin size of 1024 it was observed that the Monte Carlo estimate is at least 90% of the exact value. This average bin size, 1024, was used in the subsequent estimates.

Figure 6 summarizes the characteristics found in the simulation. The cases simulated are given in Table 2 where the weight w is about m/10.

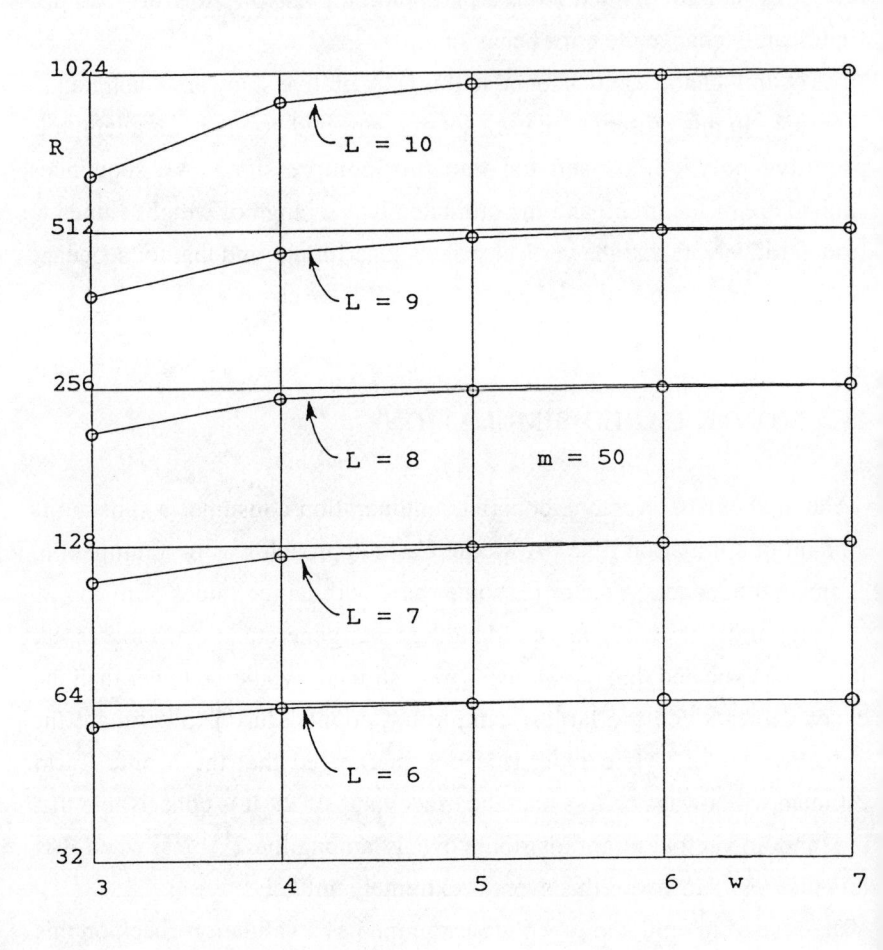

Fig. 5. R for Sequence Length 50 for Primitive Polynomials of Degree L

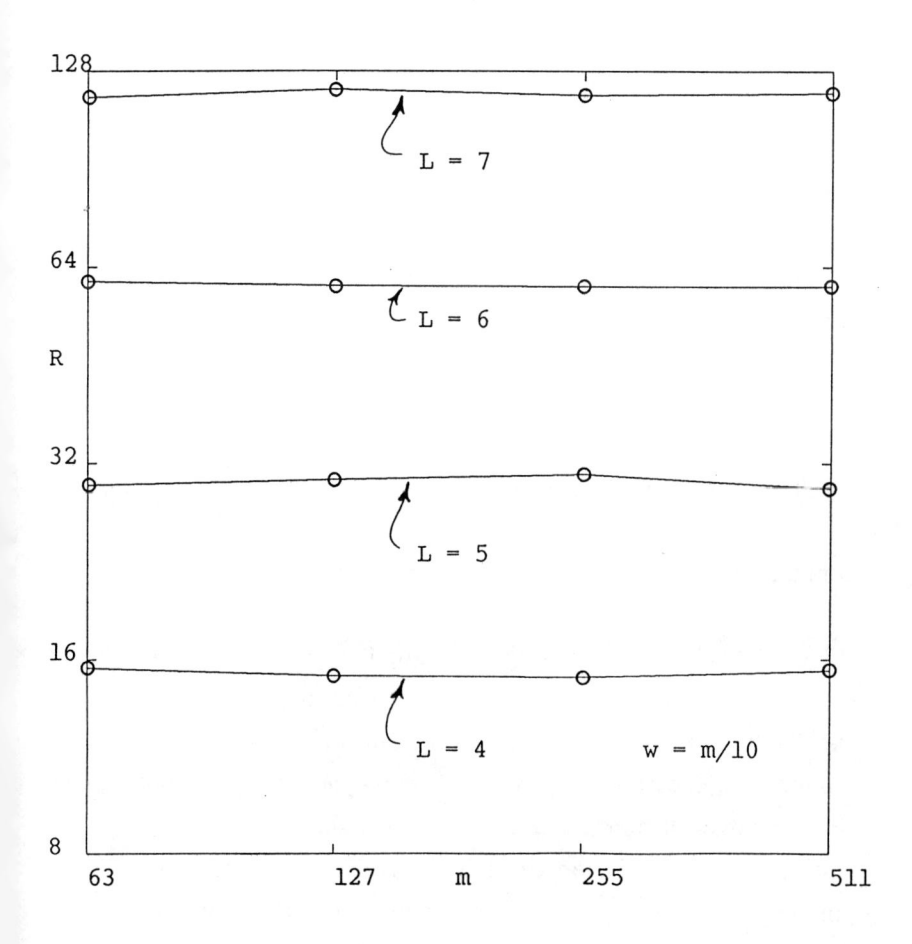

Fig. 6. Monte Carlo Estimates of R for L = 4,5,6,7

TABLE 2.

Cases for Figure 6

m, sequence length	63	127	255	511
w, weight	7	13	26	51

Figure 6 reports cases where the sequence length m is mostly longer than 2^L. Again it is seen that the reduction factor R is close to 2^L. Additional cases for L up to 10 and for a range of weights w were also examined and were found to give the same behavior.

V. CONCLUSIONS

From the results of the earlier sections we conclude:

1. Parallel compression using a length L signature with syndrome compression will reduce the syndrome error ambiguity by a factor of nearly 2^L, even if the test length is larger than 2^L.
2. In the parallel case above, the signature polynomial should not have an even number of nonzero terms, i.e. $1 + x$ should not be a factor. A factor of $(1 + x)$ will approximately double the number of aliases.
3. In the parallel case above, a primitive feedback polynomial (with about half nonzero coefficients) seems optimal. The differences between such a primitive polynomial and any polynomial without $(1 + x)$ as a factor are only noticeable for small count totals (w less than 6). As w becomes larger the reduction factor approaches 2^L.
4. Items 1 to 3 above also hold for transition counting, Walsh Spectral coefficients, or any linear operation followed by counting. Hence any of

these counter based methods can have their error ambiguity reduced using parallel signature compression. From an error count view, the particular counter based method having the smallest count total would result in the smallest ambiguity.

VI. REFERENCES

1. T. W. Williams and K. P. Parker (Jan. 1982)."Design for Testability -- A Survey," *Proceedings of the IEEE*, pp. 98-112.

2. T. W. Williams (Oct. 1984)."VLSI Testing," *COMPUTER*, pp. 126-136.

3. E. J. McCluskey (April 1985)."Built-In Self-Test Techniques," *IEEE Design and Test of Computers*, **Vol. 2, No. 2**, pp. 21-28.

4. N. Benowitz, et al. (May 1975)."An Advanced Fault Tolerant System for Digital Logic," *IEEE Trans. on Computer*, **Vol. C-24, No. 5**, pp. 489-497.

5. J. E. Smith (June 1980)."Measures of the Effectiveness of Fault Signature Analysis," *IEEE Trans. on Computers*, **Vol. C-29**, pp. 510-514.

6. J. Savir (June 1980)."Syndrome Testable Design of Combinational Circuits," *IEEE Trans. on Computers*, **Vol. C-29**, pp. 442-550.

7. A. K. Susskind (Feb. 1983)."Testing by Verifying Walsh Coefficients," *IEEE Trans. on Computers*, **Vol. C-32**, pp. 198-201.

8. D. M. Miller and J. C. Muzio (Aug. 1984)."Spectral Fault Signatures for Single Stuck-At-Faults in Combinational Networks," *IEEE Trans. on Computers*, **Vol. C-33**, pp. 765-769.

9. W. W. Peterson and E. J. Weldon, Jr. (1972). "Error-Correcting Codes", MIT Press, Cambridge, MA.

10. T. Kasami, T. Fujinara and S. Lin (Nov. 1985)."An Approximation to the Weight Distribution of Binary Linear Codes," *IEEE Trans. Information Thy.*, **IT-31, No. 6**, pp. 769-780.

ALIASING PROBABILITIES OF SOME DATA COMPRESSION TECHNIQUES

J.C. Muzio, F. Ruskey, R.C. Aitken, M. Serra

Fault Detection Research Group
Department of Computer Science
University of Victoria
Victoria, B.C., Canada

ABSTRACT

The aliasing probability of a number of common data compression techniques are investigated. Initially a general theoretical framework for the analysis is developed leading to general results for worst case and average case aliasing for any such data compression scheme. The results are applied to the cases of linear feedback shift register testing, syndrome testing, accumulator testing and spectral coefficient testing. The second half of the chapter is an in depth evaluation of the practical behaviour of a set of 41 single and multiple output circuits under the various data compression methods to determine coverage against all single stuck-at faults.

I. INTRODUCTION AND NOTATION.

In a classical testing scheme, a test set has to be generated at design time for use at test time. The approach is very time consuming, and especially complex when dealing with large VLSI circuits, since its complexity increases exponentially with the number of gates in the circuit. This expense occurs only once, and may be considered part of the design cost of the circuit. The main disadvantage of test set methods, however, is the need to store the test set vectors, and then load them from an external file when they have to be applied at test time. This cost occurs each time a circuit is tested. Consequently, test set schemes are not applicable to built-in testing because of their high storage requirement.

In order to avoid the expense of test set generation, research has proceeded primarily in two directions. The first is random test generation, the second is data compression. In random test generation, the test vectors are generated pseudo-randomly, typically using a feedback shift register, and it is hoped that the test patterns generated will cover the vast majority of all possible faults. More information on these methods can be found in (9). Random test generation alone is unsuitable for built-in testing, because of the continued need to store correct output values. For a successful built-in test, some form of data compression is required. Such data compression schemes form the subject of this chapter.

Data compression is the label given to a set of techniques which use encoding schemes to reduce the length of the output vector to a reasonable sized entity called a signature. Formally, a signature consists of a set of attributes of the output vector. For such a signature to be useful in testing, a fault in the circuit should cause the signature to change. The most well-known data compression technique uses a linear feedback shift register (LFSR) to obtain a signature which is equivalent to the remainder of the polynomial division over GF(2) of the input sequence by a fixed primitive polynomial which is determined by the feedback points of the LFSR. Many other works (see for example (12)) have deduced measures of the effectiveness of LFSR testing.

Definition 1: Let $f(x_1,...,x_n)$ be a function of n inputs. The syndrome is defined as $S(f) = weight(f)/2^n$, where weight(f) is just the number of ones in the output vector for f. The range of the syndrome is thus [0,1].

The syndrome is simply the weight normalized by a factor of 2^n. In practice any syndrome tester just tests the weight of the function, all the functions being formally considered as functions of n inputs so that the normalizing factor is identical for all of them and can consequently be ignored.

The range of values for the weight of the non-constant functions is $(0,2^n)$, and thus it takes at most n bits to store its maximum value. Since the storage required for a syndrome counter grows linearly with the number of inputs, it will exceed that required by a standard 16-bit LFSR for n>16 (although for n<16 the storage required is less).

A syndrome counter is a simple device consisting of a binary up-counter (n bit for n-input function), n bits of storage (for the correct value) and a comparator. Of course, the n bits of storage may be on a piece of paper somewhere, and the comparator may be the person who is performing the test. The complete set of hardware is required for a built-in self test module on a chip, although the correct value may be hardwired into the comparator and need not be stored in a register.

The linear feedback shift register is also a simple construction. In the standard HP model (2), taps are taken on positions 7, 9, 12 and 16, the values in these positions are summed modulo 2 (exclusive-or'ed) and the result is exclusive-or'ed with the present value of the output stream. Each successive output value causes a shift of the LFSR, and hence a new sum is computed. The result is an essentially random sequence in the register after the circuit has been exhaustively tested.

Of course, with 2^{2^n} possible output combinations compressed into either 2^n (syndrome counting) or 2^{16} (LFSR) signatures, there will be cases where more than one output stream produces a given signature. In this situation the signature fails to detect the fault and it is the analysis of the frequency of this occurrence which is the substance of the rest of this chapter.

Definition 2: When a fault free and a faulty circuit have the same signature, "aliasing" is said to occur.

The purpose of this chapter is to give a detail analysis of the underlying theory behind the data compression methods and to derive some results for the worst case aliasing and average case aliasing of the methods. These theoretical results are derived on a probabilistic basis, which require a number of assumptions to be made. In particular we assume a uniform distribution of the error patterns resulting from a faulty circuit, that is, in the presence of a fault all other bit patterns are equally likely. In practice the set of possible faults in a circuit only produces a tiny fraction of all possible bit patterns. Consequently the assumption of uniform distribution depends on these faulty patterns being randomly distributed among all the possible bit patterns. The assumption is simplistic and, as was shown in (12), it can produce some unexpected and clearly unreasonable results in extreme cases. However it does provide us with a basis for comparison of the methods.

After giving the detailed theoretical results for linear feedback shift registers and for syndrome testing, we consider the more general signatures based on several spectral coefficients. Recently Saxena (11) has proposed a new approach based on accumulator testing combined with the syndrome. The theoretical results for this method follow from our general theory and are also included.

While it is possible to compare the methods based on their theoretical performance, this only gives some feel for their average behaviour and does not necessarily indicate their relative merits in a practical application. The second half of the chapter looks in detail at a set of 39 single output circuits together with two larger multiple output circuits and gives complete details of the performance of all of the methods and their aliasing probabilities for each of the circuits. In a number of cases the actual results differ widely from the theoretical average behaviour. This is not surprising since some types of circuits are known to be difficult to test.

In the following discussion β will denote a sequence of base b digits of length N. Our applications will be primarily for cases when N is a power of b, with usually b = 2. The function t is the circuit testing function describing the chosen method of data compression, so that, for example, in syndrome testing t just evaluates the weight of the function realized by the circuit. t takes a digit sequence β and returns an integer. The number of digit sequences for which t returns the value K is denoted by a_K. More formally,

$$\beta \in \{0,1,...,b-1\}^N$$

$$t : \{0,1,...,b-1\}^N \to Z$$

For convenience we restrict the range of t to be the non-negative integers.

$$a_K = |\{\beta \in \{0,1,...,b-1\}^N \mid t(\beta) = K\}|$$

i.e. a_K denotes the number of β such that $t(\beta) = K$.

II. A GENERAL APPROACH TO ALIASING PROBABILITIES

In this section we consider two distinct problems. The first is an analysis of the worst case probability of aliasing, and then the more interesting situation which is the average case. Initially the former problem can be stated as follows:

The Worst Case Aliasing Problem: Let K_0 be the value of K that maximizes a_k. Given β chosen uniformly at random, determine the *worst-*

case aliasing probability W(t), defined as

$$W(t) = \text{Prob}(\, t(\beta) = K_0\,) = b^{-N} a_{K_0} \tag{1}$$

While the results given by W(t) are useful in instituting an upper bound, the choice of a particular data compression function t must also be based on its overall average performance in aliasing. The evaluation of this average probability is based on the following statement of the problem:

The Average Case Aliasing Problem: Given β and $\hat{\beta}$ chosen uniformly at random, determine the *average-case aliasing probability A(t)*, defined as

$$A(t) = \text{Prob}(\, t(\beta) = t(\hat{\beta})) \tag{2}$$

Within these two aliasing problems there are at least two different aspects that need consideration. One is a practical approach to finding functions t that minimize A(t) and W(t) subject to reasonable constraints. The constraints should somehow mirror the difficulty of the implementation. In the sequel we attempt to give a crude measure of the cost of implementation by using the maximum number of digits required to store $t(\beta)$. The other side of the problems is to determine A(t) and W(t) exactly if possible, or to obtain asymptotic estimates for simple functions t. For example, it is important to determine (2) for some functions that are easy to implement in hardware. These include such functions as syndrome counting and LFSR's.

It might be remarked that, instead of (2), one usually considers the conditional probability

$$\text{Prob}(\, t(\beta) = t(\hat{\beta}) \mid \beta \neq \hat{\beta}\,)$$

Since $\text{Prob}(\, \beta \neq \hat{\beta}\,)$ is independent of t, if we minimize (2) then we minimize the conditional probability as well. If the conditional probability is desired then it can be computed from the following expression:

$$\text{Prob}(\,t(\beta) = t(\hat{\beta}) \mid \beta \neq \hat{\beta}\,)$$

$$= \frac{\text{Prob}(\,t(\beta) = t(\hat{\beta}) \text{ and } \beta \neq \hat{\beta}\,)}{\text{Prob}(\,\beta \neq \hat{\beta}\,)}$$

$$= \frac{\text{Prob}(\,t(\beta) = t(\hat{\beta})) - b^{-N}}{1 - b^{-N}}$$

$$= \frac{b^N \text{Prob}(\,t(\beta) = t(\hat{\beta})) - 1}{b^N - 1}$$

Prior to attacking the specific problems of the average and worst case aliasing, we give some observations listing some properties which will be useful later. For the detailed analysis a number of specific conditions are imposed upon t. These conditions are intended to reflect some of the practical requirements of data compression functions, together with assisting the analysis of the aliasing probabilities.

Observation 1:

$$\sum_K a_K = b^N \qquad (3)$$

This result is obvious in the sense that each possible β sequence is counted once on the left hand side and there are b^N such sequences.

Observation 2:

$$A(t) = b^{-2N} \sum_K a_K^2 \qquad (4)$$

A(t) is just the probability that $t(\beta)$ and $t(\hat{\beta})$ take the same value. This probability, for a specific value of K, is just $\left[\dfrac{a_K}{b^N}\right]^2$. Hence A(t) is as above.

We introduce the *range size* of a function to be the number of distinct values assumable by t. That is, $R(t) = |\{K\}|$ where $K = t(\beta)$ for some β and we exclude all those K for which $a_K = 0$. R(t) is the number of possible values that t can actually assume and so it is a measure of the comparative complexity of the storage of the relevant function. Of course the number of digits of storage required is $\lceil \log R(t) \rceil$, where the logarithm is to the base b.

Observation 3: Among all functions t with a given range size R(t), both probabilities A(t) and W(t) are minimized when the a_K are equal for all K in the range. In this case, $A(t) = W(t) = R(t)^{-1}$.

This can only occur when $R(t) | b^N$, that is R(t) is a divisor of b^N, in particular $a \times R(t) = b^N$ where $a_K = a$. If t is constrained only by some range restriction such as $0 \le t(\beta) < b^M$, then (2) is minimized if t is taken as the first M digits, treated as a base b number: $t(\beta_1 \cdots \beta_N) = (\beta_1 \cdots \beta_M)_b$. From a testing point of view this is not a very reasonable function since it ignores the remaining digits. (Note that Smith (12) makes a similar point observing that the theoretical coverage of a signature composed of the first r digits is the same as that of a signature resulting from an r-bit LFSR.) Since it seems unreasonable to make practical use of such functions we exclude them under the weak covering condition below.

Before proceeding with the analysis we list some desirable conditions for the function t to satisfy. The covering conditions are desirable from a testing point of view. The other conditions simplify the mathematical analysis. We state all five conditions prior to giving a discussion of their implications.

Weak Covering Condition: This condition states that no digit position is ignored. Formally, if for all i, $1 \le i \le N$ and $d_1 \ne d_2$ there exist sequences $\beta = \beta_1 \cdots \beta_{i-1} d_1 \beta_{i+1} \cdots \beta_N$ and $\hat{\beta} = \beta_1 \cdots \beta_{i-1} d_2 \beta_{i+1} \cdots \beta_N$ such that $t(\beta) \ne t(\hat{\beta})$, then t satisfies the weak covering condition.

Strong Covering Condition: This condition states that changes in a single position produce a change in t. Formally, if for all i, $1 \leq i \leq N$, and $d_1 \neq d_2$ and all sequences $\beta = \beta_1 \cdots \beta_{i-1} d_1 \beta_{i+1} \cdots \beta_N$ and $\hat{\beta} = \beta_1 \cdots \beta_{i-1} d_2 \beta_{i+1} \cdots \beta_N$ and we have $t(\beta) \neq t(\hat{\beta})$, then t satisfies the strong covering condition. A function satisfying the strong covering condition also satisfies the weak covering condition.

Independence Condition: This condition states that the function t treats each digit independently of the other digits. It is not given as a formal definition of independence, but left as an intuitive notion.

Linearity Condition: This condition states that the function t is a linear combination of the digits. Any function that satisfies the linearity condition also satisfies the independence condition. If the coefficients are all positive then a linear function also satisfies the strong covering condition.

Symmetry Condition: This condition states that the distribution of the a_K is symmetric about its mean. In other words, if m is twice the mean of the a_K then t satisfies the symmetry condition if $a_K = a_{m-K}$ for all K. For example in the binary case (b=2) and for n=2, we have $a_0 = a_4$, $a_1 = a_3$. It can be verified that any function that satisfies the linearity condition also satisfies the symmetry condition.

The purpose of the weak covering condition is to exclude functions t which, for example, choose some fixed subset of the digits from β. The strong covering condition ensures that no single digit error can cause aliasing, which appears to be a reasonable practical goal. It is easily seen that syndrome testing satisfies all the conditions. Accumulator testing (see below and (11)) satisfies the covering conditions; it requires careful thought to be convinced that it satisfies the independence condition. As is shown below (section 3.4) it even satisfies the linearity condition.

Intuitively, classic LFSR testing is not independent. Theorem 2 of (12) shows that LFSR testing satisfies the strong covering condition if at least two of the coefficients of the LFSR polynomial are non-zero.

As mentioned above one of the interesting aspects of the problems is the asymptotic evaluation of A(t) for some specific functions t. The following bound (4) is valuable for this exercise:

$$\frac{2^{2n}}{\sqrt{\pi}(n^2+n/2+1/8)^{1/4}} < \binom{2n}{n} < \frac{2^{2n}}{\sqrt{\pi}(n^2+n/2+3/32)^{1/4}} \quad (5a)$$

This bound is very good even for small values of n. An asymptotic expression follows from the bound and can be stated as:

$$\binom{2n}{n} \sim \frac{2^{2n}}{\sqrt{\pi n}} \quad (5b)$$

III. THE GENERATING FUNCTION APPROACH

In this section we use the concepts of an ordinary generating function for linear functions to assist in the derivation of results for A(t) and W(t). Readers who are not familiar with these ideas are referred to any standard text on the subject such as (6). Let $g(x) = \sum_K a_K x^K$ denote the ordinary generating function of the numbers a_K:

$$g(x) = \sum_{K \geq 0} a_K x^K$$

If t satisfies the symmetry condition then $a_K = a_{m-K}$, where m is twice the mean of the a_K; hence

$$\sum_{K \geq 0} a_K^2 = \sum_{K \geq 0} a_K a_{m-K}$$

Let < > denote "coefficient of" in g(x). Then

$$A(t) = b^{-2N} \sum_K a_K^2 \quad \text{(observation 2)}$$

$$= b^{-2N} <x^m> g(x)^2$$

If t satisfies the linearity condition then for some set of constants c_i:

$$t(\beta) = \sum_{i=1}^{N} c_i \beta_i$$

Any function that satisfies the linearity condition also satisfies the symmetry condition with $m = \sum_{i=1}^{N}(b-1)c_i$. Note that m is twice the mean of the a_K. It can be shown that, in this case, the generating function has the form:

$$g(x) = \prod_{i=1}^{N}(1+x^{c_i}+\cdots+x^{(b-1)c_i})$$

By substitution this gives

$$A(t) = b^{-2N} <x^m> \prod_{i=1}^{N}(1+x^{c_i}+\cdots+x^{(b-1)c_i})^2 \tag{6}$$

which is the general solution for A(t). Also from (1):

$$W(t) = b^{-N} \max_K \{a_K\} \tag{7}$$

$$= b^{-N} \max_K <x^K> \prod_{i=1}^{N}(1+x^{c_i}+\cdots+x^{(b-1)c_i})$$

which is the general solution for W(t). Since our major application is to exhaustive binary (b=2, N= 2^n) testing we restate the results in this restricted situation.

$$A(t) = 2^{-2N} <x^m> \prod_{i=1}^{N}(1+x^{c_i})^2$$

$$W(t) = 2^{-N} \max_{K} <x^K> \prod_{i=1}^{N}(1+x^{c_i})$$

For example if n=2, b=2 (a two-input binary function that is to be tested) and suppose that $(c_1,c_2,c_3,c_4) = (1,2,3,4)$ then we have $t(\beta) = \beta_1 + 2\beta_2 + 3\beta_3 + 4\beta_4$, explicitly listed here:

| circuit outputs |||| |
β_1	β_2	β_3	β_4	$K=t(\beta)$
0	0	0	0	0
0	0	0	1	4
0	0	1	0	3
0	0	1	1	7
0	1	0	0	2
0	1	0	1	6
0	1	1	0	5
0	1	1	1	9
1	0	0	0	1
1	0	0	1	5
1	0	1	0	4
1	0	1	1	8
1	1	0	0	3
1	1	0	1	7
1	1	1	0	6
1	1	1	1	10

The values of K, a_K are given below.

K	0	1	2	3	4	5	6	7	8	9	10
a_K	1	1	1	2	2	2	2	2	1	1	1

Note that $R(t) = 11$.

$$g(x) = \sum_{K=0}^{10} a_K x^K = 1 + x + x^2 + 2x^3 + 2x^4 + \cdots + x^{10}$$

$g(x)$ can also be written in the form:

$$g(x) = (1 + x^{c_0})(1 + x^{c_1})(1 + x^{c_2})(1 + x^{c_3})$$

$$= (1 + x)(1 + x^2)(1 + x^3)(1 + x^4).$$

Hence

$$A(t) = 2^{-8} <x^{10}> g(x)^2 = 2^{-8} \times 26$$

and $W(t) = 2^{-4} \times 2$

Observe that the maximum in the worst case does not necessarily occur at $K = m/2$, although it does in the examples of the next section. For example, if $N = 3$, $b = 2$ and $c_1, c_2, c_3 = 1, 1, 4$, then we have

K	0	1	2	3	4	5	6
a_K	1	2	1	0	1	2	1

Note that the mean is 3 but $a_3 = 0$ and the worst case aliasing is for a_1 and a_5. Also $R(t) = 6$. There are a number of instances in the literature where the authors have observed the symmetry of their functions and then inferred that the maximum occurs at the mean. Such reasoning is incorrect unless some additional property (such as convexity or unimodality) is also shown. In most practical situations however the worst case is at the mean.

IV. APPLICATIONS

In this section we present some practical applications of the use of the expressions derived in the previous section. The last two examples are not linear but are included for completeness and since the values of $A(t)$ and $W(t)$ are known for them. The major discussion of most of the practical approaches is reserved for the next section when we consider the performance of the methods for a set of practical circuits.

A. Syndrome counting.

To get the syndrome set $b = 2$, $N = 2^n$ and $c_i = 1$ for all i. The syndrome can take on any value between 0 and N. It follows that $R(t) = N+1$. The distribution of weights for functions of a fixed number of inputs can readily be seen to be binomial. The total number of functions with weight K ($0 \le K \le N$) is given by: $a_K = \binom{N}{K}$

Because of the symmetric nature of the weight distribution, it is apparent that the greatest amount of aliasing occurs for the mid-range value, where the weight is N/2. Hence

$$W(t) = \frac{\binom{N}{N/2}}{2^N} \tag{8}$$

Using the asymptotic expression given earlier in (5b) we deduce that

$$W(t) \sim (\pi N/2)^{-1/2} \tag{9}$$

For the average computations, equation (7) becomes

$$A(t) = 2^{-2N} <x^N> (1+x)^{2N} = 2^{-2N} \binom{2N}{N} \sim (\pi N)^{-1/2} \tag{10}$$

This agrees with Savir & McAnney in (10), but was derived independently. The major point worth noting here is that $W(t) \sim \sqrt{2} \, A(t)$.

B. LFSR testing.

Set $b = 2$ and $N = 2^n$. The standard aliasing probability for a LFSR of length D (in bits) is given by Smith in (12) as :

$$\frac{2^{N-D}-1}{2^N-1} \tag{11}$$

For large n this simplifies to 2^{-D}, and given the standard 16-bit LFSR size the aliasing probability is approximately 2^{-16}. The range size is $R(t) \leq 2^D$. The average and worst case aliasing probabilities are identical.

$$W(t) = A(t) = 2^{-D} \tag{12}$$

In a certain sense LFSR testing is the optimal way to test since it partitions the set of all possible outcomes into equal sized equivalence classes.

Carter (1) has given the aliasing probability as $\leq 2 N^{-1}$ with several provisos. First, the shift register must be longer than n bits; second, the feedback connections must be made randomly; and finally, the inputs must be presented in random order. Carter also adds that it is his personal belief that the probability is $\leq N^{-1}$, but that he is unable to show this.

With respect to the first restriction, if the shift register is shorter than n bits, the aliasing probability will be greater than N^{-1}. For example, suppose a 16-bit LFSR is used for a function with n>16. There are only 2^{16} possible signatures for such a LFSR, so a method using the LFSR which gives aliasing of less than 2^{-16} on average is somewhat unlikely. The degree to which the other two requirements are satisfied by a genuine LFSR scheme is best left to other works, although it seems unlikely that cycling through inputs in order would satisfy the random order restriction.

We have chosen to give Carter's supposition the benefit of the doubt and present the aliasing probability of a D-bit LFSR with an exhaustive test using n inputs as:

$$\max \left[\frac{2^{N-D}-1}{2^N-1}, N^{-1} \right] \tag{13a}$$

or, in the case of a 16-bit LFSR,

$$\max(2^{-16}, N^{-1}) \tag{13b}$$

C. Generalized counting techniques

Case 1: This example is a generalization of syndrome counting. Assume that $b = 2$ and the bits have been grouped into L *groups* each of which contains M bits. The function t is linear and the coefficients are the same for all bits in a given group; that is t is a linear sum of the weights of each of the groups. Abusing the previous notation slightly,

$$t(\beta) = \sum_{i=1}^{L} c_i \times m_i(\beta) \tag{14}$$

where $m_i(\beta)$ denotes the number of 1's in the ith group of β. Let $C = \sum c_i$. Clearly, $R(t) = CM+1$, $m = CM$. From (7) we obtain

$$A(t) = 2^{-2N} <x^{CM}> \prod_{i=1}^{L}(1+x^{c_i})^{2M} = 2^{-2N} \sum \prod_{j=1}^{L} \begin{bmatrix} 2M \\ k_j \end{bmatrix} \tag{15}$$

where the sum is over all k_i, each $k_i \leq 2M$, such that

$$\sum_{i=1}^{L} c_i k_i = CM \qquad (16)$$

Note that we obtain syndrome counting if $L = 1$.

Case 2: If (16) has the unique solution $k_i = M$ then

$$A(t) = 2^{-2N} \begin{bmatrix} 2M \\ M \end{bmatrix}^L \sim (\pi M)^{-L/2} \qquad (17)$$

This gives us a lower bound on (15). The unique solution corresponds to computing L distinct numbers; one number from each group. Thus the worst case aliasing is obtained by raising the worst case aliasing within a group to the Lth power.

$$W(t) = 2^{-N} \begin{bmatrix} M \\ \lfloor M/2 \rfloor \end{bmatrix}^L \sim (\pi M/2)^{-L/2} \qquad (18)$$

Case 3: It has been suggested that taking the syndrome in addition to one other spectral coefficient (7) would be an attractive alternative to syndrome testing. This corresponds to the case where $L = 2$. Then

$$W(t) \sim \frac{4}{\pi N} \quad \text{and} \quad A(t) \sim \frac{2}{\pi N}$$

Some empiric results for this are presented and discussed later.

Case 4: The case where $L = 4$ is discussed later (as an alternative to using both the accumulator and the syndrome for testing). There

$$W(t) \sim \frac{64}{\pi^2 N^2} \quad \text{and} \quad A(t) \sim \frac{16}{\pi^2 N^2}$$

D. Accumulator testing

When computing the syndrome of a sequence β, we can denote by $\alpha_1\alpha_2\cdots\alpha_N$ the successive contents of the counter used. Obviously α_N is the resulting syndrome. At any time i the syndrome counter has a content

$$\alpha_i = \sum_{j=1}^{i} \beta_j \tag{19}$$

The "accumulator s" is defined as the sum of the partial syndromes:

$$s = \sum_{i=1}^{N} \alpha_i \tag{20}$$

From (19) and (20) we have

$$s = \sum_{i=1}^{N} (N+1-i)\, \beta_i$$

Thus the result of the accumulator compression function is to evaluate a weighted sum of all the bits, where the weights are distinct and vary from 1 to N. For more details see (11). As an example, for b=2, N=8, β=(01101110), we have that $(\alpha_1,\alpha_2,\cdots,\alpha_8)$=(01223455) and s=22.

Case 1: Accumulator testing is a special case of Case 1 of section C, in which L = N, M = 1, c_i = N+1−i, C = N(N+1)/2, and R(t) = 1 + N(N+1)/2. For the average case, (15) becomes

$$A(t) = 2^{-2N} \sum \prod_{j=1}^{N} \binom{2}{k_j}$$

where the sum is over all k_i such that $\sum_{i=1}^{N} ik_i = N(N+1)/2$

Note that there is no contribution to A(t) unless k_i = 0, 1, or 2. For the worst case, from (6),

$$W(t) = 2^{-N} <x^K> \max_K \prod_{j=1}^{N}(1+x^j) = 2^{-N} \max_K \sum \prod_{j=1}^{N} \begin{bmatrix} 1 \\ k_j \end{bmatrix}$$

where the sum is over all k_i such that $\sum_{i=1}^{N} ik_i = K$

This can also be written as

$$W(t) = 2^{-N} \max_K D(K,N)$$

where $D(K,N)$ is the number of partitions of K into distinct parts such that the maximum part is at most N. For fixed N the sequence $D(K,N)$ is symmetric and unimodal. The term *unimodal* simply means that the values are non-decreasing and then non-increasing as K varies between 0 and $N(N+1)/2$. Symmetry follows from the symmetry of t. The unimodality has been observed experimentally, but has not been proven. The maximum therefore occurs at $K = \lfloor N(N+1)/4 \rfloor$. Thus,

$$W(t) = 2^{-N} D(\lfloor N(N+1)/4 \rfloor, N)$$

Case 2: In the ACT testing of (11) where the signature contains two attributes, namely both the syndrome and the accumulator, we can take $c_i = 1 + (N+1)(N+1-i)$. Then $t(\beta) = s + (N+1)a$ where s is the syndrome and a is the accumulator. The range size is $R(t) = (N+1)[1+N(N+1)/2]$. In reference (11) it is shown that

$$W(t) = 2^{-N} P(\lfloor N/2 \rfloor, \lfloor N/2 \rfloor, \frac{N^2}{8})$$

$$\sim 2^{-N} \begin{bmatrix} N \\ N/2 \end{bmatrix} \sqrt{\frac{24}{\pi N^2(N+1)}} \sim \frac{4\sqrt{3}}{\pi N^2}$$

where $P(n,r,t)$ denotes the number of partitions of t into at most n parts such that each part is not greater than r. However the derivation of the asymptotic result in (11) is not rigorous.

E. Evaluation of Theoretical Results

Table 1 presents the computed evaluation of A(t) and W(t) according to the theoretical results stated in equations (6) and (7). In the various columns the expected values for LFSR, syndrome (L=1), and the groupings for L=2 and L=4 are shown for both W(t) and A(t). The results in each case give the probability of aliasing in exponential notation so that a number is written normalized in the form d.ddd–e meaning d.ddd 10^{-e}.

V. EMPIRICAL RESULTS

In this section we present the empirical performance of the signatures for the set of sample circuits and discuss the results in comparison to the theoretical expectations. The faults modeled are single stuck-at faults at the gate level; we attempt to deduce some comparisons of the effectiveness of the different signatures.

The sample consists of 44 single output circuits and 2 multiple output ALU's. The single output circuits vary from 3 to 12 inputs, collected by the authors from various locations including papers, books, and actual chips; only 39 of them are listed in the tables as some were similar realizations of duplicate functions. The first ALU is the standard SN74181 chip, with 14 inputs and 8 outputs, the second one is a bit-slice implementation from (13, p.85), with 9 inputs and 6 outputs. Because of the large amount of computing resources required for simulation, a limit on the number of inputs had to be imposed.

The resulting 39 circuits are not very large, but quite varied in type and size. A catalogue of them can be found in the Appendix: their sources and appropriate references can be obtained from the authors.

n	LFSR	syndrome W(t)	syndrome A(t)	L = 2 W(t)	L = 2 A(t)	L = 4 W(t)	L = 4 A(t)
4	6.250 -02	1.995 -01	1.410 -01	7.958 -02	3.979 -02	2.533 -02	6.333 -03
5	3.125 -02	1.410 -01	9.974 -02	3.979 -02	1.989 -02	6.333 -03	1.583 -03
6	1.562 -02	9.974 -02	7.052 -02	1.989 -02	9.947 -03	1.583 -03	3.958 -04
7	7.812 -03	7.052 -02	4.987 -02	9.947 -03	4.974 -03	3.958 -04	9.895 -05
8	3.906 -03	4.987 -02	3.526 -02	4.974 -03	2.487 -03	9.895 -05	2.474 -05
9	1.953 -03	3.526 -02	2.493 -02	2.487 -03	1.243 -03	2.474 -05	6.184 -06
10	9.766 -04	2.493 -02	1.763 -02	1.243 -03	6.217 -04	6.184 -06	1.546 -06
11	4.883 -04	1.763 -02	1.247 -02	6.217 -04	3.108 -04	1.546 -06	3.865 -07
12	2.441 -04	1.247 -02	8.816 -03	3.108 -04	1.554 -04	3.865 -07	9.663 -08
13	1.221 -04	8.815 -03	6.233 -03	1.554 -04	7.771 -05	9.663 -08	2.416 -08
14	6.104 -05	6.233 -03	4.408 -03	7.771 -05	3.886 -05	2.416 -08	6.039 -09
15	3.052 -05	4.408 -03	3.117 -03	3.886 -05	1.943 -05	6.039 -09	1.510 -09
16	1.526 -05	3.117 -03	2.204 -03	1.943 -05	9.714 -06	1.510 -09	3.775 -10
17	1.526 -05	2.204 -03	1.558 -03	9.714 -06	4.857 -06	3.775 -10	9.436 -11
18	1.526 -05	1.558 -03	1.102 -03	4.857 -06	2.429 -06	9.436 -11	2.359 -11
19	1.526 -05	1.110 -03	7.792 -04	2.429 -06	1.214 -06	2.359 -11	5.898 -12
20	1.526 -05	7.792 -04	5.510 -04	1.214 -06	6.071 -07	5.898 -12	1.474 -12

TABLE 1 : Theoretical Results.

The empirical aliasing results given here may not be numerically valid as "average cases" for "typical" circuits, because of the small number of circuits and their size. However, we claim that most of the circuits presented are practical circuits with the exception of a few chosen specifically because they are syndrome-untestable.

The linear feedback shift registers simulated were of n bits, where n is the number of inputs, with the feedback tap on the last bit, implementing the simple polynomial $x^n + 1$. Other polynomials would have performed quite differently, but no effort was made to find the best such structure. This is to emphasize that there is no logical criterion connecting a fault effect with an aliasing LFSR signature. At equal length of bits with a syndrome tester, there are up to 2^{n-1} possible simulations to be done by an LFSR in order to find the best polynomial for a particular circuit. The only exception is the SN74181 ALU (with 14 inputs), where the standard 16-bit Hewlett-Packard LFSR was used.

Analysis was performed both with and without fault collapsing (the process of eliminating equivalent faults). The aliasing was computed as a percentage by dividing the number of faults which aliased by the total number of faults simulated. Different testing methods were computed to reflect the theoretical analysis. For the first set of comparisons, the signatures of a syndrome tester and of the n-bit LFSR described above were computed for all possible faults on the circuits, as for A and B in section IV. The second method used accumulator and ACT testing as described recently in (11) and presented in section D. Thirdly we evaluated spectral signatures comprised of the syndrome and one first-order spectral coefficient: it follows case 3 in section C. Lastly a signature which partitioned the output vector into 4 parts, computing the weight of each part, was simulated, as in case 4, section C.

Table 2a presents the empirical results for the 39 single output circuits. They are grouped by number of inputs. Column 3 gives the total number of line segments (including roots and stems of fan-out branches

no.	inputs	lines	undetectable	faults	SYN	LFSR
1	3	11	0	22	0.091	0.091
2	3	19	0	38	0	0
3	3	27	0	54	0.1111	0.1111
4	3	13	0	26	0.2308	0.0385
5	3	12	0	24	0.3333	0.2083
6	3	13	0	26	0.2308	0.0385
7	3	9	0	18	0	0
9	4	11	0	22	0.2727	0.2727
10	4	20	22	18	0	0.175
11	4	20	0	40	0.25	0.2
12	4	17	0	34	0.1765	0.5588
13	4	26	4	48	0.1667	0.0833
14	4	12	0	24	0	0.4166
15	4	31	0	62	0.0322	0.0645
16	4	22	0	44	0.1818	0.5909
17	4	18	8	28	0	0
18	4	41	18	64	0	0.0156
19	5	16	2	30	0	0
20	5	13	1	25	0	0
21	5	8	0	16	0.25	0
22	5	28	0	56	0.0357	0.0178
23	5	16	0	32	0	0
24	6	21	12	30	0	0.4333
25	6	17	4	30	0	0.0666
26	6	41	38	44	0	0.0682
27	6	23	8	38	0	0.1579
28	6	18	13	23	0	0.4783
29	6	27	0	54	0.1481	0.5185
31	7	14	0	28	0	0
32	7	11	0	22	0	0
33	7	12	0	24	0	0
36	8	18	0	36	0	0.0555
37	8	13	0	26	0	0
38	8	15	0	30	0	0
40	9	23	0	46	0	0
41	9	23	0	46	0	0.1304
42	10	21	0	42	0.9047	0.0238
43	12	29	0	58	0	0.0344
44	12	24	0	48	0	0.0416

TABLE 2a : Single Output Circuits - No Fault Collapsing.

no.	inputs	lines	dependent	undetectable	faults	SYN	LFSR
74181							
out 1	14	240	244	0	244	0.0328	0
out 2	14	240	304	0	304	0.0263	0
out 3	14	240	162	0	162	0.0494	0
out 4	14	240	308	0	288	0.1215	0
out 5	14	240	230	0	230	0.2565	0
out 6	14	240	390	30	360	0.0222	0
out 7	14	240	158	0	158	0.2025	0
out 8	14	240	92	0	92	0.2717	0
all	14	240	480	0	480	0.01875	0
ALU2							
out 1	9	113	74	0	74	0	0
out 2	9	113	110	0	110	0	0
out 3	9	113	114	0	114	0.9825	0
out 4	9	113	74	0	74	0.973	0
out 5	9	113	42	0	42	0.9524	0
out 6	9	113	16	0	16	0.875	0
all	9	113	226	0	226	0.3893	0

TABLE 2b : ALU's - No Fault Collapsing.

separately). Column 4 shows the number of undetectable faults. (By undetectable faults we mean a stuck-at-fault on a line segment which produces no change in the output vector). The total number of detectable faults in column 5 is thus calculated as twice the number of lines minus the number of undetectable faults.

Table 2b presents the same results for the two ALU's. The aliasing percentage is given for each output as tested separately, and for the outputs as tested in parallel (the row labelled "all"). Note that line segments do not usually lead to all outputs, and thus in column 4 ("dependent") we count the actual number of faults on line segments which do depend on that output.

Table 3a and table 3b present similar results after fault collapsing was done. We only considered two-way equivalent faults and the collapsing was applied as follows:

AND gate	input/0	collapsed to	output/0
OR gate	input/1	collapsed to	output/1
NAND gate	input/0	collapsed to	output/1
NOR gate	input/1	collapsed to	output/0
inverter	input/0 or /1	collapsed to	output/1 or /0

Column 5 of table 3a gives the actual number of simulated faults after collapsing. In table 3b the number of faults on dependent line segments, undetectable faults and collapsed faults are explicitly shown in columns 4, 5 and 6 respectively, while the number of actual simulated faults is in column 7.

In table 4 those circuits which showed any aliasing by accumulator testing and ACT are listed: none of the remaining circuits had any aliasing. Columns 4-6 are for all detectable faults, columns 7-9 are after fault collapsing.

In table 5 we consider signatures obtained by partitioning the output vector in two or four parts and evaluating the syndromes of each part separately. This corresponds to selecting two or four spectral coefficients. The columns for L=2 show the aliasing for the choice of

no.	inputs	lines	undetectable	faults	SYN	LFSR
1	3	11	0	16	0.125	0.125
2	3	19	0	22	0	0
3	3	27	0	36	0.1666	0.1666
4	3	13	0	14	0.2857	0.0714
5	3	12	0	17	0.4706	0.2941
6	3	13	0	14	0.2857	0.0714
7	3	9	0	10	0	0
9	4	11	0	12	0.3333	0.1666
10	4	20	22	10	0	0.5
11	4	20	0	28	0.2142	0.2857
12	4	17	0	22	0.1818	0.5454
13	4	26	4	30	0.2666	0.0666
14	4	12	0	12	0	0.3333
15	4	31	0	36	0.0555	0.1111
16	4	22	0	27	0.2962	0.7037
17	4	18	8	16	0	0
18	4	41	18	41	0	0.0244
19	5	16	2	18	0	0
20	5	13	1	13	0	0
21	5	8	0	11	0.3636	0
22	5	28	0	32	0.0625	0.0312
23	5	16	0	18	0	0
24	6	21	12	21	0	0.3809
25	6	17	4	18	0	0.0555
26	6	41	38	22	0	0.1363
27	6	23	8	36	0	0.1666
28	6	18	13	12	0	0.3333
29	6	27	0	28	0.2857	0.5
31	7	14	0	16	0	0
32	7	11	0	12	0	0
33	7	12	0	13	0	0
36	8	18	0	22	0	0.0909
37	8	13	0	14	0	0
38	8	15	0	16	0	0
40	9	23	0	26	0	0
41	9	23	0	27	0	0.1111
42	10	21	0	38	0.9473	0.0263
43	12	29	0	32	0	0.0625
44	12	24	0	26	0	0.0769

TABLE 3a : Single Output Circuits - Fault Collapsing.

no.	inputs	lines	dependent	undetectable	collapsed	faults	SYN	LFSR
74181								
out 1	14	240	244	0	77	167	0.0479	0
out 2	14	240	304	0	94	210	0.038	0
out 3	14	240	162	0	46	116	0.0689	0
out 4	14	240	308	0	91	217	0.129	0
out 5	14	240	230	0	69	161	0.2546	0
out 6	14	240	390	30	116	244	0.0327	0
out 7	14	240	158	0	46	112	0.2142	0
out 8	14	240	92	0	24	68	0.2647	0
all	14	240	480	0	146	334	0.0269	0
ALU 2								
out 1	9	113	74	0	12	62	0	0
out 2	9	113	110	0	27	83	0	0
out 3	9	113	114	0	25	89	0.9775	0
out 4	9	113	74	0	16	58	0.9655	0
out 5	9	113	42	0	8	34	0.9412	0
out 6	9	113	16	0	0	16	0.875	0
all	9	113	226	0	58	168	0.375	0

TABLE 3b : ALU's - Fault Collapsing.

no.	inputs	lines	faults (no coll.)	ACC	ACT	faults (coll.)	ACC	ACT
3	3	27	54	0.1111	0.1111	36	0.1666	0.1666
11	4	20	40	0.0250	0.0000	28	0.0357	0.0000
42	10	21	42	0.9047	0.9047	38	0.9473	0.9473
74181 out7	14	240	158	0.0120	0.0120	118	0.0178	0.0178
ALU2 out3	9	113	114	0.9649	0.9649	89	0.9551	0.9551
out4	9	113	74	0.9459	0.9459	58	0.931	0.931
out5	9	113	42	0.9048	0.9048	34	0.8824	0.8824
out6	9	113	16	0.75	0.75	16	0.75	0.75
all	9	113	226	0.3539	0.3539	209	0.2631	0.2631

TABLE 4 : Accumulator and ACT Testing.

no.	inputs	lines	faults	L=2 best {syn.r(i)}	{syn.height}	L=4
1	3	11	22	0	0	0
3	3	27	54	0.1111	0	0.074
4	3	13	26	0	0	0
5	3	12	24	0.0833	0.25	0
6	3	13	26	0	0	0.2307
9	4	11	22	0	0.2727	0
11	4	20	40	0.025	0	0.225
12	4	17	34	0	0	0
13	4	26	48	0.0416	0	0
15	4	31	62	0	0	0
16	4	22	44	0.0909	0	0.0454
21	5	8	16	0.125	0	0.25
22	5	28	56	0	0.0357	0.0357
29	6	27	54	0.1111	0	0.1111
42	10	21	42	0.9047	0	0.8095
74181						
out 1	14	240	244	0	0.0328	0.0327
out 2	14	240	304	0	0.0263	0.0263
out 3	14	240	162	0	0.0494	0.0493
out 4	14	240	288	0	0	0.0277
out 5	14	240	230	0.0086	0.2565	0.1478
out 6	14	240	360	0	0	0.0222
out 7	14	240	158	0.0253	0.2025	0.0949
out 8	14	240	92	0.0326	0.2717	0.2717
all	14	240	480	0	0.002	0.0187
ALU 2						
out 3	9	113	114	0.9649	0.0526	0.0877
out 4	9	113	74	0.9459	0.973	0.9729
out 5	9	113	42	0.9047	0.9524	0.9523
out 6	9	113	16	0.75	0.875	0.875
all	9	113	226	0.3539	0.23	0.2389

TABLE 5 : Generalized Counting Aliases.

syndrome (r_0) and the "best" first order spectral coefficient r_i, followed by the aliasing for the syndrome and the highest order spectral coefficient, called the "height", $r_{1,2,...,n}$. No fault collapsing was done here. The circuits listed are only those where the syndrome alone aliased in tables 2a or 2b. Obviously the remainder will not alias.

It is important to note when examining the empirical results that the aliasing encountered for actual circuits can be considerably greater than theoretical "worst-case" values for aliasing probability. This phenomenon will always result when random samples are taken from a distribution. For this reason, "worst-case" refers to the worst-case distribution which can result from fixing some parameter, rather than actual samples taken from that distribution. The mean of worst-case values from sample circuits will approach the theoretical worst-case result, provided that sample circuits represent a random sample from the set of all possible functions.

We now discuss the results presented in the tables and draw some possible conclusions.

A. Syndrome Testing

(See Tables 2 and 3) The worst syndrome aliasing occurs for circuit 42, and for the ALU of (13). Circuit 42 is the 74LS630 parity generator, and consists entirely of exclusive-or gates. Every line but the final output is a candidate syndrome-untestable line, and all of these alias. It is well-known that this sort of circuit is very unsuitable for syndrome testing (see for instance, (7)). The only spectral coefficient which may be used to test this circuit is the "height", $r_{1,2,...,10}$. Of course, the circuitry required to calculate the height is equivalent to the circuit under test, since in both cases an exclusive-or tree is necessary. Notice, though, that the 10-bit LFSR had very low aliasing. If this circuit is tested with the standard 16-bit H-P LFSR, no aliasing occurs. Because shift register techniques are pseudo-random, there does not appear to be any method of predicting which circuits will have higher LFSR aliasing than others. For this reason, an LFSR is more suitable than counting methods on circuits with many candidate syndrome-untestable lines.

The other circuit with a large amount of syndrome aliasing is the second ALU. Most outputs come from exclusive-or gates, so all lines but the outputs are candidate syndrome-untestable. An interesting point to note is that the 74LS181 ALU (which was investigated by Savir in his original paper on syndrome testing (8)) is largely syndrome testable. With the addition of two gates and a single control input, Savir was able to make the circuit entirely syndrome testable. Most other functions with large syndrome aliasing probabilities contain exclusive-or sections. Not all such circuits alias, though. Circuit number 27 (from (3)), for example, consists entirely of exclusive-or and exclusive-nor gates, yet has no syndrome aliasing at all.

The first conclusion which may be drawn from the results is that the syndrome appears to either not alias at all, or alias too much to be of use, for circuits with more than four inputs. For this reason, if a circuit is to be tested with a syndrome counter, it is undoubtedly important to use design for testability criteria, starting with the marking algorithm of (7). This deterministic analysis which can be applied in time linear in the number of lines, will give a precise indication of the maximum syndrome aliasing. One may then check the candidate syndrome untestable lines in various manners: by exhaustive simulation as for an LFSR, by random sample or accurately by theoretical spectral analysis as in (5). If the syndrome aliasing determined in this fashion is too high, another signature should be chosen. The fact that this deterministic analysis is possible, without the large amount of simulation required by other methods, is a distinct advantage of the technique.

B. LFSR Signatures

Much work has been done on both the theoretical and empirical performance of the LFSR. These latest results merely corroborate the previous findings. Some aliasing is observed for the circuits, including some very large values (0.5909 for circuit 16, and 0.5588 for circuit 12, for instance). All aliasing appears to be removable through the use of a longer shift register, however. In fact, the standard 16-bit HP LFSR was also simulated for all circuits and no aliasing was observed.

C. Accumulator Testing

(See Table 4). The aliasing of the accumulator testing or of the combined ACT testing of (11) was in most cases zero, but given the size of the tester this is not particularly surprising. What is remarkable, however, is its terrible failure on several circuits (42 and the ALU). In these cases, the accumulator performs only marginally, if at all, better than the syndrome tester. For a testing method to be considered an improvement over syndrome testing, it must be able to provide an effective test on circuits with many candidate syndrome-untestable lines, in particular exclusive-or networks.

Despite the significant increase in tester size, as well as the reduced performance resulting from the need to perform additions instead of merely counting, no improvement is seen in pathologically syndrome untestable circuits. For this reason, it appears that while accumulator testing is theoretically interesting, in practice it is likely of little value.

D. Generalized Counting

(See Table 5). While the theoretical expectations for the partitions of L=4 of the output truth vector using the partial syndromes as a signature are quite good, the empirical results are not much superior to the L=2 partitions for the problem circuits. This is not particularly surprising since the signature is merely augmented in size, but the randomization process performed by adding extra spectral coefficient is based on the same principles.

The more interesting aspect is that in a number of cases the L=4 partitions gave results comparable or even better than the accumulator testing. This is an advantage since the test circuitry required for built-in-test is much smaller and can be optimized in the design in various ways. These last figures emphasize once more that a design for testability approach and a selection of testing method appropriate to the circuit are the important aspects. Any a priori knowledge about the circuit under test can be exploited in the more deterministic counting methods: the LFSR randomization can be used alternatively when no such knowledge is available or when no design analysis can be applied.

REFERENCES

1. Carter, J. (May 1982). "The theory of signature testing for VLSI". *Proc. 14th ACM Symposium on Theory of Computing.* pp. 111-113, Santa Clara, CA.

2. Frohwerk, R.(May 1977). "Signature analysis: a new digital field service method". *Hewlett-Packard Journal.*

3. Goel, P. (October 1980). "An implicit enumeration algorithm to generate tests for combinational logic circuits". *Proc. Symposium on Fault Tolerant Computer Systems.* **10,** 145-151.

4. Grosswald, E. (January 1977). "Solution of advanced problem 6019". *American Math. Monthly.* 63-65.

5. Hurst, S.L., Miller, D.M. and Muzio, J.C. (1985). "Spectral Techniques in Digital Logic". Academic Press, London and New York.

6. Knuth, D.E. (1969). "The Art of Computer Programming". Addison-Wesley.

7. Miller, D.M. and Muzio, J.C. (August 1984). "Spectral fault signatures for single stuck-at-faults in combinational networks". *IEEE Transac. Computers.* **C-33, 8,** 765-769.

8. Savir, J. (June 1980). "Syndrome testable design of combinational networks". *IEEE Transac. Computers.* **C-29, 6,** 442-451.

9. Savir, J., Ditlow, G. and Bardell, P. (June 1983). "Random pattern testability". *Proc. Symposium on Fault Tolerant Computer Systems.* **13,** 80-89.

10. Savir, J. and McAnney, W.H. (November 1985). "On the masking probability with one's count and transition count". *IEEE Int. Conference on Computer-Aided Design.* pp. 111-113, Santa Clara, CA.

11. Saxena, N.R. and Robinson, J.P. (April 1986). "Accumulator compression testing". *IEEE Transac. Computers.* **C-35, 4,** 317-321.

12. Smith, J.E. (June 1980). "Measures of effectiveness of fault signature analysis". *IEEE Transac. Computers.* **C-29, 6,** 510-514.

13. Waser, S. and Flynn, M.J. (1982). "Introduction to arithmetic for digital system designers". Holt, Rinehart and Winston, New York, NY.

APPENDIX

C 1

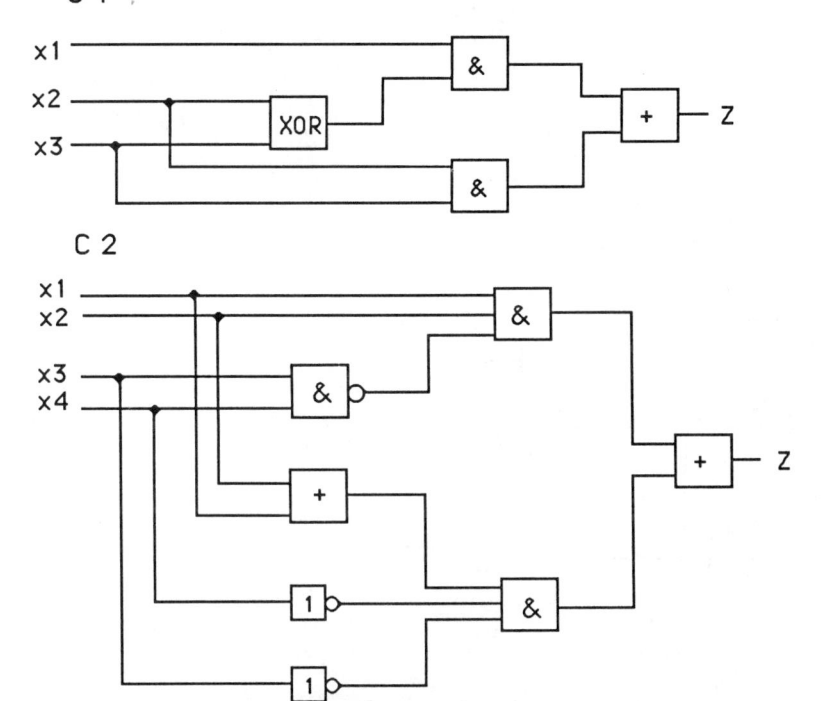

C 2

C 3

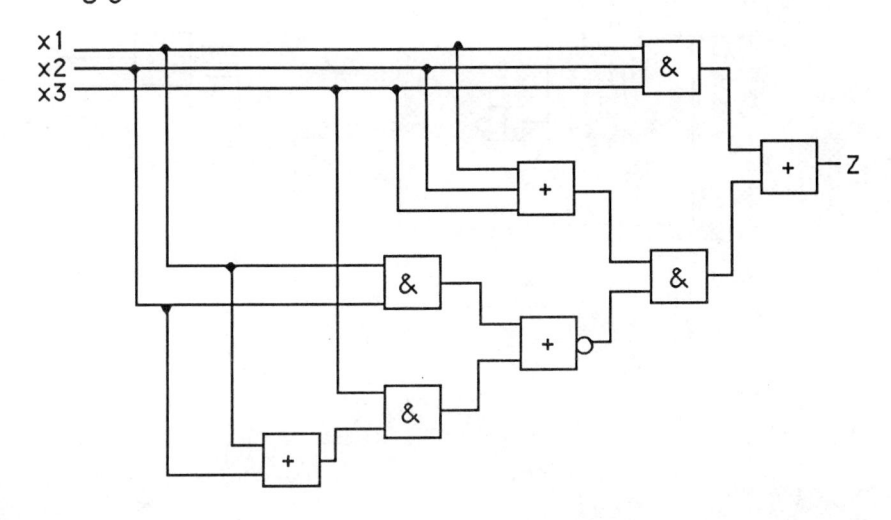

C4

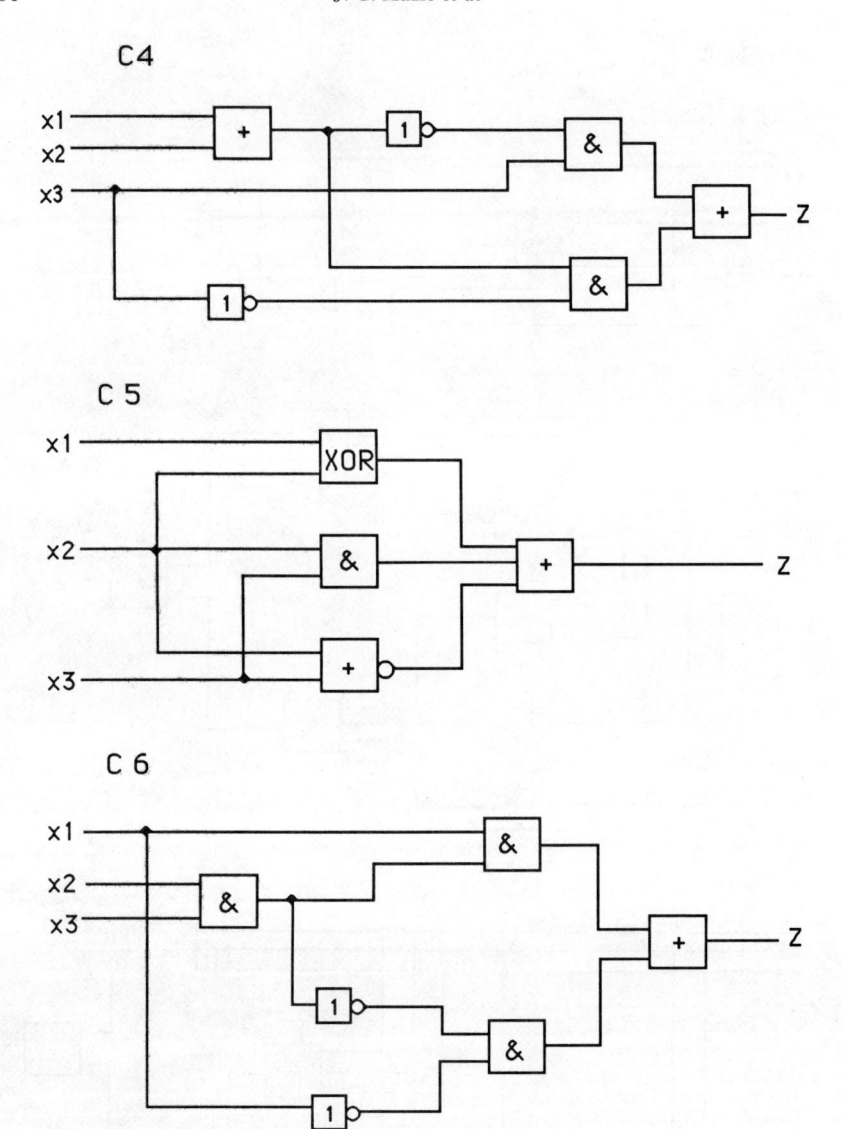

C5

C6

Aliasing Probabilities

C 7

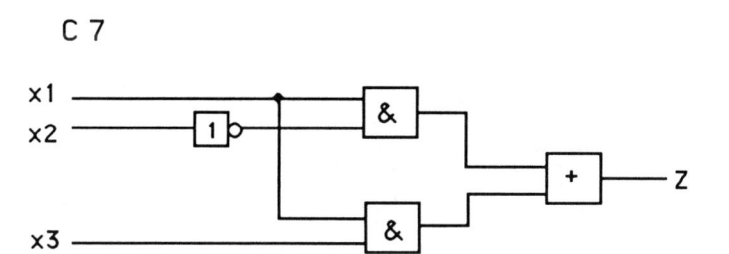

C 9

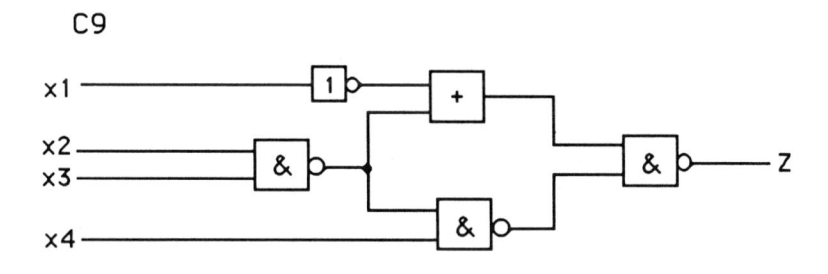

C 10

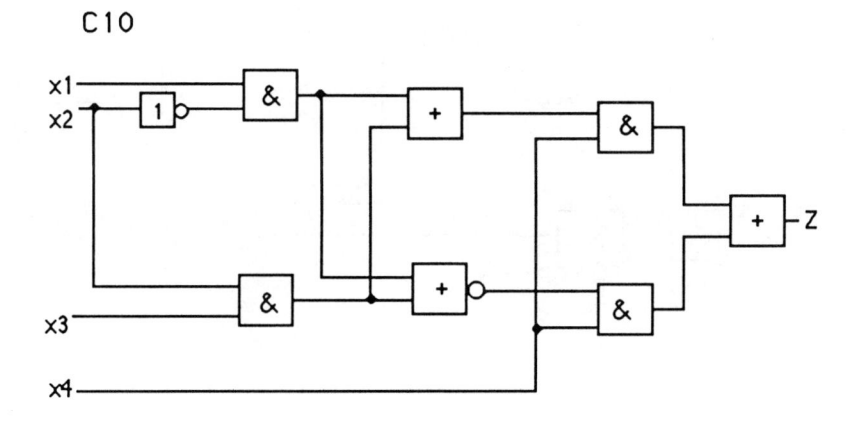

C11

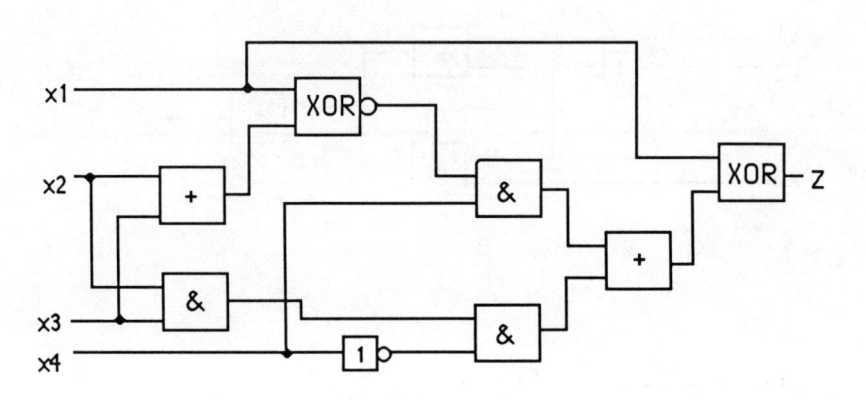

C12

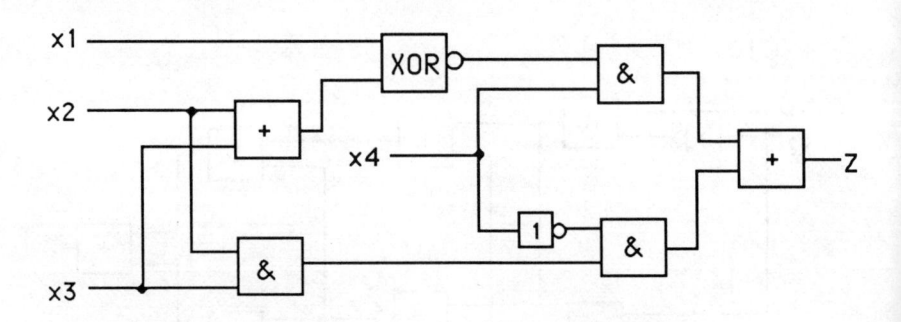

C13

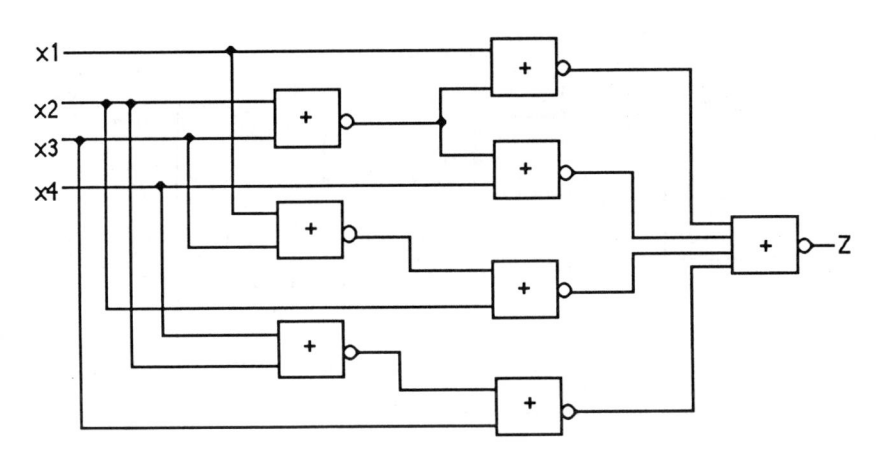

C14

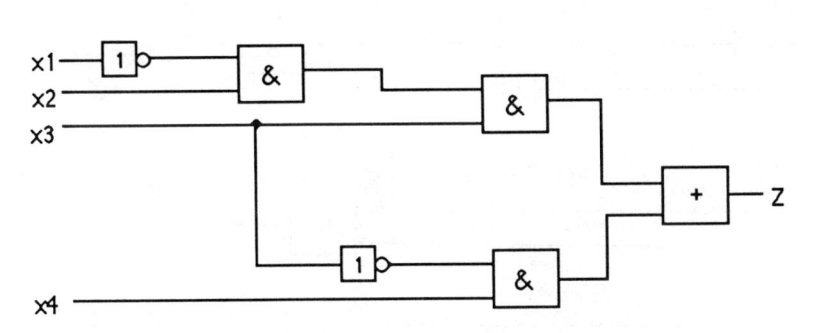

C15

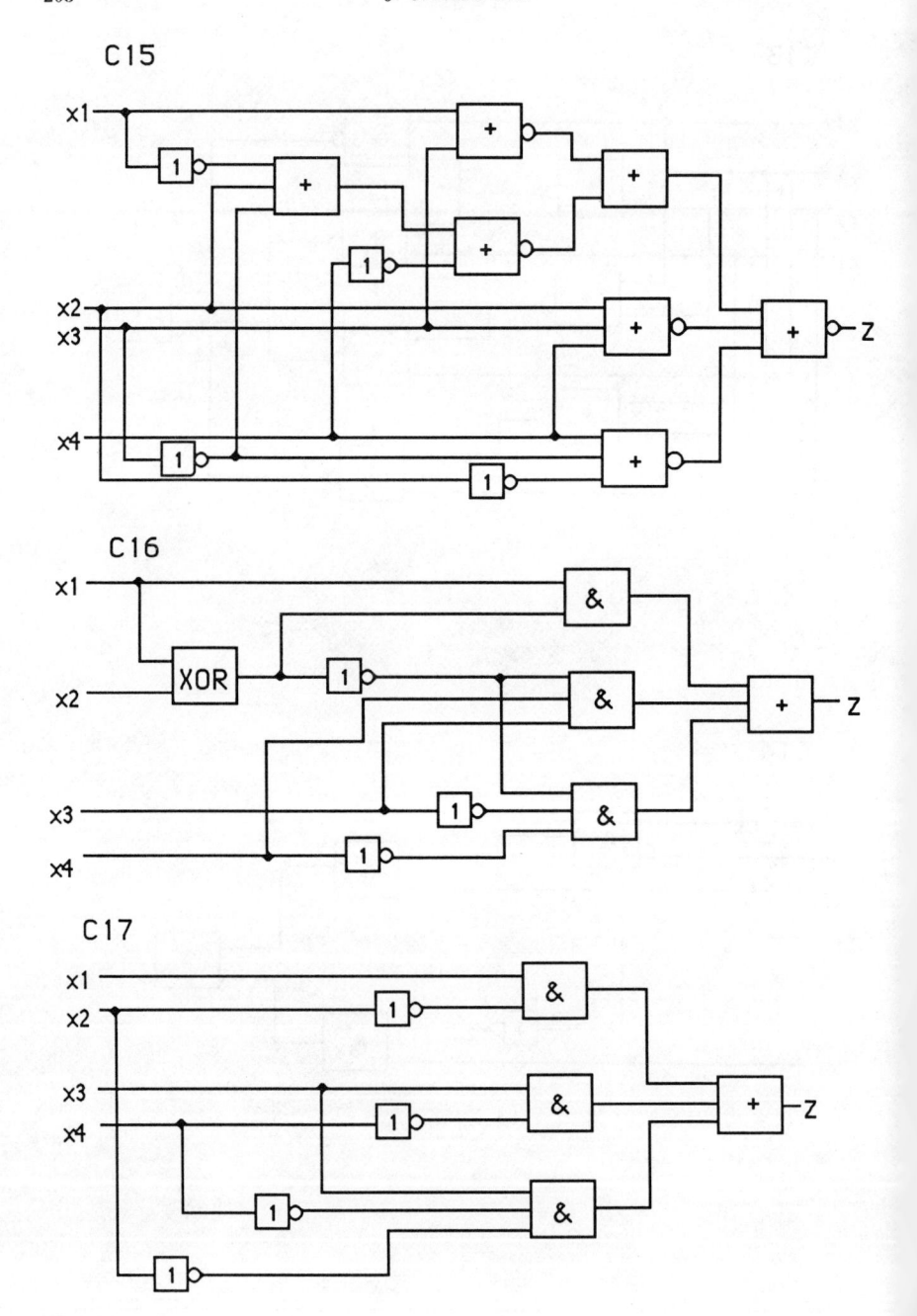

C16

C17

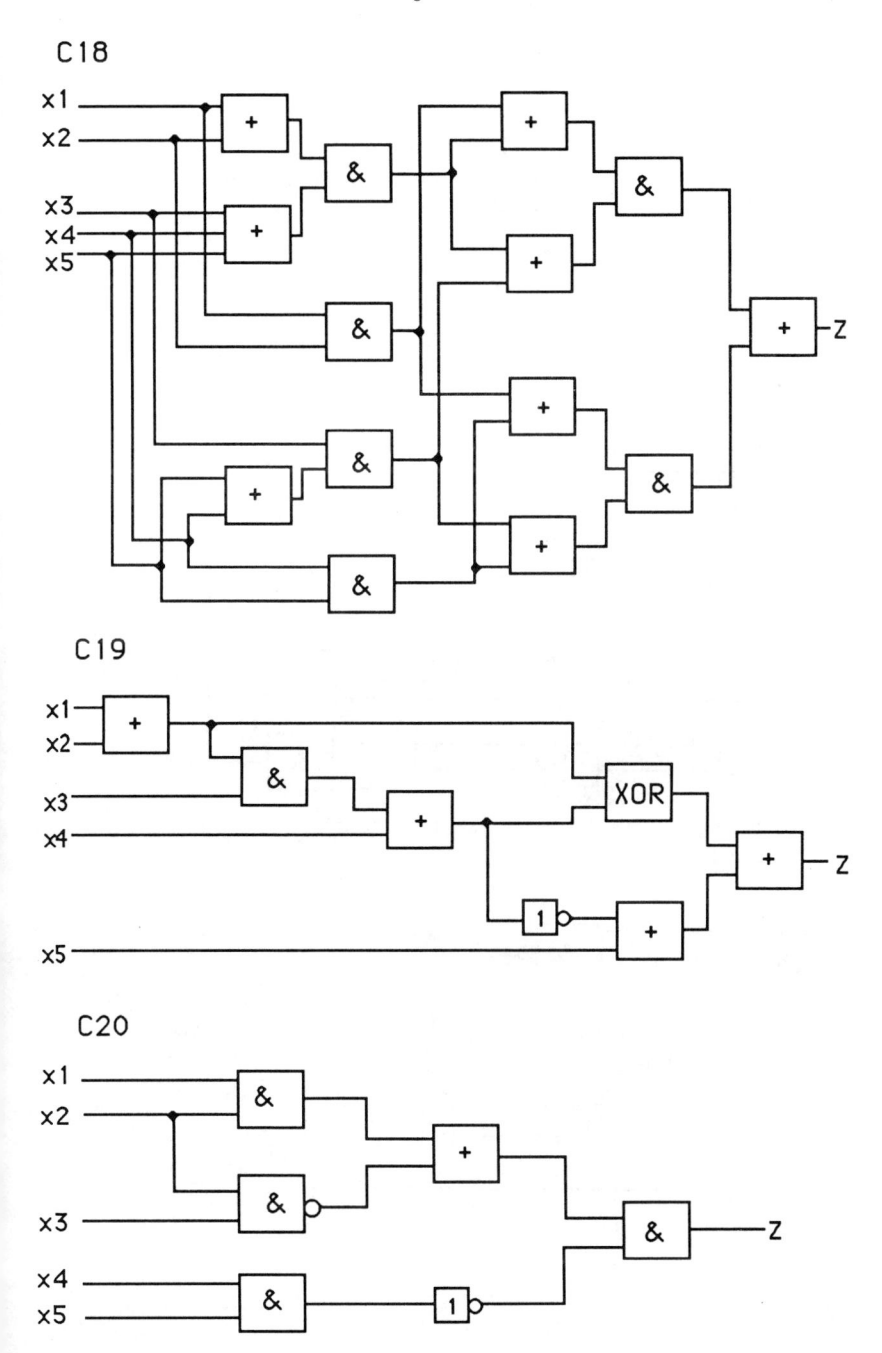

Aliasing Probabilities

C18

C19

C20

C21

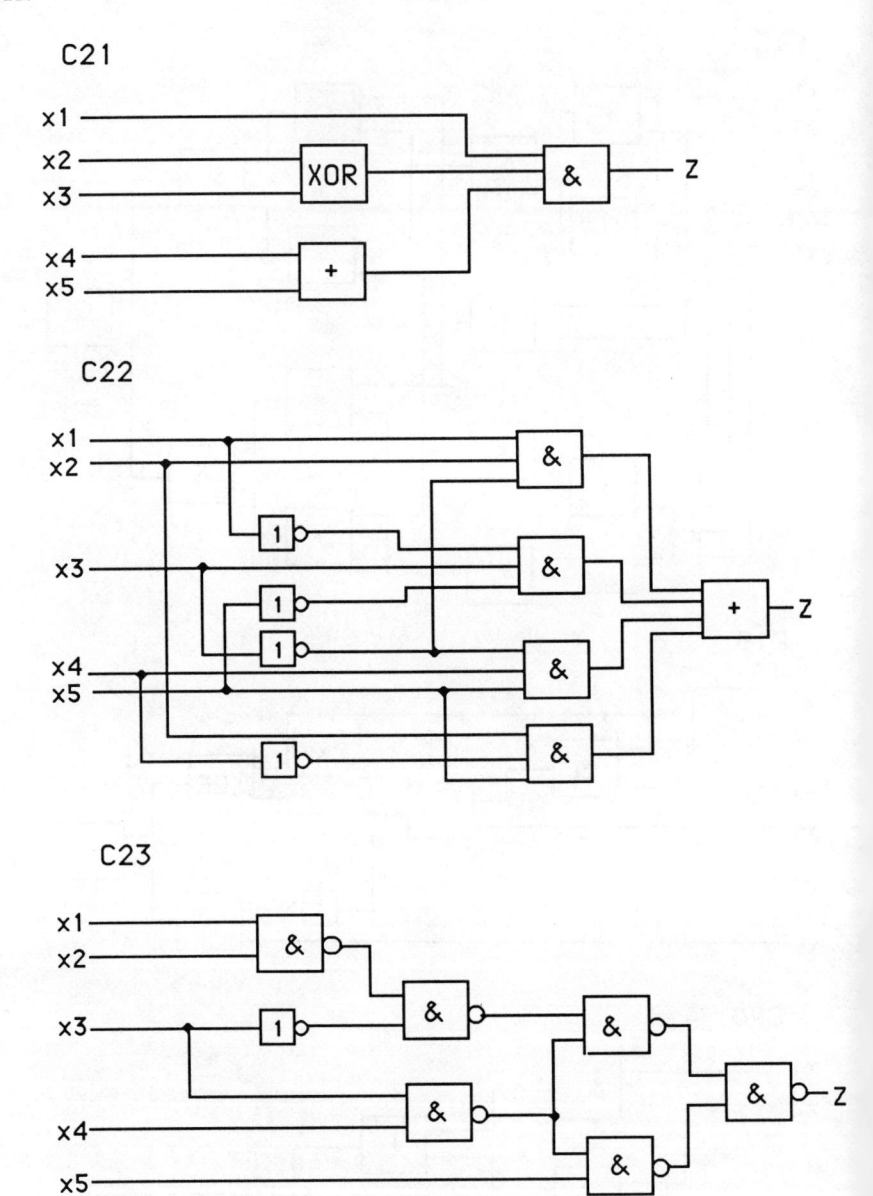

C22

C23

Aliasing Probabilities

C24

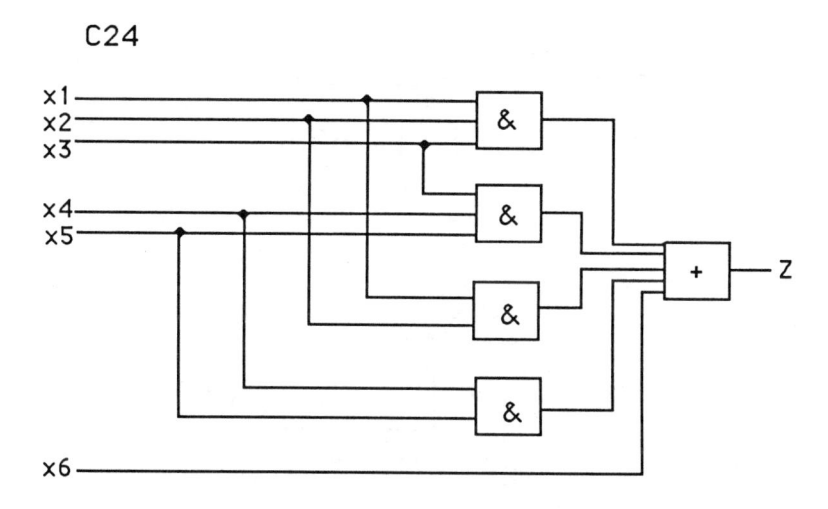

C25

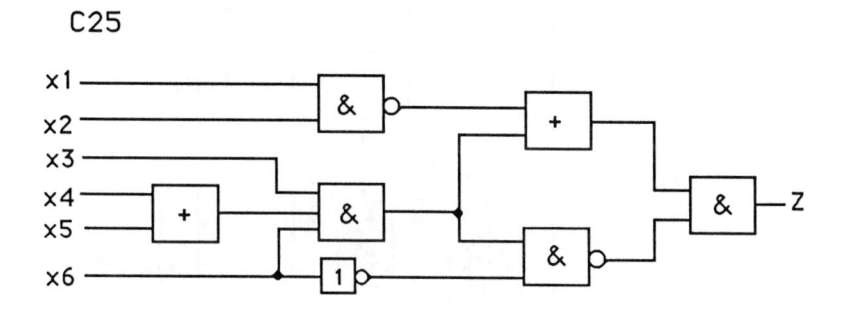

C26

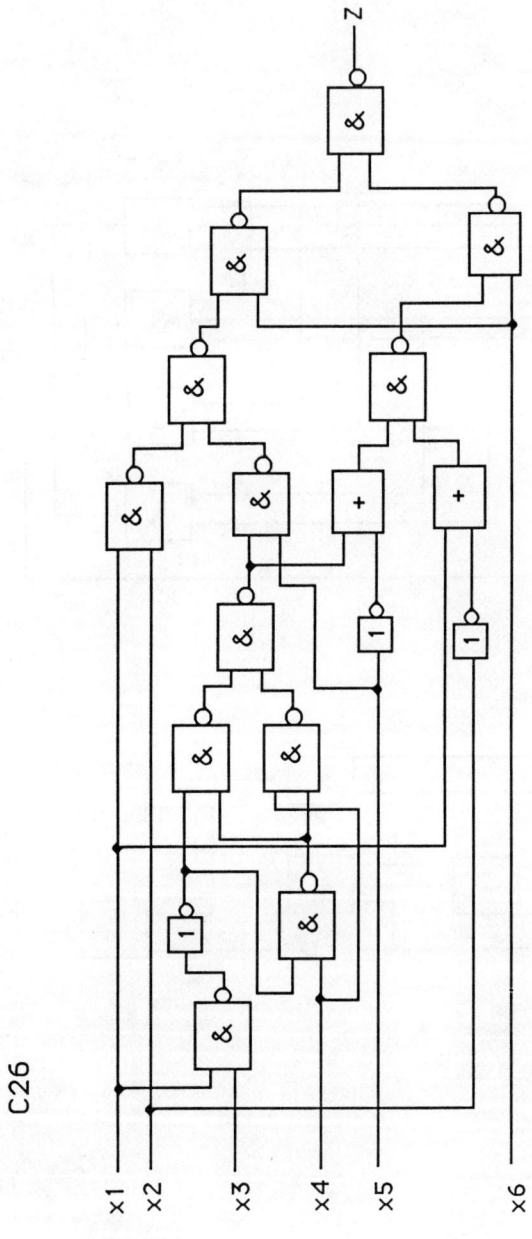

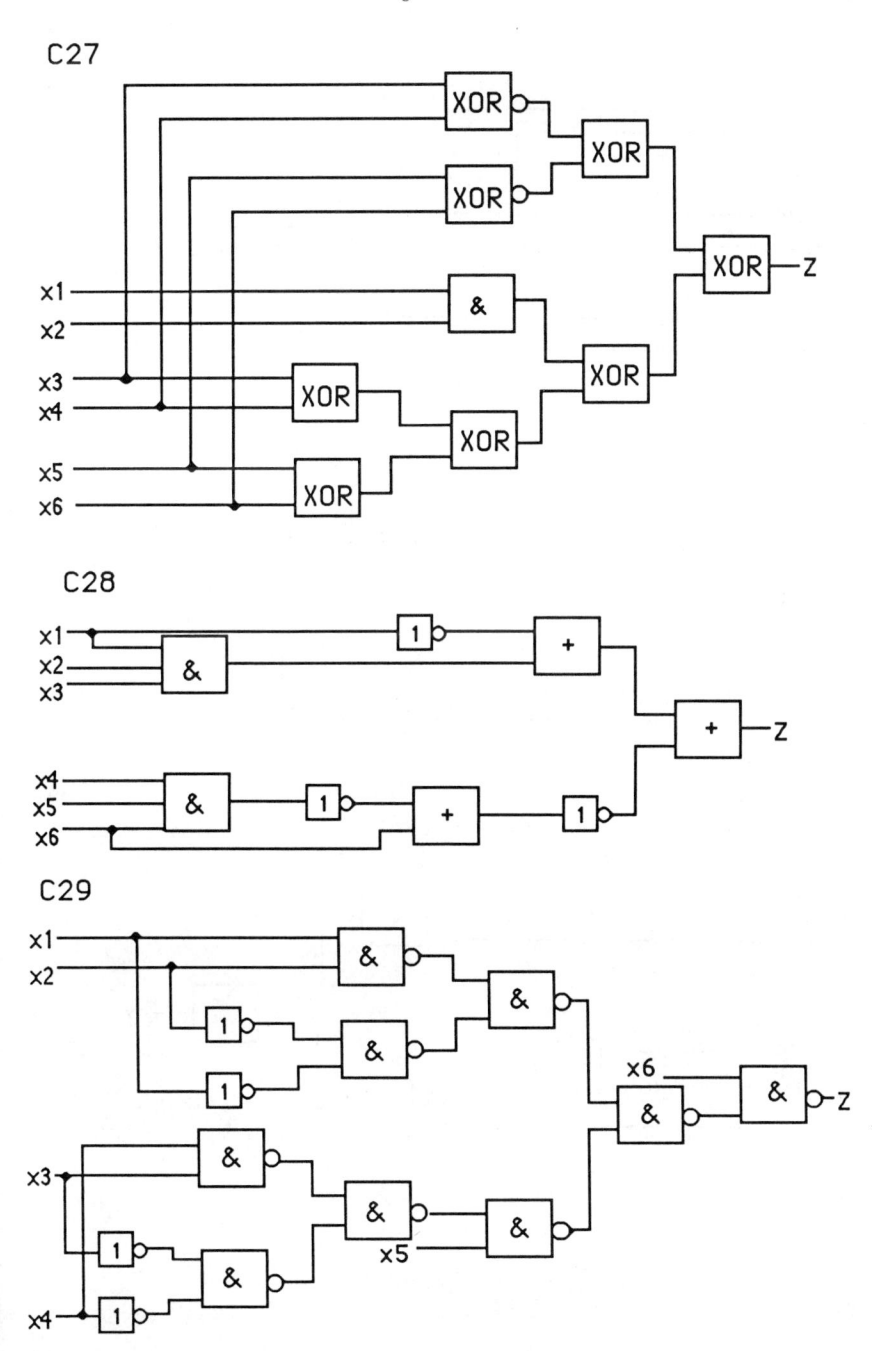

Aliasing Probabilities

C27

C28

C29

C31

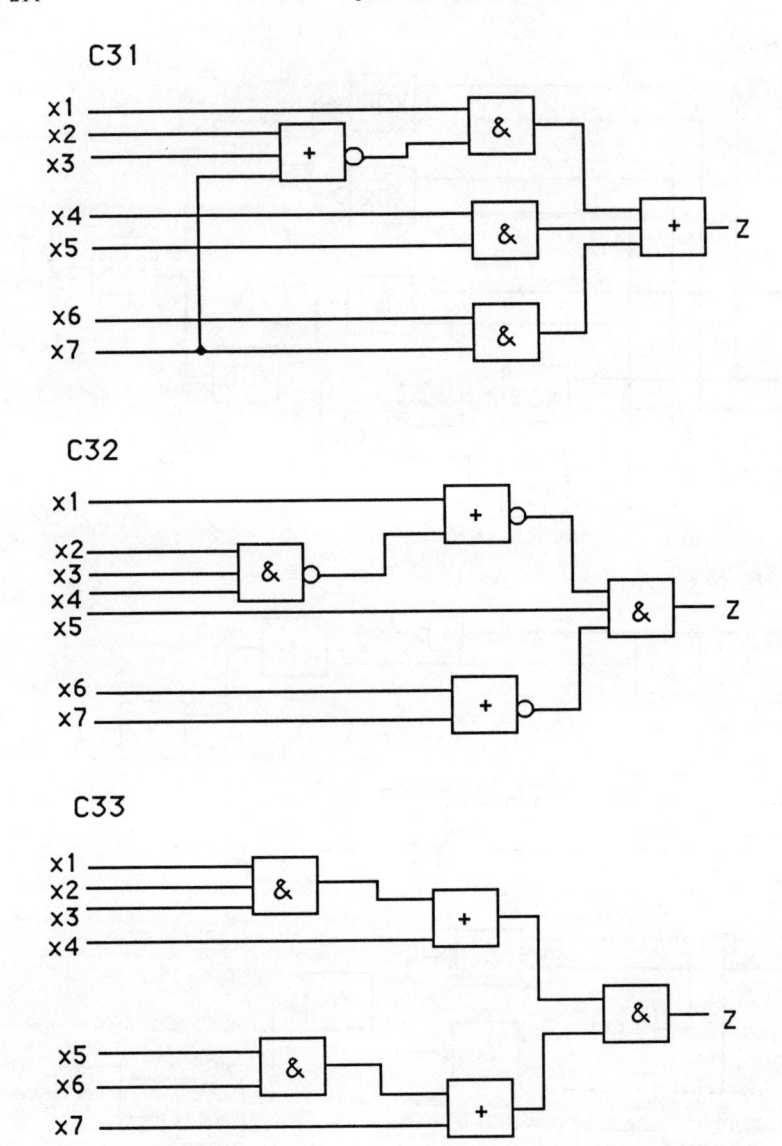

C32

C33

C36

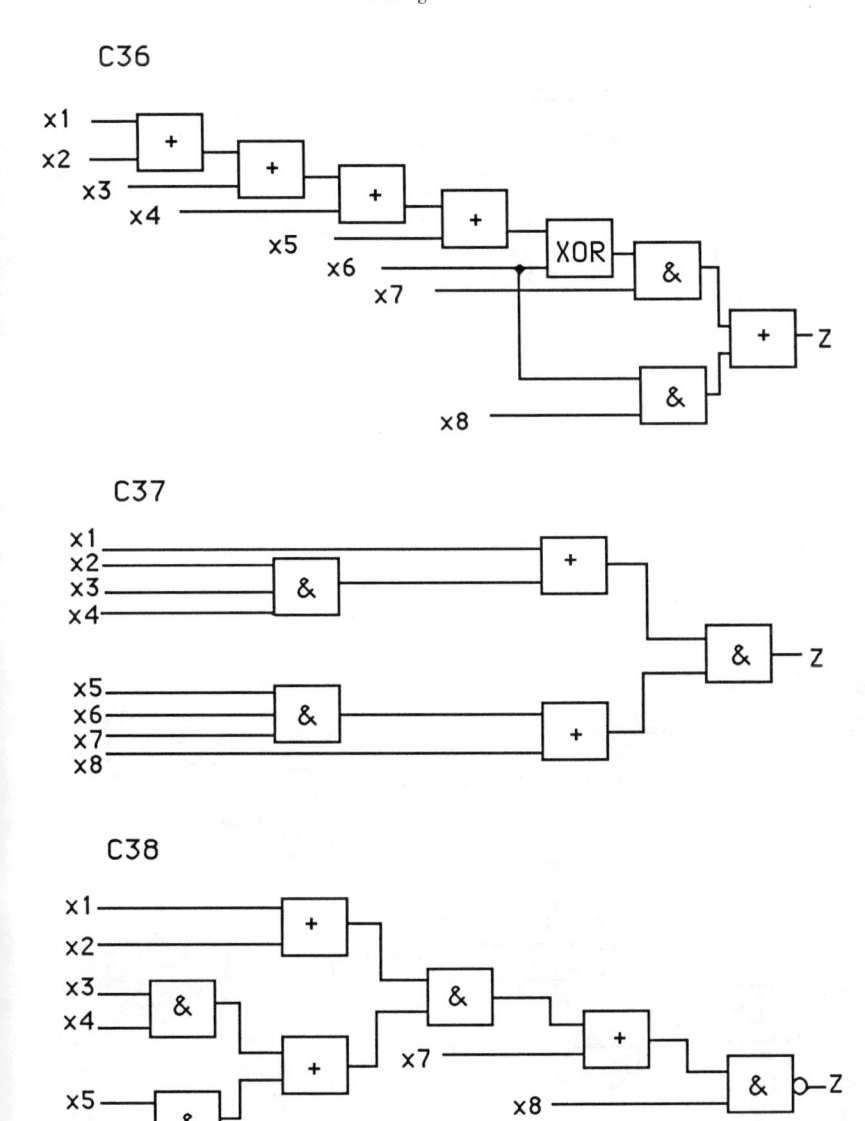

C37

C38

C40

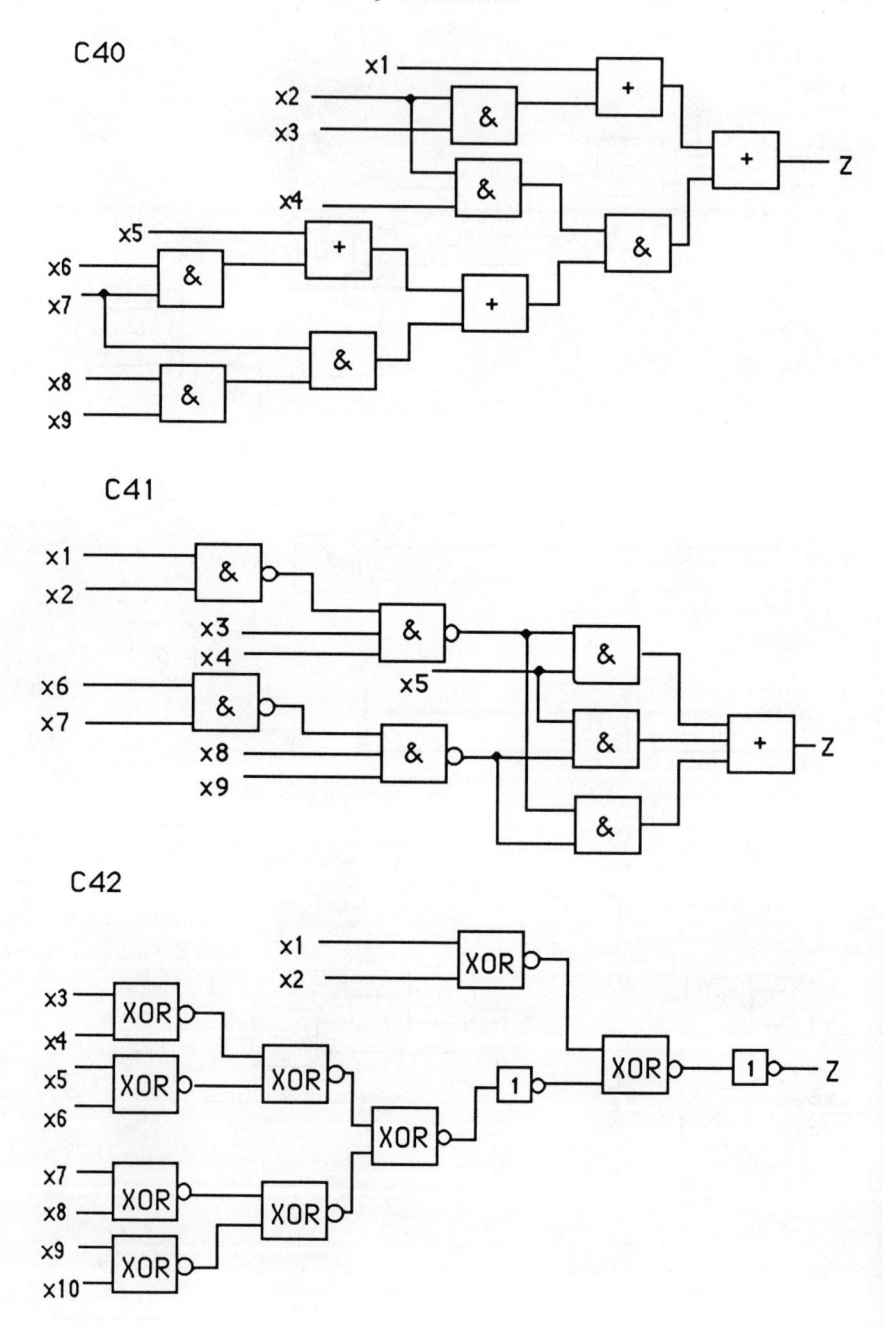

C41

C42

C43

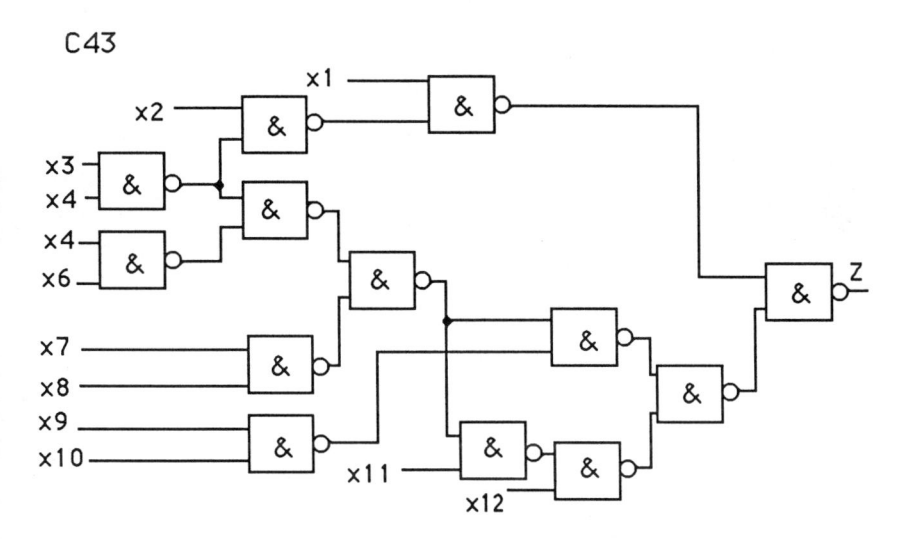

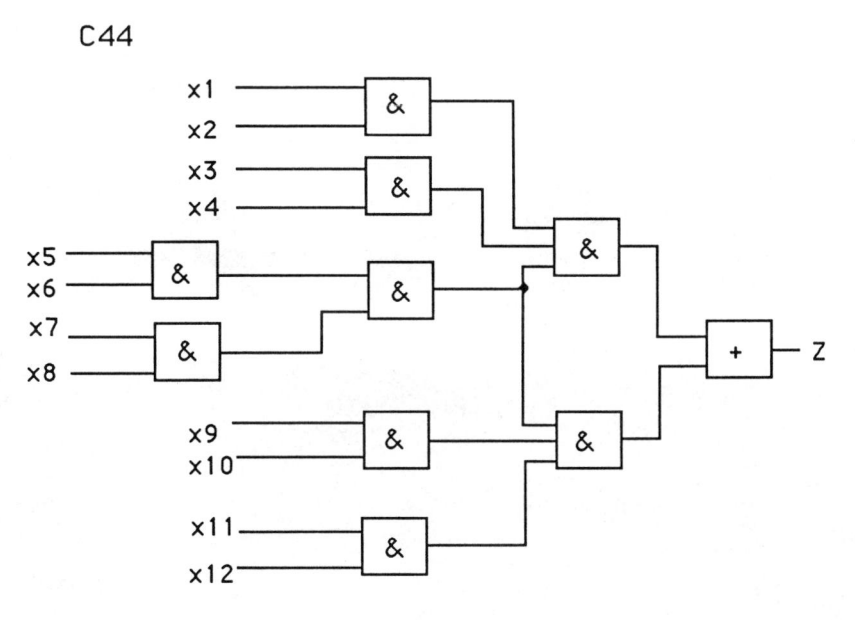

C44

AN INTRODUCTION TO AN OUTPUT DATA MODIFICATION SCHEME

Vinod K.Agarwal and Yervant Zorian

VLSI Design Laboratory
Department of Electrical Engineering
McGill University
Montreal, CANADA

I. INTRODUCTION

The emergence of built-in self-test $(BIST)$ as an alternative to the conventional form of external testing of $VLSI$ circuits is posing very challenging problems (23). One of these problems is to increase the quality of fault coverage of a $BIST$ implementation without incurring a large overhead. In particular, the loss of information in the output data compressor, which is typically a multi-input linear feedback shift register$(MISR)$, is a major cause of concern (2),(4),(19),(20) and (24). Recently, a new scheme, (1),(25),(26),(27),(29), has been proposed which appears to provide the best improvement in the quality of $BIST$ over all the other schemes. This scheme, called Output Data Modification (ODM), exploits the knowledge of the functionality of the circuit under test (CUT) to improve the quality of standard $BIST$. In this chapter, the ODM concept, modifier sequence generation methods as well as some implementational aspects will be considered, followed by some simulation results. It will be seen that the ODM scheme can indeed serve a very useful role in providing high fault coverage through $BIST$, with a minimal increase in the area overhead.

The remainder of the chapter is organized as follows. The error masking problem and the ODM concept are introduced in section II. A specific element of the ODM scheme is the modifier sequence which depends on the functionality of the CUT. Various modifier sequence generation methods are presented in section III. The realization of the scheme as a $BIST$ design along with the additional hardware requirements are considered in section IV. The implementation of ODM scheme and the design of the *Modifier* block is detailed in section V. Finally, some simulation results and conclusions are presented in section VI.

II. OUTPUT DATA MODIFICATION SCHEME:

In this chapter we describe the (ODM) scheme for a CUT which is being tested by l input patterns, and where output is being compressed on m-output lines. The input generation is either pseudo-random or exhaustive, generated by an $LFSR$ or a *Counter*. The assumption of scan design assures that the CUT under consideration is a combinational block (8).

A. Error Masking Problem:

The output data matrix compression in any $BIST$ approach can be thought of as a many-to-one function f, which maps $l \times m$ bit matrices into k-bit signatures, where $k \ll l \times m$. This reduction in data is bound to result in a loss of information. A typical measure of this loss is in terms of error coverage (and not fault coverage), which is determined by a probabilistic approach (19). To be more specific let A_0 represent the $l \times m$ matrix of the particular fault-free CUT which is being considered, and let **A** be the the set of all possible $l \times m$ matrices (exluding A_0) such that for each $A_i \in $ **A**, the signature $f(A_i)$ equals the error-free signature $f(A_0)$.

The number of elements in this set **A** is defined as the deception volume of A_0 with respect to f and denoted as $dv(A_0, f)$. If $dv(A_0, f) = 0$, there is no error information loss, but if $dv(A_0, f) > 0$, then there are $dv(A_0, f)$ possible erroneous matrices which are undetectable by the testing scheme under consideration.

If each of the total number $T = 2^{l \times m} - 1$ of all possible $l \times m$ matrices has the same probability of occurance as a result of faults in the CUT, then the ratio:

$$pr = \frac{dv(A_0, f)}{T}$$

is referred to as the *error masking probability* of f, given A_0. A small pr is considered as an indicator of an efficient f and/or A_0. It is known((5) that no CUT might be able to satisfy such an equi-likely assumption; however, until more specific results are known in this area, one has to live with the conventional wisdom.

Nontheless, this definition of pr allows simple calculations to determine its value. For instance, if the output compression is done using an $MISR$, then $dv(A_0, f) = 2^{l \times m - k} - 1$ and pr is simply 2^{-k} for large l and m. If $k = 16$, 2^{-k} provides a reasonable masking probability. On the other hand, the realization that the denominator T used in calculating pr is an extremely large and unverified number, it appears reasonable to expect that for large and complex CUT's, pr should be a much smaller number. In the following discussions, we will use the two expressions *decrease error masking* and *increase error coverage* without any distinction. Moreover, wherever, the compression function f is obvious, we will write $dv(A_0)$ instead of $dv(A_0, f)$.

B. Improving Error Coverage:

Some attempts to reduce error masking have been reported in the literature. The simplest scheme is to increase the size of the signature register from k to say $2k$. With this the probability of error masking now is 2^{-2k}, which however is not significant in terms of reducing the deception volume. For instance, when $l = 1000$ and $m = 50$, the deception volume with $k = 16$ is $2^{50000-16}$ and with $k = 32$ is $2^{50000-32}$. Another approach (12) is based on applying the same test set more than once but in a different order. The resulting signatures so obtained can be shown to reduce the deception volume by the similar factor as above. Yet another approach (2) suggests using different type of compressors to reduce the deception volume. Here again, the reduction however is no better than that obtained in the previous two approaches. Moreover, some improvement in the reduction of deception volume is obtainable in the Accumulator Compressor approach (18). However, in this case, considerable additional hardware is required due to the adoption of an extra adder-accumulator.

C. ODM Concept:

In the ODM scheme, the probability pr is decreased by a very significant amount. To see how this scheme works, notice first that if a compression scheme produces equal deception volumes to each of its input matrices $(l \times m)$, it is said to have a uniform deception volume. For instance Fig.1, where each point on its x-axis refers to a specific $l \times m$ matrix, shows the constant deception volume with respect to

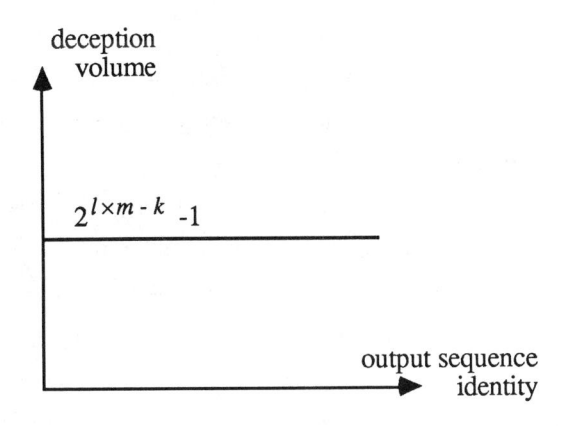

[Fig.1. Uniform deception volume ($MISR$), showing deception volume at $2^{l \times m - k} - 1$ versus output sequence identity]

Fig.1. Uniform deception volume ($MISR$)

the identity of the output for polynomial based compression scheme (7) and (9).

On the other hand, the count based schemes(10),(13), and (17), like One's Counting, Transition Counting and Edge Counting, have a non-uniform deception volume as shown in Fig.2, where the deception volume is a function of weight, number of transitions and number of edges, respectively. For instance if One's Counting is applied on a single output CUT, then the resulting signature is the weight of its

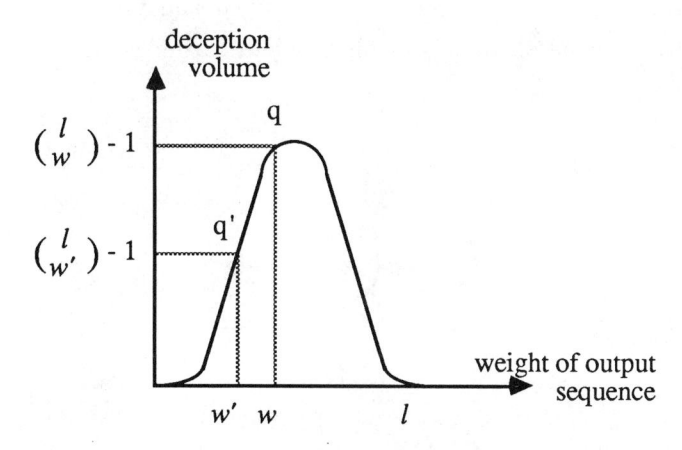

[Fig.2. Non-uniform deception volume graph with $\binom{l}{w}-1$ at peak q and $\binom{l}{w'}-1$ at q', plotted against weight of output sequence, with w', w, l marked on the x-axis]

Fig.2. Non-uniform deception volume (One's Counting)

l-bit long output string. Since this weight is generally close to $l/2$, its deception volume, as seen in Fig.2, will be $\binom{l}{l/2}$. However, if the weight happend to be, either close to zero or one, then the deception volume will be much smaller.

The purpose of the output data modification concept is to take the advantage of this non-uniform distribution, so as to reduce the deception volume of the output sequence. Fig.3 shows that this reduction in deception volume is realized by modifying the output sequence **q**, by using a modifier sequence **s**, so that the resulting sequence **q'** has a deception volume which is much smaller than that of **q**, $dv(\mathbf{q'}) << dv(\mathbf{q})$. This is done with no loss, since the original string can be fully recoverd from the modified one by performing the bit-wise Exclusive-OR with **s** again. However, this scheme is practical only if the modifier string **s** can be generated easily and by inexpensive hardware.

In fact the ideal value for the modifier string **s** is the original string **q** itself, so that after the bit-wise Exclusive-OR is performed, **q'** would be a string of all zeros; that is $dv(\mathbf{q'}) = 0$. Since this is not an inexpensive possibility, **s** has to be as close as possible to the original string **q**, in the case of One's Counting scheme. Because the closer the provided **s** to **q** is, the smaller the deception volume that is obtained.

An application of this concept is developed and presented next as a new *BIST* general model, which provides higher certainty of error coverage.

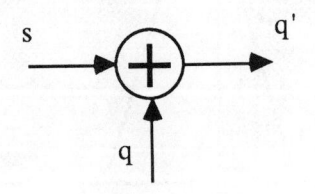

Fig.3. Modification Operation

D. BIST General Model:

The realization of such a model is shown by the setup in Fig.4. Here the dotted box containing CUT shows the standard form of $BIST$ scheme. This general model adopts, either a pseudo-random or an exhaustive input pattern generator (IPG). Two additional blocks appear in Fig.4. The block on the left is called the *Modifier* and the one on the right is called the *Improved Compressor*. This second block receives as input two l-bit long single bit streams: s from the *Modifier* and the quotient bit stream q produced by the $MISR$. The two streams of data are Exclusively-ORed to produce a modified q denoted by q'. Finally, q' is compressed by a count based scheme. For instance, if One's Counting is adopted then the deception volume is changed from $\binom{l}{w} - 1$ to $\binom{l}{w'} - 1$, where the number of one's in the q is w and that in q' is w'. Correspondingly, the error masking probability is reduced from $(\binom{l}{w} - 1)/(2^l - 1)$ to $(\binom{l}{w'} - 1)/(2^l - 1)$. For $l = 4000$, $w = 1900$ and $w' = 700$, the two making probabilities respectively are 2^{-21} and 2^{-1400}. A detailed description of these design blocks is presented in section IV.

In order to generate the closest possible s which uses a very small silicon area dedicated to the *Modifier* block, various modifier sequence

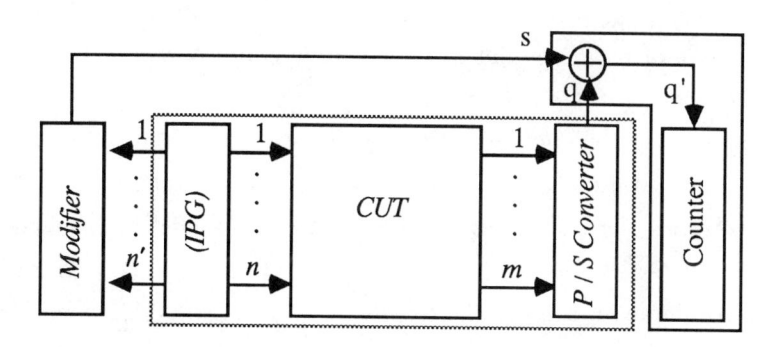

Fig.4. *ODM* Scheme as a *BIST* General Model

generation methods are proposed in section III. In the remainder of this chapter, we will assume that the *Improved Compressor* adopts the One's Counting scheme. However, our discussions can be easily extended to other count based schemes.

III. MODIFIER SEQUENCE GENERATION

Since the *Modifier* block is a part of the total overhead, the amount of hardware assigned to implement it should be small. For this reason, several methods have been developed to generate the modifier sequence inexpensively. In addition to the hardware constraint, another aspect to be considered is the capability of a method to generate a large set of modifier sequences, so that the possibility of finding a closer sequence to the original q is increased. The l-bit long sequences generated by $O(n')$ gates, where $n' = log(l)$, will be refered to as easily generatable sequences (EGS). In a sense, an EGS here has a regular structure which allows it to be generated so easily. If such a sequence s is used as a modifier sequence, then it provides a new perspective to the concept of compression. Consider $q \oplus s = q'$, then $q = q' \oplus s$. Since the Exclusive-OR operation is simply addition modulo-2, we can see that q consists of two parts: a regular part and an irregular part. Thus, if we wish to compress q, then simply compressing the irregular part is enough, because the compression of the regular part can be taken care of by a modifier string. Thus the aim of our ODM scheme can be seen as the extraction of as much regularity of a sequence as possible. Without any loss of generality we will in the following deal with only those q's wherein $w \leq l/2$.

Two types of modifier sequence generation methods are developed to produce EGS's. The methods which generate a sequence by segments are considered in section III.A, whereas a method which generates a sequence by totality is described in section III.B .

A. Methods to Generate a Sequence by Segments:

A method belongs to this type if it forms a sequence by generating its segments and then by concatenating them, using combinational circuits. Two methods, the Exponentially Increasing Segments (EIS) method and the Equal Size Segments (ESS) method, to generate EGS's are introduced below.

1. EIS Modifier Sequence Generation Methods:

(a). Single Input Method (1):
A sequence s^i of length 2^i will be said to consist of exponentially increasing segments if s^i is recursively obtained by concatenating two subsequences of identical length, 2^{i-1}. Each sequence is generated by starting with a subsequence s^0 of length 1 and recursively building subsequences of exponentially increasing lengths. More specifically s^1 consists of s^0 as one half and a sequence r^0 of length 1 as the other half, where r^0 is s^0, $\overline{s^0}$, Z^0 or 1^0, and

$$s^1 = r^0.s^0$$

Here Z^i (1^i), $0 \le i \le n'$, refers to a sequence of length 2^i consisting of all zeros (all 1's). For convenience, we assume that $l = 2^{n'}$, where $n' \le n$. Thus the more general form can be given as

$$s^{i+1} = r^i . s^i$$

where $r^i = s^i$, $\overline{s}^i$, Z^i or 1^i, $0 < i < n' - 1$.

Recall now an $s^{n'}$ of length l is to be generated in our method, which is as close as possible to a given q. Thus, the generation of $s^{n'}$ is governed by the makeup of q. In other words, at each stage i of generation, we make the choice of r^i from the four possible options based on what the best choice for a given q is.

Lemma 1: The total number of EGS's of length $l = 2^{n'}$ is $O(2^{2n'})$ and thus their ratio to the total number of sequences of length l tends to 0 for large l, that is
$$\frac{2^{2n'}}{2^l} \to 0 \quad \text{for } l \to \infty$$
Proof: Let $N(n')$ denote the total number of EGSs of length $2^{n'}$. Then from the above definition of EGS, we have $N(n') \simeq 4N(n'-1)$. A simple iteration shows that
$$N(n'+1) \simeq 4^{n'+1} N(0) \simeq 4^{n'+1} . 2 \simeq 2^{2n'+3}$$
Thus $N(n') = O(2^{2n'})$.

Given a sequence **q**, we then have to determine an optimal EGS, such that the weight of $\mathbf{q} \oplus \mathbf{s}$ is minimal over all possible choices of $O(2^{2n'})$ sequences. Due to the exponential size of the choices to be made, an exhaustive search can become very expensive. In a typical application, l may be equal to 2^{20} where 20 is the number of input lines to the CUT. The total number of EGS's then would be $2^{40} \simeq 10^{12}$. In the following we thus propose a heuristic algorithm of $O(n')$ to determine an **s** for a given sequence **q** of length $l = 2^{n'}$.

Algorithm(1):
begin
let $\mathbf{s}^0$, the most-significant-bit of **s**, be the same as $\mathbf{q}^0$,
   which is the most-significant bit of **q**;
for $i = 0$ **to** $n' - 1$ **do**
   begin
   let $\mathbf{s}^i$ be the *segment* generated so far;
   let $\mathbf{q}^i$ be $\mathbf{s}^i$'s corresponding *segment* on **q**;
   select $\mathbf{r}^i$ among $[\mathbf{s}^i, \overline{\mathbf{s}^i}, \mathbf{Z}^i, 1^i]$ so as:
      $\mathbf{r}^i \oplus \mathbf{q}_l^i$ has minimal weight
   (where $\mathbf{q}_l^i$ represents the 2^i-bit long segment to the left of $\mathbf{q}^i$);
   $\mathbf{s}^{i+1} := \mathbf{r}^i . \mathbf{s}^i$
   end;
$\mathbf{s} := \mathbf{s}^{n'}$
end.

Example(1): Let q be as given below:

$$\begin{align}
\mathbf{q} &= 1010\ 0100\ 1001\ 0010 \\
\mathbf{s}^0 &= 0 \qquad \mathbf{r}^0 = \overline{\mathbf{s}^0} \\
\mathbf{s}^1 &= 10 \qquad \mathbf{r}^1 = \mathbf{Z}^1 \\
\mathbf{s}^2 &= 0010 \qquad \mathbf{r}^2 = \overline{\mathbf{s}^2} \\
\mathbf{s}^3 &= 1101\ 0010 \qquad \mathbf{r}^3 = \mathbf{Z}^3 \\
\mathbf{s}^4 &= 0000\ 0000\ 1101\ 0010 \\
\mathbf{q'} &= 1010\ 0100\ 0100\ 0010
\end{align}$$

Clearly the weight of **q** is $w = 6$, whereas that of **q'** becomes $w' = 4$. Notice that in choosing $\mathbf{r}^3$, there were two options; the other being 0010 1101 . There is no restriction in making a choice in such situations.

Lemma 2: The number of bit-operations in (Algorithm 1), is $O(l)$ and the number of bit-vector operations is $O(log(l))$, where $l = 2^{n'}$.

Proof: For $i = 0$ to $n' - 1$, we notice that a constant number of bit-vector operations are performed to determine each $\mathbf{s}^i$. However, since each bit-vector operation corresponds to an increasing number of such bit-operations, the total number is $O(2^{n'})$.

Lemma 3: The weight w' of $\mathbf{q'} = \mathbf{q} \oplus \mathbf{s}$ is always less than or equal to w, wherein w is the weight of **q**, and **s** is obtained from (Algorithm 1).

Proof: Note that at every iteration step for $i = 0, 1 \ldots n' - 1$ in (Algorithm 1), there are four choices one of which is $\mathbf{Z}^i$. Therefore, in the worst case when $\mathbf{s}^i$ can not decrease the weight of $\mathbf{q'}$ at all, then the choice for $\mathbf{s}^i$ is $\mathbf{Z}^i$. Therefore $w' \leq w$.

Suppose now that $w - w' = u$ is the reduction in the weight of **q**. This reduction will lead to a very large reduction in the modified sequence. More specifically, let $u/w = d$ and $l/w = b$. If we assume that $d \geq .07$, then it is possible to prove the following:

Theorem 2: The deception volume of **q'** is at least $2^{.1u}$ times smaller then the deception volume of **q**. In other words:

$$dv(\mathbf{q}) \geq 2^{.1u}.dv(\mathbf{q'})$$

Proof: The proof is largely combinatorial in nature and relies on Sterling's formula for factorial approximation. First of all, we have that

$$dv(\mathbf{q}) = \binom{l}{w} = \frac{l!}{w!(l-w)!}$$

$$= (\frac{l}{e})^l \sqrt{2\pi l} \frac{1}{(\frac{w}{e})^w \sqrt{2\pi w} (\frac{l-w}{e})^{l-w} \sqrt{2\pi(l-w)}}$$

$$= \frac{l^l}{w^w (l-w)^{l-w}} \sqrt{\frac{2\pi l}{2\pi w \; 2\pi(l-w)}} \cdots \quad (1)$$

Similarly

$$dv(\mathbf{q}') = \binom{l}{w'} = \binom{l}{w-u}$$

$$= \frac{l^l}{(w-u)^{w-u}(l-(w-u))^{l-(w-u)}} \sqrt{\frac{2\pi l}{2\pi(w-u) \; 2\pi(l-(w-u))}}$$

$$\quad (2)$$

Notice that in eqn.(2) if $w - u$ is very small, then the use of the Sterling's formula will be invalidated. Keeping that restriction in mind, we next calculate the ratio

$$\frac{dv(\mathbf{q})}{dv(\mathbf{q}')} = \frac{w-u^{w-u}(l-(w-u))^{l-(w-u)}}{w^w (l-w)^{l-w}} \sqrt{\frac{(w-u)(l-(w-u))}{w(l-w)}}$$

$$= (1 + \frac{u}{l-w})^{l-w} . (1 - \frac{u}{w})^w . (\frac{l}{w-u} - 1)^u . R \quad (3)$$

wherein R is the square root expression. Substituting now $u = dw$ and $l = bw$ and rearranging the terms in eqn.(3) we get

$$= (\frac{(b-1+d)^{b-1+d} (1-d)^{1-d}}{(b-1)^{b-1}})^w . R$$

If we now take the log_2 on each side, it follows that

Output Data Modification

$$\log_2 \frac{dv(\mathbf{q})}{dv(\mathbf{q'})} = w((b-1+d)\log_2(b-1+d) + (1-d)\log_2(1-d)$$
$$- (b-1)\log_2(b-1)) + \log_2 R \qquad (4)$$

It can be easily shown that this expression is an increasing function for increasing values of d and b. For the minimum value of $b = 2.0$ (since $w \leq l/2$), we have that

$$\log_2 \frac{dv(\mathbf{q})}{dv(\mathbf{q'})} = w((1+d)\log_2(1-d) + (1-d)\log_2(1-d)) + \log_2 R \qquad (5)$$

where $R = \sqrt{(1-d^2)}$. Thus $\log_2 R = 1/2\log_2(1+d) + \frac{1}{2}\log_2(1-d)$. Since, w is bound to be a very large quantity in $BIST$, the influence of $\log R$ in eqn.(5) can be ignored. Finally, it can be shown that for $d \geq .07$, $(1+d)\log(1-d) + (1-d)\log(1-d) \geq .1d$. Therefore, it follows that

$$\log_2 \frac{dv(\mathbf{q})}{dv(\mathbf{q'})} \geq .1dw = .1u$$

or

$$dv(\mathbf{q}) \geq 2^{.1u}.dv(\mathbf{q'})$$

Example(2): Consider a multiplier circuit for any 8 bit by 8 bit multiplications of positive integers. If we now apply $l = 2^{16}$ input vectors to the circuit, the weight w of the most significant bit sequence $\mathbf{q}$ is calculated to be 9918. Therefore,

$$dv(\mathbf{q}) = \binom{65536}{9918}$$

and

$$pr = \frac{\binom{65536}{9918}}{2^{65536}}$$

However, if (Algorithm 1) is used to determine a modifier sequence and $\mathbf{q}$ is modified to $\mathbf{q'}$ then the weight w' of $\mathbf{q'}$ is 4036, leading to $u = w - w' = 5612$. Therefore

$$dv(\mathbf{q'}) \leq 2^{-.1u} dv(\mathbf{q}) = 2^{-561} dv(\mathbf{q})$$

and similarly

$$pr' \leq 2^{-561} pr$$

Since only $O(n') = O(16)$ gates were used, the reduction of 2^{-561} is very significant. In fact, since $d = u/w = .566$ and $b = l/w = 6.60$, if we use equation (4), we get that

$$log \frac{dv(\mathbf{q})}{dv(\mathbf{q'})} = 9918(6.17 \, log(6.17) + .434 \, log(.434)$$

$$-5.60 \, log(5.60) + \frac{1}{2} log(.48) \simeq 17262$$

Therefore, to be exact $dv(\mathbf{q'}) = 2^{-17262} dv(\mathbf{q})$.

It follows from *Example(2)* that *Theorem(2)* provides only a lower bound on the reduction. For d significantly greater than .07, the reduction is bound to be much higher than $2^{-.1u}$.

Multi-Seed Possibility (25):

It is possible to extend the single input method to reach larger choices of EGS's by selecting the very first bit $\mathbf{s}^0$ from any specific bit position out of the l possible positions of $\mathbf{q}$. We will refer to such a position and any given value (0 or 1) for that position as a seed. Starting from a seed $\mathbf{s}^0$, we start building $\mathbf{s}^i$, $i = 1, ...n'$, by choosing an $\mathbf{r}^{i-1}$ which is either to the left or to the right of $\mathbf{s}^{i-1}$, in a general form this can be given as

$$\mathbf{s}^i = \mathbf{s}^{i-1}.\mathbf{r}^{i-1}$$

or

$$\mathbf{s}^i = \mathbf{r}^{i-1}.\mathbf{s}^{i-1}$$

This is demonstrated in the following example.

Example(3): Let **q** be as given below and the seed be the thirteenth bit from the left.

$$\begin{aligned}
\mathbf{q} &= 1101\ 0011\ 1000\ 0101\ 1001\ 0110\ 1100\ 0001 \\
\mathbf{s}^0 &= 0 \\
\mathbf{s}^1 &= 01 & \mathbf{r}^0 &= \overline{\mathbf{s}^0} \\
\mathbf{s}^2 &= 0101 & \mathbf{r}^1 &= \mathbf{s}^1 \\
\mathbf{s}^3 &= 0000\ 0101 & \mathbf{r}^2 &= \overline{\mathbf{Z}^2} \\
\mathbf{s}^4 &= 1111\ 1010\ 0000\ 0101 & \mathbf{r}^3 &= \overline{\mathbf{s}^3} \\
\mathbf{s}^5 &= 1111\ 1010\ 0000\ 0101\ 1111\ 1010\ 0000\ 0101 & \mathbf{r}^4 &= \mathbf{s}^4 \\
\mathbf{q}' &= 0010\ 1001\ 1000\ 0000\ 0110\ 1100\ 1100\ 0100 \\
& w = 15 \quad w' = 11
\end{aligned}$$

Since building EGS's has an exponentially increasing nature, we need n' generation stages to obtain **s** of length l. A hardware realization of such a *Modifier* block is shown in Fig.5, wherein each generation stage i ($i = 1, 2, ...n'$) is represented by a functional box G^{i+1}, realized by one simple gate with two inputs. The first input provides the previous subsequence $\mathbf{s}^i$, and the second, a clock input c^i which is the i-th output of an n'-bit counter. Since another counter of the same length is regarded in the case of exhaustive input pattern generation (where $n' = n$), a fanout of its outputs can provide the c^i inputs required for each gate G^{i+1}, as seen in Fig.5, thus eliminating the necessity of an extra counter in the *Modifier* block. The formal algorithm for this scheme is presented in (Algorithm 2).

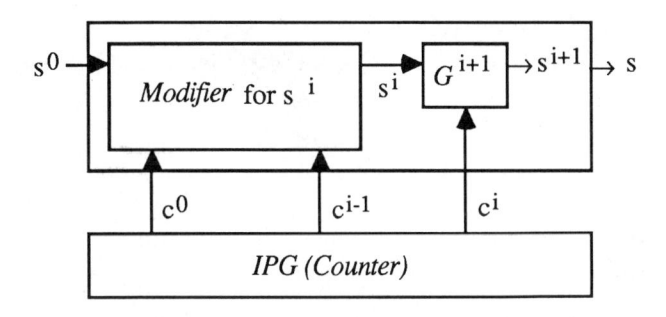

Fig. 5. Single Input Method

Moreover (Table 1) helps to adopt the simple function g^{i+1} realized by gate G^{i+1} corresponding to every stage i. The total number of EGS's of length $2^{n'}$ obtained by this scheme with multi-seeds can be shown to be $O(6^{n'})$ by using this approach.

Table 1.

Modifier Sequence Generation Functions

$\mathbf{s}^{i+1}$	$g^{i+1}(\mathbf{s}^i, \mathbf{c}^i)$
$\mathbf{s}^i.\mathbf{s}^i$	$\mathbf{s}^i$
$\overline{\mathbf{s}^i}.\mathbf{s}^i$	$\mathbf{s}^i\ XOR\ \mathbf{c}^i$
$\mathbf{1}^i.\mathbf{s}^i$	$\mathbf{s}^i\ OR\ \mathbf{c}^i$
$\mathbf{s}^i.\overline{\mathbf{s}^i}$	$\mathbf{s}^i\ XOR\ \mathbf{c}^i$
$\mathbf{s}^i.\mathbf{Z}^i$	$\mathbf{s}^i\ AND\ \mathbf{c}^i$
$\mathbf{s}^i.\mathbf{1}^i$	$\mathbf{s}^i\ OR\ \overline{\mathbf{c}^i}$

Algorithm 2:
begin
for each selected seed $\mathbf{q}^0$ **do**
    begin
    let $\mathbf{s}^0$ be equal to $\mathbf{q}^0$;
    for $i = 0$ to $n' - 1$ **do**
        begin
        let $\mathbf{s}^i$ be the *segment* generated so far;
        let $\mathbf{q}^i$ be $\mathbf{s}^i$'s corresponding *segment* on $\mathbf{q}$;
        select $\mathbf{r}^i$ among $[\mathbf{s}^i,\ \overline{\mathbf{s}^i},\ \mathbf{Z}^i,\ \mathbf{1}^i]$ so as:
            $\mathbf{r}^i \oplus \mathbf{q}_h^i$ has minimal weight
        (where $\mathbf{q}_h^i$ represents the symetric half of $\mathbf{q}^i$);
        if ($\mathbf{r}^i$ is the right-half of $\mathbf{s}^{i+1}$)
            then $\mathbf{s}^{i+1} := \mathbf{s}^i.\mathbf{r}^i$
            else $\mathbf{s}^{i+1} := \mathbf{r}^i.\mathbf{s}^i$
    end;
    let $\mathbf{s}^{n'}$ be the *sequence* which provides minimal $\mathbf{s}^{n'} \oplus \mathbf{q}$
    end;
$\mathbf{s} := \mathbf{s}^{n'}$
end.

Note that any bit in **q** could be a seed. However, it is unrealistic to generate **s**'s based on each seed, and then to select the best one. A problem of selecting the efficient seed thus needs to be solved. Nonetheless, it may be noticed here that, the number of possible EIS's is highest, when the previously generated sequence $\mathbf{s}^i$ consists of alternative 0's or 1's, instead of containing a string of all 0's or all 1's, (for instance, if the sequence $\mathbf{s}^i$ consists of all 0's then $\mathbf{s}^{i+1}$ could be also all 0's, resulting in a low number of EGS possibilities). Thus, the subsequences in **q**, where a number of transitions are concentrated should be first considered to choose a seed from.

(b). Double Input Method (25):

In this method, we provide the capability of generating higher number of EGS's by adding a second (auxiliary) input sequence $\mathbf{p}^i$, for each generating stage i where $i = 1, ...n'$, thus providing two distinct sequences $\mathbf{s}^i$ and $\mathbf{p}^i$ as bases to generate $\mathbf{s}^{i+1}$ and $\mathbf{p}^{i+1}$. However, this additional sequence necessitates an auxiliary line of gates $(G^1, G^2, ...G^{n'})$, besides the main one $(F^1, F^2, ...F^{n'})$, as shown in Fig.6.

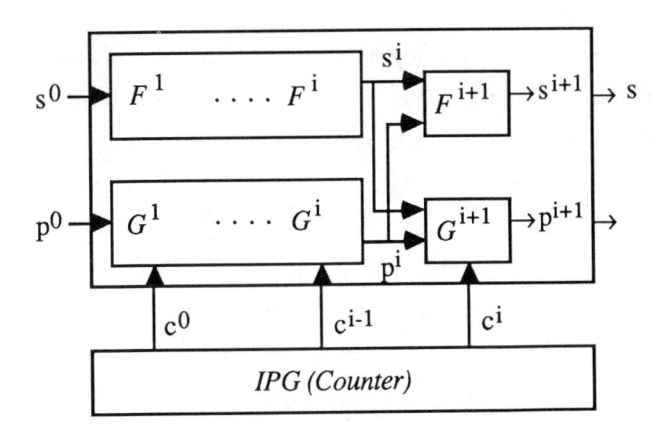

Fig.6. Double Input Method

The advantage of having two distinct sequences is in providing the possibility of obtaining up to 16 (i.e. 2^{2^2}) choices of functions at each stage. Notice that, in contrast, in the previous method only four (i.e. 2^{2^1}) functions were possible. This causes an improvment in the total number of EGS's, thus increasing the availability of an s closer to the original q. Each sequence $\mathbf{s}^{i+1}$ consists of two subsequences of identical length:

- One of them is the exponentially generated subsequence $\mathbf{s}^i$.
- The other is the auxiliary subsequence $\mathbf{p}^i$, which consists of two independent halves: $\mathbf{p}_l^{i-1}$ and $\mathbf{p}_r^{i-1}$. These two halves are generated based on both previous subsequences $\mathbf{s}^{i-1}$ and $\mathbf{p}^{i-1}$ by using the functions g_l^{i-1} and g_r^{i-1}, as stated below:

$$\mathbf{p}_l^{i-1} = g_l^{i-1}(\mathbf{s}^{i-1}, \mathbf{p}^{i-1})$$
$$\mathbf{p}_r^{i-1} = g_r^{i-1}(\mathbf{s}^{i-1}, \mathbf{p}^{i-1})$$

To determine the auxiliary functions g_l^{i-1} and g_r^{i-1} and therefore the structure of G^i, we have to use the two subsequences $\mathbf{s}^{i-1}$ and $\mathbf{p}^{i-1}$, with each one of the obtainable 16 possible functions, the 16 distinct half $\mathbf{p}^i$'s ($\mathbf{p}_l^{i-1}$, $\mathbf{p}_r^{i-1}$), and then for each half $\mathbf{p}^i$ we have to select the closest possible choice to its corresponding part on the original sequence q. These steps are shown in (Algorithm 3) and applied on (Example 3).

The use of double sequences in each stage results in a higher number of obtainable functions. The total number of EGS's in this method can be shown to be $O(256^{n'})$. If the number of input subsequences, entering each stage i is increased, then one can achieve even higher number of generatable sequences: $2^{2^{(\# \ of \ input \ sequences)}}$. However, the complexity of (Algorithm 3) will be increased in the same order, because this complexity depends on the number of generatable sequences, as it tries every possible function and then selects the appropriate one. Moreover, the silicon area requirement is bound to increase as well.

Algorithm 3:
begin
  for each selected seed $\mathbf{q}^0$ do
    begin
      let $\mathbf{s}^0$ be equal to $\mathbf{q}^0$;
      let $\mathbf{p}^0$ be the symetric bit to $\mathbf{q}^0$ on $\mathbf{q}$;
      for $i = 0$ to $n' - 1$ do
        begin
          let $\mathbf{s}^i$ be the *segment* generated so far;
          let $\mathbf{s}^{i-1}$ and $\mathbf{p}^{i-1}$ be the two *segments* of $\mathbf{s}^i$;
          let $\mathbf{q}^i$ be $\mathbf{s}^i$'s corresponding *segment* on $\mathbf{q}$;
          select the function g_l^{i-1} among the 16 possible functions:
$$\mathbf{p}_l^{i-1} = g_l^{i-1}(\mathbf{s}^{i-1}, \mathbf{p}^{i-1})$$
          so as: $\mathbf{p}_l^{i-1} \oplus \mathbf{q}_l^{i-1}$ has minimal weight;
          select the function g_r^{i-1} among the 16 possible functions:
$$\mathbf{p}_r^{i-1} = g_r^{i-1}(\mathbf{s}^{i-1}, \mathbf{p}^{i-1})$$
          so as: $\mathbf{p}_r^{i-1} \oplus \mathbf{q}_r^{i-1}$ has minimal weight;
          $\mathbf{p}^i := \mathbf{p}_l^{i-1}.\mathbf{p}_r^{i-1}$;
          if ($\mathbf{p}^i$ is the right-half of $\mathbf{s}^{i+1}$)
            then $\mathbf{s}^{i+1} := \mathbf{s}^i.\mathbf{p}^i$
            else $\mathbf{s}^{i+1} := \mathbf{p}^i.\mathbf{s}^i$
        end;
      let $\mathbf{s}^{n'}$ be the *sequence* which provides minimal $\mathbf{s}^{n'} \oplus \mathbf{q}$
    end;
  $\mathbf{s} := \mathbf{s}^{n'}$
end.

Example (4): The Double Input Method is applied on the previous example:

$\mathbf{q} = $ 1101 0011 1000 0101 1001 0110 1100 0001
$\mathbf{s}^0 =$ 0 $\mathbf{p}^0 = \overline{\mathbf{s}^0}$
$\mathbf{p}^0 =$ 1
$\mathbf{s}^1 =$ 01 $\mathbf{p}^1 = \mathbf{s}^0.\mathbf{s}^0$
$\mathbf{p}^1 =$ 00
$\mathbf{s}^2 =$ 0001 $\mathbf{p}^2 = \overline{\mathbf{p}^1.\mathbf{p}^1}$
$\mathbf{p}^2 =$ 1100
$\mathbf{s}^3 =$ 1100 0001 $\mathbf{p}^3 = (\mathbf{p}^2 + \mathbf{s}^2).\overline{(\mathbf{p}^2 + \mathbf{s}^2)}$
$\mathbf{p}^3 =$ 1101 0110
$\mathbf{s}^4 =$ 1101 0010 1100 0001 $\mathbf{p}^4 = (\mathbf{p}^3 + \mathbf{s}^3).\mathbf{s}^3$
$\mathbf{p}^4 =$ 1101 0011 1100 0001

$s^5 =$ 1101 0011 1100 0001 1101 0010 1100 0001
$q' =$ 0000 0000 0100 0100 0100 0100 0000 0000

$$w=15 \qquad w'=4$$

(c). Further EIS possibilities (1):

The modifier sequence obtained through the previous methods may not always result in drastic reductions in deception volume. In fact, any one method may never be suitable for all sequences of interest. Here, to enhance the scope of modifier sequences, some new types of EIS modifier generation methods will be considered.

If the modifier sequence s produced by (Algorithm 3) results in no significant weight reduction. Another possibility, in such a situation, will be to try a count based scheme other than the One's Counting, such as Transition Counting or Edge Counting. In such a case, same EIS methods can be adopted but by using different algorithms in order to generate new modifier sequences, which provide reductions in the number of transition or the number of edges respectively.

Another particularly interesting scheme in this regard is the simple Transition Counting (0-to-1 and 1-to-0) which is the weight of the sequence c, where the x-th bit of c is: $c(x) = q(x) \oplus q(x-1)$ and $c(0) = 0$, as shown in (Example 5).

Example(5): For instance if
$$q = \quad 0\ 1\ 0\ 1\ 1\ 0\ 1\ 0$$
$$\text{Shifted-}q = 0\ 1\ 0\ 1\ 1\ 0\ 1\ 0$$
$$c = 0\ 1\ 1\ 1\ 0\ 1\ 1\ 1\ 0$$

Thus the number of transitions in q is 6. Here, c is treated as the modified q itself, then One's Counting is applied on it:

$$dv(\mathbf{c}) = \binom{8}{6} = 28 \; < \; dv(\mathbf{q}) = \binom{8}{4} = 70$$

Notice that c so obtianed will always have one extra bit than q. However, we will without any loss of generality ignore the most significant bit. Interestingly, c has the recovery property in that q can be obtained from c using the following rules:

$$\mathbf{q}(0) = \mathbf{c}(0)$$

$$\mathbf{q}(x) = \mathbf{c}(x) \oplus \mathbf{q}(x-1)$$

Thus, c has the same fault coverage information as q does.

Another possibility is to use a two level modification scheme. For instance, c can be treated further by (Algorithm 3) to determine a modifier string s for c so that the resulting $\mathbf{c}' = \mathbf{c} \oplus \mathbf{s}$ has lesser weight than c. In (Example 5), c is so regular, that an s = 1110 1110 can be generated by (Algorithm 3) leading to $\mathbf{c}' = \mathbf{Z}^3$ consisting of all zeros. The deception volume of c' is of course simply 1.

One can easily increase the set of EGS by employing some rather simple techniques. The two-level scheme is a case in point. The c obtained from q and shifted-q in fact becomes an EGS in the sense that it can be easily generated with $O(n')$ hardware.

Another class of sequences is the set of strings which are already available on the $BIST$ chip as a part of some other subsystem. For instance, if an $LFSR$ is used as an IPG to perform testing on a circuit, then the input sequences applied to the circuit can also be used to generate modifier sequences. To be more specific, we will refer to such sequences as available sequences (AS).

A combination of EGS and AS can be used to generate modifier sequences which do not belong to either. A method called the ESS method, described in the following paragraph, is totally based on the ASs of the IPG.

2. ESS Modifier Sequence Generation Method (25):

One major disadvantage of EIS schemes is that for each stage i, a gate G^i is used to produce a segment of s which is of the same length as another segment produced by all the previous gates $(G^1, G^2, ...G^{n'})$

put together. This leads to lesser control on how close s can be generated to a given q. The ESS method described here is designed to solve this problem.

We noticed in the Double Input Method that a larger number of EGS's can be generated if more input sequences are available in each stage. In particular, if say four input sequences are used for each segment of q, then we can generate up to 2^{2^4} sequences for that segment alone. Thus, the ESS method is based on dividing s into equal segments of size, say $n' = log(l)$, each of which of length l/n' . Then a subsequent s_i $1 \leq i \leq n'$, is generated for each of the n' segments using some available sequences on the chip. To be more specific, the available sequences used in this method are the output sequences of the IPG. All the s_i's are then concatenated by a multiplexor to generate the required s. This is shown in Fig.7 .

Example(6): the following table lists the error masking probability with output data modification for the most significant bit output of an 8 × 8-bit multiplier, for each of the three previous methods.

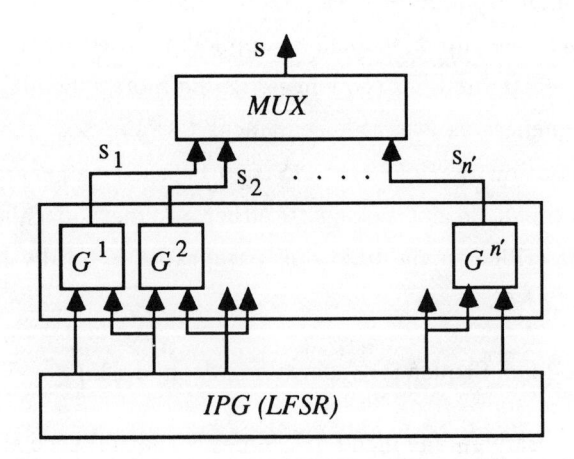

Fig.7. Equal Size Segments Method

Table 2.

8×8 multiplier

Single Input Method	Double Input Method	Equal Size Segments
2^{-23265}	2^{-24875}	2^{-9360}

A. Method to Generate a Sequence by Totality

The methods which generate a sequence by segments and proposed in the previous sections are not very useful if one wanted to reduce the deception volume to any predetermined value. Moreover these methods are not guaranteed to work effectively for all circuits of interest. Both of these problems are overcome in a method described in the following.

A convenient way to describe this method, which generates the modifier string in totality, is to think of the *Modifier* block as a $n' = log(l)$ input and a single output combinational circuit, which when applied with l input patterns produces l bits on the output line. In particular, when the l patterns are applied in a certain order, then the resulting l-bits would form the modifier string s. In order to decide the functionality of the *Modifier* block, suppose that the l-bit long string q as produced by the $MISR$ corresponds to the output column of the truth-table, whereas the input patterns for the *Modifier* block as generated by the IPG correspond to the n' variable of this table. Based on this, it is possible to write a minimal sum-of-products form expression for this function (6).

Let t be the number of products in this expression, since the *Modifier* is a part of the overhead, the amount of hardware assigned to implement it should be as small as possible. Consequently only a limited number of products say p might be allowed to be implemented.

Clearly if $t <= p$, then all the product terms could be included in the *Modifier*, and the exact sequence **q** could be generated by the *Modifier* block resulting in the deception volume of 0. However, if $t > p$, then the expected function is not entirely realized, and a subset consisting of p product terms out of t, will have to be selected to form the modifier function which generates **s**. This selection will of course be based on covering the largest number of 1-vertices in **q**.

It has to be noted that the total number of sequences that can possibly be generated by the *Modifier* block depends on p. Therefore if a higher number of products is made available the number of such sequences should be increased to any desired length. Thus, this resulting flexibility creates a trade off between the overhead required to implement the method and the number of generatable sequences. This is considered as an important advantage of using this method. However, for the method to be useful, it has to be ensured that for any average function created by sequence **q**, a significant reduction in deception volume will occur with a given p. A formal proof of this statement is provided in (26). In the following section, complete implementation of the *ODM* scheme with sequence generation by totality is described.

IV. REALIZATION OF ODM AS A BIST DESIGN

We have so far described the motivation behind our scheme, the modification concept used to realize it, and various methods to generate the modifier sequence. The realization of the scheme as a BIST design is introduced in this secton, where the functionality and the design of each one of its blocks, shown in Fig.4, are described.

A. The Standard BIST Facilities:

The dotted box in Fig.4, where the standard *BIST* facilities are placed, contains besides the CUT the IPG and the signature register which is used in this scheme as a parallel to serial converter P/S Converter.

1. Input Pattern Generator:

The actual task of this block is to generate two sets of l distinct patterns in parallel. The first set, like in any standard BIST scheme, will consist of pseudo-randomly generated test patterns and be applied on the n-input lines of the CUT; whereas, the second set will be composed of $n' = log(l)$ bit wide patterns (where $n' \leq n$) and be utilized as the $2^{n'}$ exhaustive input combinations to the *Modifier* which generates s. We will refer to the first set as the test set for the CUT and the second set as the input set for the *Modifier*.

Since an $LFSR$, whose biggest appeal lies in its capability to generate pseudo-random patterns, is so far the most efficient type of test pattern generator for $BIST$, the following two designs, which are proposed to realize the IPG block, are based on $LFSR$'s and realize both of the above mentioned functions:

a. The first design proposed is of an IPG consisting of two independent $LFSR$'s: The first a special maximum length $LFSR$ of size n' (i.e. $2^{n'} - 1$ distinct patterns) (16) with all its outputs connected directly to the first n' inputs of the CUT, as well as to all the inputs of the *Modifier*, whereas the other $LFSR$ is an $n - n'$ bit long one with its outputs applied on the remaining lines of the CUT. In this case, the test set for CUT is generated by both $LFSR$'s. It has been shown (3), that such an implementation will not affect the pseudo-random nature of the input patterns for the CUT.

b. The second proposed is based on various techniques suggested in the literature (21) and (11), to design $LFSR$'s of size n, that have the ability to generate exhaustive patterns on n' of its outputs. One such technique (22) can very conveniently be used in realizing our model. This technique shows that an $LFSR$ of size n, with a characteristic polynomial $f(x) = g(x).p(x)$ over modulo-2 can generate $2^{n'} - 1$ distinct patterns, provided that $p(x)$ is a maximum length (primitive) polynomial of degree n', $g(x)$ is any monic polynomial of degree $n - n'$, and the initial state of the LFSR is divisible by $g(x)$.

2. P/S Converter:

If the modification concept were to be applied to every single line of a circuit with m output lines, then we will require m *Modifier* blocks. However, for large values of m, such an approach will be very expensive. To solve this problem, we propose the use of a *P/S Converter* which will transform every m-bit output vector into a single bit. The selection of a proper *P/S Converter* is however crucial to the desired effectiveness of error coverage.

There are various possibilities which can be used to select a proper *P/S Converter*. In the following, we describe three such schemes with their advantages and disadvantages:

- The simplest P/S conversion scheme is to use any one output line, to create the serial string and to ignore all the others. However, this is clearly inappropriate because many faults which are localized in certain areas of the IC chip may give completely correct results on the selected output line, even though the chip is faulty.

- Another P/S conversion scheme, which performs sampling on the output lines to create a single stream of bits is also inappropriate. Since certain parts of an IC chip operate only when control variables switch them on and switch other parts off, it is possible to have bursts of incorrect values on certain output lines when these sections are switched on. This burst of erroneous bits may occur between the times that the multiplexor takes a bit from the other output lines, and therefore may be absent from the serial stream.

- Since an $MISR$ converts each output vector, during each clock cycle, into a single quotient bit (the most significant bit of $MISR$) by utilizing all the m-bits, it seems to be a better choice for a proper P/S converter than the above two schemes. Thus, in our scheme we use an $MISR$ as a P/S Converter and monitor the quotient bit, so that the reconstructed serial string **q** of length l could be used in the following *Improved Compression* stage.

The loss of information, for this stage, is in mapping the $l \times m$ bit matrix to an l-bit serial string. Since this mapping involves only modulo-2 additions, it is uniform. Therefore, there are equal numbers of $l \times m$ matrices, out of T that map to the same serial string; this number is $T/2^l$. If we assume each of the T possible matrices is equally likely, the error masking probability for this stage becomes 2^{-l}. Since l is usually in thousands or millions, the loss of error coverage at this stage is not very significant.

Another advantage of using an $MISR$ as a P/S Converter is that it allows the implementation of our $BIST$ model on any existing $MISR$ based $BIST$ scheme, by simply adding the *Modifier* and the *Improved Compressor* blocks. Thus, the ODM scheme proposed in this chapter should be used only when a very high error coverage is essential and can be implemented by some simple additional circuitry to the existing scheme.

B. Modifier:

As was mentioned earlier the *Modifier* consists of a predetermined number of product terms p and generates the modifier string **s**. The cubical notation will be used to represent product terms. Since the

Modifier realizes a single output function, a very convenient implementation for it could be the functional block model used in (15).

It must be indicated here that the size of this block is flexible and depends simply on p and n'. Moreover, layout estimates show that a *Modifier* block which can implement $p = 150$ cubes and $n' = 16$ input variables will require a silicon area which corresponds to the size of a 16-bit $MISR$. Thus, a large number of s sequences can be generated with a relatively small area of the *Modifier*.

C. Improved Compressor:

Two data streams, the modifier string **s** and the quotient-bit string **q**, are Ex-ORed in this block to produce the l-bit long modified string **q'**. The new string **q'** enters the Counter to be compressed into an n'-bit long signature. This compression must be done by using a non-uniform deception volume scheme. Otherwise, the modification will not serve any purpose. The use of the best scheme will of course be judged by the decrease in the deception volume. For instance, if the compression is done by the One's Counting scheme, and if the number of ones in **q** is w and in **q'** is w', then the improvement in deception volume is given by the ratio:

$$\frac{pr'}{pr} = \frac{\binom{l}{w'}}{\binom{l}{w}}$$

where $pr'(pr)$ is the error masking probability obtained by compressing **q'** (respectvely **q**). Similarly, if the compression is done by Transition Counting, and there are t and t' transitions in **q** and **q'** respectively, then the improvement ratio is

$$\frac{pr'}{pr} = \frac{\binom{l-1}{t'}}{\binom{l-1}{t}}$$

Similar formulaes can be obtained for other count based schemes.

If it is possible to replace the *Counter* by an $LFSR$ to perform the count based compression, a gain in terms of hardware overhead can be obtained, since an $LFSR$ design requires typically less area than a counter. This is realizable in the ODM model because the signature in our *Counter* does not necessarily have to be the *expected number of counts* itself. In fact, it suffices to have any signature provided that it indicates that the *Counter* has performed the *expected number of counts*. Thus, we will adopt the use of an $LFSR$ as a *Counter* in the *Improved Compression* block. This is realized in our design, by adopting an autonomous $LFSR$ (i.e. with no external inputs), where the string under count q' is applied as a control signal, which enables the $LFSR$ clocks to change a state for each count.

Moreover, notice that it is possible as well to force such an $LFSR$, with no external inputs, to any known final state given the *expected number of counts* and the characteristic polynomial (14). If the predetermined final state is assigned to be a regular sequence, like for instance a string of alternative 1's and 0's, then the important benefit obtained from this fact is in the possiblity of eliminating the *REFERENCE* block used to store the fault-free signature.

V. IMPLEMENTATION OF ODM SCHEME

In general, the implementation of any $BIST$ scheme requires a considerable amount of information about the CUT and its functionality. This information is required for instance to determine the number of pseudo-random patterns which need to be applied to the CUT, to identify the characteristics of both IPG and $MISR$, and to perform a functional simulation to find out the fault-free signature. To implement the ODM scheme, one needs to carry out all the above activities except that instead of determining the fault-free signature in the $MISR$, the fault-free quotient bit stream **q** of the $MISR$ is needed. In addition, some extra work needed to implement the ODM scheme is described in the following steps:

1. Apply Boolean Function Minimization coupled with Cube Selection:

   • Given **q** and the corresponding input set of patterns for the *Modifier*, find out the functionality of the *Modifier* block and its minimal sum-of-products form expression. Notice that since l is often not much larger than say 1 million, n' is less then 20. An exact minimization is possible for such functions with a very efficient algorithm (6).

   • Given the set of products in the previous expression and the size p of *Modifier*, perform the selection of the proper p cubes if $t > p$, by using a special *cube select algorithm* (28), which carries out the selection process based on covering the largest number of 1-vertices in the expected function F(**q**).

2. Form *Modifier* Block:

   • Design the Modifier Functional Block (15) of p cubes, and determine the modifier string **s** generated by this block.

3. Determine the specifications of *Counter*:

   • Form an $LFSR$-counter of size n', with maximum length characteristic polynomial.

   • Based on **q** and **s**, compute w', the weight of the modified string.

   • Based on w' and the characteristic polynomial of the *Counter*, find out the initial state, for a string of alternative 1's and 0's to be the final state (i.e. the signature).

Example(7): The following example serves to illustrate the above description of the implementation steps.

In order to maintain the simplicity of the example, the predetermined test length has been limited to $l = 64$; thus the number of variables required to generate the l-bit long modifier string s, will become $n' = 6$. Based on the previous information, functional simulation is performed to determine the serial string q. To be more explicit, some bits in q associated with their corresponding n'-bits wide input pattern to the *Modifier*, are shown in (Table 3).

Table 3.

Modifier Inputs and Quotient bit string

IPG's input set for Modifier	Quotient bit string
f e d c b a	F(q)
0 0 0 1 0 1	1
0 0 1 0 1 0	1
0 1 0 1 0 0	1
1 0 1 0 0 0	1
0 1 0 0 0 1	1
1 0 1 1 1 0	0
0 1 1 1 0 1	1
1 1 0 1 1 0	0
. . . . . .	.
. . . . . .	.
. . . . . .	.

Let us suppose that, the following sum-of-product expression is obtained as a result of applying boolean minimization to the modifier function.

$$F(\mathbf{q}) = \overbrace{\overline{a}\overline{b}}^{4-cube} + \overbrace{\overline{bd}f + \overline{c}\overline{d}f}^{3-cubes} + \overbrace{b\overline{c}d\overline{e} + a\overline{d}ef}^{2-cubes}$$
$$+ \overbrace{abc\overline{d}\overline{e} + abc\overline{d}f + a\overline{c}de\overline{f} + \overline{a}b\overline{c}ef}^{1-cubes} + \overbrace{\overline{a}bc\overline{d}ef}^{0-cubes}$$

The set of cubes derived from this expression, consists of 10 cubes presented in (Table 4) after being sorted on the basis of cube size.

Table 4.

The List of Cubes

$\underline{a\ b\ c\ d\ e\ f}$

$4 - cubes \{\ 0\ 0\ x\ x\ x\ x$

$3 - cubes \begin{cases} x\ 0\ x\ 0\ x\ 0 \\ x\ x\ 0\ 0\ x\ 1 \end{cases}$

$2 - cubes \begin{cases} x\ 1\ 0\ 1\ 0\ x \\ 1\ x\ x\ 0\ 1\ 1 \end{cases}$

$1 - cubes \begin{cases} 1\ 1\ 1\ 0\ 0\ x \\ 0\ 1\ 1\ 1\ x\ 0 \\ 1\ x\ 0\ 1\ 1\ 0 \\ 0\ 1\ 0\ x\ 1\ 1 \end{cases}$

$0 - cubes \{\ 0\ 1\ 1\ 0\ 1\ 0$

It was mentioned that p is decided according to either the area dedicated to the Modifier block or the required amount of deception volume. However, in this example, let us consider different cases by assigning p to several values, say 3, 6, 8 and 10. The selection step is

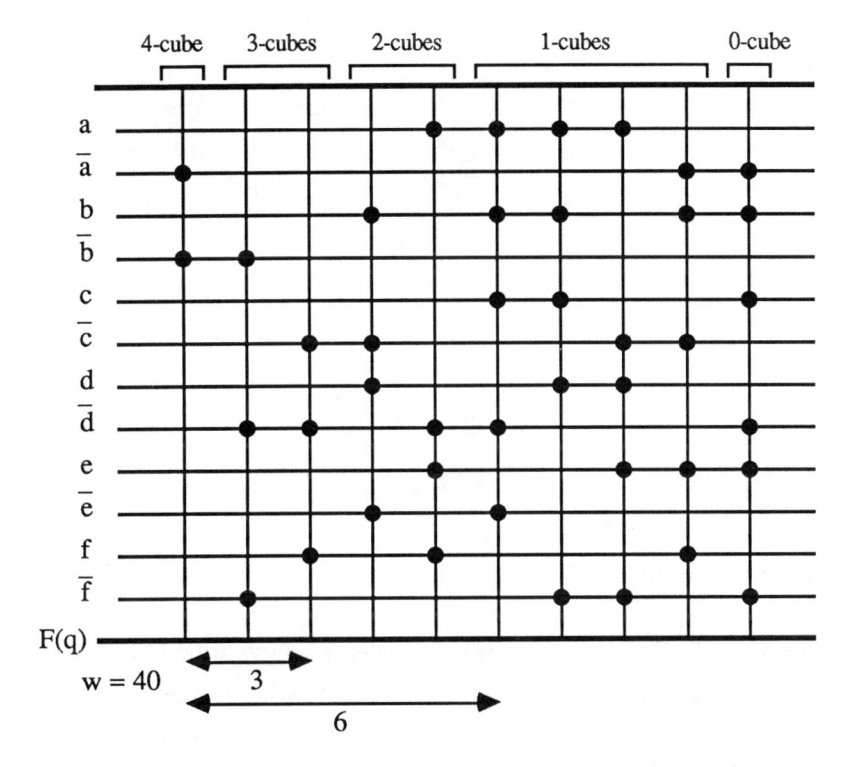

Fig.8. The Hardware Realization of the Modifier Block

performed by utilizing the *cube select algorithm* (28). The contents of the *Modifier* block is eventually determined during this step. However a convenient way to visualize its hardware realization is to present it in its totality Fig.8, where each cube is distinguished by its hardware programmed points.

Table 5 demonstrates the change in weight and deception volume for the different values assigned to p. For instance, if the number of cubes is limited to 6 the weight of the modified string **q'** will be reduced from 40 to 6, whereas the deception volume will go down from 1.4×10^{19} to 5.2×10^8.

Table 5.

# of cubes	weight	Deception Volume
0	40	1.4×10^{19}
3	14	1.8×10^{15}
6	6	5.2×10^{8}
8	2	2×10^{3}

Furthermore this simplified example, apart from its basic illustrative objective, confirms the flexibility and the effectiveness of the ODM scheme, when the modifier sequence is generated by totality.

VI. CONCLUSIONS

It is proven in (26) that in general the deception volume of an average sequence of length l and any specific weight w is reduced by a significant amount as a result of applying output data modification. This reduction can be controlled by the predetermined number of cubes p. Since the higher the p is, the larger the subset of selected cubes will become; and therefore the reduction in weight and deception volume will increase. This provides a trade off between the size of the *Modifier* block and the reduction in error masking desired.

On the other hand, if our objective is to reach a specific deception volume, then the required amount of p (i.e. the size of the *Modifier* block), can be obtained from the deception volume curves which are provided as above-mentioned, for average sequences of specific lengths.

Various lengths of average sequences with different amount of p's have been under consideration. Their deception volume curves attest a significant amount of reduction in error masking. Fig.9 demonstrates a case study where the strings of length $l = 2^{12}$ with various values

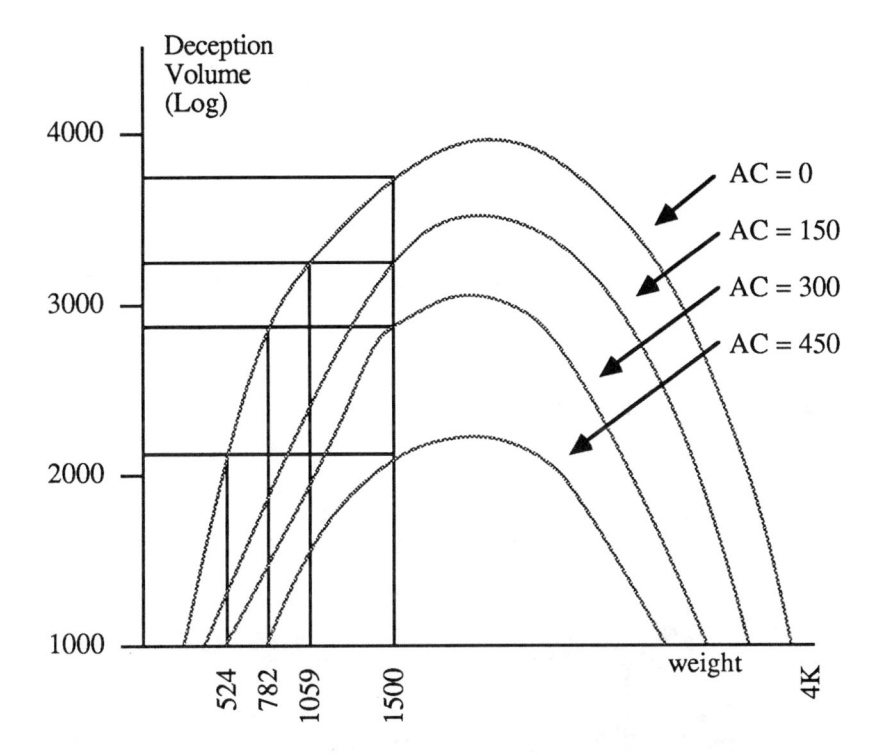

Fig. 9. Trade Off Curves

of p's are taken into account. For instance if three different sizes of *Modifier* blocks are considered $p = 150, 300$ and 450, an average string of weight $w = 1500$ is reduced respectively to 1059, 782 and 524. Thus the reduction in deception volume is more then $2^{500}, 2^{1000}$, and 2^{1600} times.

This demonstrates the interesting fact that a simple trade off curve allows one to decide the optimal solution between the reduction desired versus silicon area overhead. No other published scheme in the literature can provide such a small overhead.

Although this probabilistic approach, adopted to estimate the loss of information, leads to considerable amounts of reduction in the deception volume due to the adoption of ODM scheme, it does not provide any estimate in terms of fault coverage. In the following we will present the results of actual fault simulations, in order to show the tangible effects of ODM scheme. Numerous simulations were run on several example networks, where the fault coverage of each network was computed, first in the case where no output compression is performed, and then in both cases where the standard $BIST$ approach and the ODM scheme are adopted. Table 6 shows the degree of fault coverage in the previous three cases. The coverages here present average value s over various runs, using different IPG's and $MISR$'s.

Table 6.

Simulation Results

Tested IC	# of patterns	No Cmprs.	Strd BIST	ODM
alu	256	100%	98.73%	100%
mc432	1024	99.24%	97.7%	99.24%
mc880	4096	99.36%	98.3%	99.36%
mc1355	4096	99.49%	98.53%	99.49%

In (Table 6) the No Compression column demonstrates the coverage of the test set which consists of l pseudo-random patterns. The second column shows the amount of reduction in fault coverage due to error masking in the standard $BIST$ approach, whereas the third column reveals the effectiveness of the ODM scheme in eliminating the error masking effect. In these cases the *Modifiers* were selected to consist of $p = 50$ cubes. These experimental results confirm the significant benefits, of adopting the ODM scheme, in terms of improved fault coverage, as well as to justify the low cost of its implementation in terms of additional hardware.

The design of each of the four blocks, that have been presented in section III, is noticed to have two important characteristics. The first is that the scheme is, in general, application-independent, and the second is that each block is composed of identical cells with uniform interconnections along one dimension. Moreover, it has to be noted that the CUT function dependent steps, presented in section IV, have the capability to be automated. Therefore it is very realistic and promising to consider developing a CAD tool in order to provide automated implementation of the ODM scheme.

VII. REFERENCES

1. Agarwal, V.K. (1983). *Proc. FTCS*, **13**, 227-234.
2. Bhavsar, D.K. and Krishnamurthy, B. (1984). *Proc. IEEE ITC*, 134-139.
3. Bhavsar, D.K. (1985). *Proc. IEEE ITC*, 88-93.
4. Carter J.L. (1982). *ACM Symposium on Theory of Computing*, **14**, 1982.
5. Carter, W.C. (1982). *Proc. FTCS*, **12**, 289-296.
6. Dagenais, M.R., Agarwal, V.K., and Rumin, N.C. (1985). *Proc. Design Automation Conference*, **22**, 667-673.
7. David R. (1978). *Proc. FTCS*, **8**, 103-107.
8. Eichelberger E.B. and Williams T.W. (1977). *Proc. Design Automation Conference*, **14**.
9. Frohwerk R.A. (1977). *Hewlett Packard Journal*, 2-8.
10. Fujiwara, H. and Kinoshita, K. (1978). *Proc. FTCS*, **8**, 108-113.
11. Golan, P., Nocak, O. and Hilavcka, J. (1985). *Proc. FTCS*, **16**, 404-409.
12. Hassan, S.Z. and McCluskey, E.J.(1984). *Proc. FTCS*, **14**, 354-359.
13. Hayes, J.P. (1976). *IEEE Transactions on Computers*, **C-25** 613-620.
14. McAnney W.H. and Savir J. (1986). *Proc. IEEE ITC*, 54-58.
15. Mead, C. and Conway, L. (1980). "Introduction to VLSI Systems". Addison-Wesley Publishing Company.

16. Peterson, W.W. and Weldon, Jr. (1972). "Error Correcting Codes" 2nd edition. MIT Press, Cambridge Mass. and London.
17. Savir, J.E. (1980). *IEEE Transactions on Compters*, **C-29**, 442-550.
18. Saxena, N.R. and J.P. Robinson (1985). *Proc. FTCS*, **15**, 300-305.
19. Smith, J.E. (1980). *Transactions on Computers*, **C-29**, 510-514.
20. Sridhar, T., Ho, D.S., Powell, T.J. and Thatte S.M. (1982). *Proc. IEEE ITC*, 656-661.
21. Tang, D.T. and Chen, S.L. (1983). *Proc. FTCS*, **13**, 222-226.
22. Wang L.-T. and McCluskey E.J. (1986). *IEEE Transactions on Computers*, **C-35**, 367-370.
23. Williams T.W., and Parker, K.P. (1982). *IEEE Transactions on Computers*, **C-31**, 2-15.
24. Williams T.W., Daehn W., Gruetzner M. and Starke C.W. (1986). *Proc. IEEE ITC*, 282-288.
25. Zorian, Y. and Agarwal, V,K. (1984). *Proc. IEEE ITC*, 140-147.
26. Zorian Y. and Agarwal V.K. (1986). *Proc. FTCS*, **16**, 410-415.
27. Zorian, Y. and Agarwal V.K. (1986). *Proc. of Technical Workshop: New Directions for IC Testing*, 2/1-2/14.
28. Zorian Y. and Agarwal V.K. (1985). *VLSI Design Laboratory*, TR **85-36R**, McGill University, Montreal.
29. Zorian, Y. and Agarwal V.K. (1986). *Proc. International Conference on Custom and Semicustom IC's*, **6**.

EFFECTIVENESS MEASURES FOR DATA COMPRESSION TECHNIQUES UNDER UNEQUALLY LIKELY ERRORS

Nirmal Raj Saxena

Hewlett-Packard
19111 *Pruneridge Avenue*
Mail Stop 44UX
Cupertino, CA 95014

ABSTRACT

In this chapter, the concept of distance distribution of errors is introduced. A heuristic method based on distance distribution is developed to compute the probability of aliasing in various compression methods when errors are not equally likely.

I. INTRODUCTION

Aliasing or error masking probability is a fundamental measure of the effectiveness of data compression methods. Most of the earlier reported results (1-5) have considered this measure in the context of *equally likely errors*. Experimental studies on the error behaviour of digital circuits (6,7) have generally shown non-conformance to the *equally likely error* (ELE) model. This measure of effectiveness would be more objective if true error situations were considered. Therefore, it is important to revisit the problem of computing aliasing probability in the context of *unequally likely errors* (ULE).

Predicting error behaviour is difficult and may require exhaustive fault simulation. Developing a mathematical apparatus to estimate aliasing probability from the predicted behaviour also appears to be difficult.. This may be the reason the ELE model has been used. Given this difficulty in computing aliasing probability, we restrict our attention to approximate algorithms. In this chapter, we present one such algorithm. Smith's dependent error model (8), a departure from ELE, provides the necessary heuristics.

We define a mapping which associates a positive integer called *ß-distance* to every error sequence. By error sequence we mean the (bitwise) sum modulo two of the faulty and fault free output sequences. We will primarily consider single-output digital circuits. The ß-distance concept is a variation of the dependent error model in (8). In the ELE model 2^m-1 length m error sequences are possible. This mapping allows us to distribute the domain of 2^m-m-1 (excluding m single bit error sequences) length m error sequences over the range of $\{1, \ldots, m-1\}$ ß-distances. The distribution function thus obtained enables us to determine how close the observed error behaviour is to the ELE ß-distance distribution. Aliasing relates to the number of undetected error sequences. With the heuristics presented in this chapter we can predict the number of undetected error sequences both for ELE and ULE. The heuristics are validated by comparing the predicted number of undetected errors with the exact number of undetected errors. Actual results show a strong correlation.

The data compression schemes that are of interest in this chapter are *signature analysis* (SA) and *syndrome testing* (ST). We present a straightforward extension of the results in (8) to ST. A testability tool called *expert gate analysis* (EGA) is presented. EGA approximately predicts the ß-distance distribution without resorting to fault simulation.

II. DEPENDENT ERRORS

Dependent errors, as defined in (8), have the property that erroneous bits in an error sequence are in a regular pattern in positions i, ..., i+bn, ..., where b is a power of two and $0 \leq i < b$. Polynomial representations of error sequences are denoted as error polynomials. For example, the error sequence 10101 corresponds to the error polynomial $x^4 + x^2 + 1$. Equivalently, *dependent* error sequences can be expressed (8) as $x^i E(x)^b$. The degree of $E(x)$ is $\leq \lfloor (m-i-1)/b \rfloor$.

It is shown in (8) that the number of error sequences missed by a signature polynomial p(x) of degree r for a length m output sequence under the dependent error model is

$$\sum_{i=0}^{b-1} (2^{\lfloor (m-d-i-1)/b \rfloor + 1} - 1) \tag{1}$$

and the probability that p(x) will not detect an error sequence is given by

$$\frac{\sum_{i=0}^{b-1} (2^{\lfloor (m-d-i-1)/b \rfloor + 1} - 1)}{\sum_{i=0}^{b-1} (2^{\lfloor (m-i-1)/b \rfloor + 1} - 1)} \tag{2}$$

where m is the length of the output sequence and d depends on the characteristics of p(x). If p(x) is irreducible, then d assumes the value br. Henceforth, we will assume irreducible polynomials and in particular primitive polynomials. Simplifying (2) for the case of irreducible polynomials we have the following theorem:

Theorem 1: (8) If SA is performed using an irreducible polynomial p(x) of degree r on length m output sequences, where m is a power of 2, then the probability of aliasing under the dependent error model is 0 for $m/b < r$ and

$$\frac{2^{\frac{m}{b} - r} - 1}{2^{\frac{m}{b}} - 1} \quad \text{for all } m/b \geq r$$

Proof: Since m is a power of 2 and d = br, expression (1) can be reduced to

$$2^{(m-br)/b} \sum_{i=0}^{b-1} 2^{\lfloor -(i+1)/b \rfloor + 1} - \sum_{i=0}^{b-1} 1$$

$\lfloor -(i+1)/b \rfloor = -1$ for $0 \leq i < b$

Simplifying further, the number of undetected error sequences is given by

$$b(2^{m/b-r} - 1)$$

In a similar way we can reduce the denominator in eqn. (2) to

$$b(2^{m/b} - 1)$$

which enumerates all possible error sequences. Thus the aliasing probability for $m/b \geq r$ is

$$\frac{2^{\frac{m}{b}-r} - 1}{2^{\frac{m}{b}} - 1} \quad (3)$$

The aliasing probability is zero when $m/b < r$. In this case, the error polynomials are of the form $x^i E(x)^b$ and the degree of $E(x)$ is less than r. Therefore $x^i E(x)^b$ will not be a multiple of $p(x)$ and all error sequences of this type are detected. Q.E.D.

An asymptotic aliasing probability function for ST under *dependent errors* is derived in the next theorem.

Theorem 2: If k ones are uniformly distributed in a length m output sequence and ST is performed, then the probability of aliasing under dependent errors is given by

$$\frac{(2^{m/b}/\sqrt{\pi m/2b})e^{-2(k-m/2)^2/bm} - 1}{2^{m/b} - 1}$$

Proof: In a length m output sequence, under dependent errors, there are at most $\lfloor m/b \rfloor$ bits that are equally likely to be in error. Assuming uniform distribution of k ones, there will be approximately $\lfloor k/b \rfloor$ ones in these $\lfloor m/b \rfloor$ bits, which, excluding the error free case, is simply

$$\binom{\lfloor m/b \rfloor}{\lfloor k/b \rfloor} - 1$$

The choice of $\lfloor m/b \rfloor$ bits corresponds to positions i, i + b, i + 2b, ..., in the output sequence. There are b ways of selecting i. Thus the number of ways syndrome k is preserved is

$$b\left(\binom{\lfloor m/b \rfloor}{\lfloor k/b \rfloor} - 1\right)$$

There are $2^{\lfloor m/b \rfloor} - 1$ possible dependent error sequences for every choice of i. This implies that there are $b(2^{\lfloor m/b \rfloor} - 1)$ possible error sequences (same as in Theorem 1). Thus the aliasing probability for ST is

$$\frac{b\left(\binom{\lfloor m/b \rfloor}{\lfloor k/b \rfloor} - 1\right)}{b(2^{\lfloor m/b \rfloor} - 1)}$$

Using Sterling's approximation and dropping the *floor* operator we have aliasing probability

$$\frac{(2^{m/b}/\sqrt{\pi m/2b})e^{-2(k-m/2)^2/bm} - 1}{2^{m/b} - 1} \tag{4}$$

Q. E. D.

For a non-uniform distribution of k ones the aliasing probability assumes the form

$$\frac{(2^{m/b}/\sqrt{\pi m/2b})\sum_{i=0}^{b-1} e^{-2(k_i b - m/2)^2/bm} - b}{b(2^{m/b} - 1)} \quad (5)$$

where k_i is the number of ones in bit positions i, i + b, ..., and $\sum k_i = k$. If k is in the neighbourhood of m/2 then expression (5) is upper-bounded by expression (4). The restriction that b be a power of 2 does not apply to Theorem 2. In ST error masking probability increases as b increases. In SA, however, the opposite trend manifests itself. More discussion on this will follow after we develop the ß-distance heuristics.

III. DISTANCE DISTRIBUTION

Error sequences due to faults in digital circuits do not necessarily conform to a single value of b. Furthermore, the spacing of erroneous bits may not be in multiples of b. In spite of these limitations, the dependent error model does provide valuable information. For instance, in terms of the dependence of aliasing on b, the model clearly distinguishes between SA and ST. As we shall see later, the performance of ST degrades as b increases.

It can be seen that the minimum distance between two successive erroneous bits in a dependent error sequence is b. In developing the ß-distance heuristic, we will use this property.

A. Notation and Definitions

Let $E = e(1)e(2)...e(m)$ represent an error sequence of length m. It follows that there exists a j such that $e(j) = 1$. Locations j where $e(j) = 1$ are precisely the locations where the output sequence is in error. The weight of E is defined as

$$W(E) = \sum_{j=1}^{m} e(j)$$

$W(E)$ is bounded by $1 \leq W(E) \leq m$.

Definition 1: Let $t_1, t_2, ..., t_s$ belonging to $\{1, ..., m\}$ be such that $e(t_1)$, $e(t_2), ..., e(t_s) = 1$, and $t_1 < t_2 < ... < t_s$. Then the ß-distance α of E is defined as follows:

$$\alpha = \min \{ t_{u+1} - t_u \} \text{ for } 1 \leq u < s \text{ and } W(E) \geq 2.$$

ß-distance cannot be defined for weight one error sequences. It is important to note that these weight one error sequences correspond to output sequences with single bit errors. Both SA and ST will detect these error sequences and therefore they do not contribute to aliasing. Next, we enumerate the number of length m error sequences that have the same ß-distance α.

Theorem 3: Let $f_\alpha(m)$ be the number of length m error sequences having ß-distance α. Then

(a) when $\alpha < \lfloor m/2 \rfloor$

$$f_\alpha(m) = f_\alpha(m-1) + f_\alpha(m-\alpha-1) + 2^{m-2\alpha} - \sum_{j=1}^{\alpha-1} f_j(m-2\alpha)$$

(b) when $\lfloor m/2 \rfloor < \alpha < m$

$$f_\alpha(m) = m - \alpha$$

(c) when $\alpha = \lfloor m/2 \rfloor$

$f_\alpha(m) = m - \alpha$ if m is even

$f_\alpha(m) = m - \alpha + 1$ if m is odd

Proof: case (a):
Here we can partition the error sequences into three disjoint types:

(i)   0 e(2) e(3) ... e(m)

(ii)   100...0 e(α+2) e(α+3) ... e(m)
The terms in the subsequence e(2) ... e(α+1) are all zero.

(iii)   100...0100...0 e(2α+1) ... e(m)
The terms in the subsequences e(2) ... e(α), e(α+2) ... e(2α) are all zero and e(1) = e(α+1) = 1.

In type (i) ß-distance α is preserved if the subsequence e(2) ... e(m) preserves α. Therefore, there will be $f_\alpha(m-1)$ of this type of error sequence.

Similarily, in type (ii) subsequences e(α+2) ... e(m) should preserve α. The length of these subsequences is m - α - 1; therefore, there will be $f_\alpha(m - \alpha - 1)$ of type (ii) error sequences.

Our counting argument is different for type (iii) sequences. Here to preserve ß-distance α, subsequence e(2α+1) ... e(m) should not have its own ß-distance less than α. Out of $2^{m-2\alpha}$ possible subsequences, we have to exclude all those subsequences that have ß-distance less than α. Thus the number of type (iii) sequences is

$$2^{m-2\alpha} - \sum_{j=1}^{\alpha-1} f_j(m-2\alpha)$$

A special case in type (iii) is when $\alpha = 1$ where the error sequences will have the form 11e(3)...e(m). There are 2^{m-2} such sequences because all 2^{m-2} subsequences (e(3)...e(m)) will preserve $\alpha = 1$. Another interpretation for this corner case is that there does not exist any sequence that has ß-distance less than one, so we can essentially retain the expression for the number of type (iii) sequences.

Adding type (i), (ii), and (iii) sequences we have for $\alpha < \lfloor m/2 \rfloor$

$$f_\alpha(m) = f_\alpha(m-1) + f_\alpha(m-\alpha-1) + 2^{m-2\alpha} - \sum_{j=1}^{\alpha-1} f_j(m-2\alpha) \quad (6)$$

Case (b): $\lfloor m/2 \rfloor < \alpha < m$

One representative sequence in this case is 100...0100...0 where the terms in e(2)...e(α) and e(α + 2)...e(m) are all zero. However, terms e(1) and e(α + 1) are both equal to one. All distinct right shifts of this sequence will span all possible sequences of this type. There are m - α possible shifts that preserve α. Thus, $f_\alpha(m) = m - \alpha$

Case (c): $\alpha = \lfloor m/2 \rfloor$ and m is even

This is similar to case (b); thus $f_\alpha(m) = m - \alpha$. When m is odd, the counting argument up to m - α is the same as that of the foregoing case. However, there is one additional sequence 100...0100...01, where e(1) = e($\lfloor m/2 \rfloor$ + 1) = e(m) = 1, which preserves α. Therefore in this case

$$f_\alpha(m) = m - \alpha + 1 \qquad \text{Q. E.D.}$$

The following boundary conditions and identities hold:

$$f_1(1) = 0 \,;\, f_1(2) = 1,\, f_2(2) = 0$$

$$f_m(m) = 0 \text{ for all possible m}$$

$$\sum_{j=1}^{m-1} f_j(m) = 2^m - m - 1 \tag{7}$$

For example, using these boundary conditions and the results in the foregoing theorem we can enumerate the ß-distance distribution for length 4 sequences. We will obtain $f_1(4) = 8$, $f_2(4) = 2$, $f_3(4) = 1$. Listed below are the corresponding sequences for each α:

(a) $\alpha = 1$; 1111, 1110, 0111, 1011, 1101, 1100, 0110, 0011
(b) $\alpha = 2$; 1010, 0101
(c) $\alpha = 3$; 1001

Identity (7) is a consequence of the fact that ß-distance distribution partitions $2^m - m - 1$ error sequences. Not included are all single bit errors (there are m such) and the all zero sequence (the error free case). We define a normalized distribution function $g_\alpha(m) = f_\alpha(m) / (2^m - m - 1)$.
Clearly

$$\sum_{j=1}^{m-1} g_j(m) = 1 \tag{8}$$

The distribution of $g_\alpha(m)$ over α characterizes the ELE model. If S is any arbitrary set of length m error sequences then we can partition S into at most m disjoint subsets $S_1,..., S_{m-1}$, O, where S_j represents error sequences having ß-distance j. Subset O corresponds to single bit error sequences. An empirical normalized distribution for S could be obtained by dividing the cardinality of each S_α by the cardinality of S - O. If the empirical distribution obtained closely mimics the distribution predicted by $g_\alpha(m)$, then the set S is said to have the characteristics of the ELE model. Note that for S to mimic the ELE model its cardinality need not be $2^m - 1$.

IV. HEURISTICS

Thus far we have gathered enough results to develop the heuristics. The ß-distance concept as stated before, is an abstraction of the minimum distance property of the dependent error model. Going a step further, we assume that the aliasing probability functions for the class of error sequences having ß-distance α will *approximately* track the aliasing probability functions (3) and (4) for SA and ST respectively, when parameter b is replaced by α.

Let $P_{st}(m,k,\alpha)$ denote the aliasing probability function for ST, if a length m output sequence having k ones is compressed and the error sequences have ß-distance α. Similarly, we can define $P_{sa}(m,r,\alpha)$ for SA. Stating the heuristics in terms of the assumptions we have

$$P_{st}(m,k,\alpha) \cong \frac{(2^{m/\alpha}/\sqrt{\pi m/2\alpha})\,e^{-2(k-m/2)^2/\alpha m} - 1}{2^{m/\alpha} - 1} \qquad (9)$$

$$P_{sa}(m,r,\alpha) \cong \frac{2^{m/\alpha-r} - 1}{2^{m/\alpha} - 1} \qquad (10)$$

In order to justify the above heuristics some validation tests are necessary. Two tests are proposed. Digressing a bit, we will discuss the effect of the dependent error model on both SA and ST before returning to these validation tests.

It was mentioned before that the dependent error model clearly distinquishes between SA and ST. The best way to illustrate this is by way of an example. Figure 1 plots the aliasing functions (3) and (4) for parameters: $m = 512$, $k = 200$ and $r = 5$. It can readily be seen that the aliasing probability for ST increases as b increases whereas aliasing for SA remains fairly flat. However, as b approaches m/r, the aliasing probability in SA starts dropping to zero.

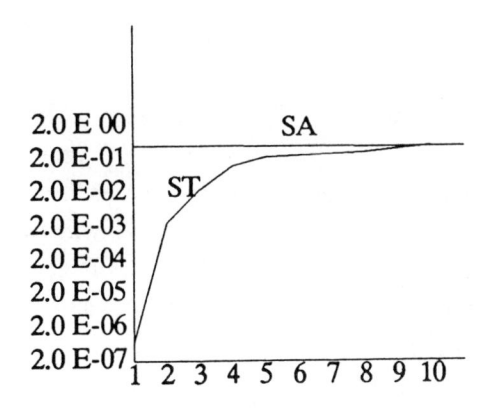

Figure 1. Aliasing vs. b for signature (SA) and syndrome (ST).

In previous papers (5,6), it has been shown that by reducing the count value k, the effectiveness of ST can be enhanced significantly. This enhanced effectiveness is guaranteed under the ELE model. The behaviour of ST under the dependent error model as illustrated by Figure 1 should be a cause for concern.

To explain this better we will introduce the parameter *equivalent length* L_e discussed in (6). L_e is defined as an equivalent length of signature register that would detect the same number of error sequences as ST would. From (6) we have

$$L_e = \frac{1}{2} \log_2 (\pi m / 2) + \frac{2}{\log_e 2} \frac{(k - m/2)^2}{m} \qquad (11)$$

For ST to be more effective than SA, L_e should be greater than r. In the case of dependent errors we will have effective length L'_e given by the equation (12).

$$L'_e = \frac{1}{2} \log_2 (\pi m / 2b) + \frac{2}{\log_e 2} \frac{(k - m/2)^2}{mb} \qquad (12)$$

It is clear from eqn. (12) that L_e degrades by almost a factor of b to L'_e. This is a significant degradation in the performance of ST. For example in Figure 1, aliasing probability in ST increases from 1.687×10^{-7} to 1.09×10^{-4} as b moves from 1 to 2. SA outperforms ST when L'_e becomes less than r. In the example, this happens when b exceeds 9.

It is conjectured that similar degradation in ST will occur if ß-distance distribution is assumed. There is experimental evidence to this effect. In fact this behaviour should follow from the heuristics and therefore validation tests will verify this conjecture.

A. Validation Tests

The approach followed in verifying the heuristics is twofold. First, assuming the ELE model we compute the expected number of aliases from the heuristics with the exact number of aliases. In SA, under the ELE model, the exact number of aliases is $2^{n-r} - 1$ and in ST it is

$$\binom{m}{k} - 1$$

Second, by performing fault simulation experiments on digital circuits the exact fault coverage for SA and ST, and for the empirical distance distribution has been obtained. The extimated fault coverage from the empirical distribution is compared with the exact coverage.

Under the ELE model the expected number of undetected error sequences by SA and ST is given by eqn. (13) and eqn. (14) respectively.

$$M_{sa} = \sum P_{sa}(m,r,\alpha) f_\alpha(m) \quad 1 \leq \alpha \leq m/r \qquad (13)$$

$$M_{st} = \sum P_{st}(m,k,\alpha) f_\alpha(m) \quad 1 \leq \alpha < m \qquad (14)$$

Tables 1 and 2 list the expected and exact values. It is clear that both M_{sa} and M_{st} approach their respective exact values asymptotically.

B. Fault Simulation Validation

Some preliminary simulation experiments have been done to study the error characteristics. A parallel fault simulator was used to obtain empirical ß-distance distribution by monitoring the exclusive-or of faulty and fault-free output sequences, for the various faults. The experimental set up is described as follows:

- Four random logic circuits (complexity 200 gates) were considered.
- Gate stuck type faults were assumed.
- Test lengths were of the order of 64.
- Multiple outputs were space compressed to a single output by an exclusive-or tree.

TABLE 1

Validation Test Results for SA

m,r	M_{sa}	Exact
16, 4	4.0639×10^3	4.095×10^3
32, 5	$1.3421716.. \times 10^8$	$1.3421772.. \times 10^8$
64, 6	$2.88230376.. \times 10^{17}$	$2.88230376.. \times 10^{17}$
128, 7	$2.6584559.. \times 10^{36}$	$2.6584559.. \times 10^{36}$

TABLE 2

Validation Test Results for ST

m,k	M_{st}	Exact
16, 8	1.33151×10^4	1.3072×10^4
32, 16	$6.0614.. \times 10^8$	$6.0580.. \times 10^8$
64, 32	$1.839818.. \times 10^{18}$	$1.839817.. \times 10^{18}$
128, 64	$2.399824.. \times 10^{36}$	$2.399824.. \times 10^{36}$

TABLE 3

SA Fault Coverage Simulation

Circuit	Estimated	Exact
A	97.43	98.20
B	97.23	97.10
C	97.00	96.93
D	97.30	97.20

TABLE 4

ST Fault Coverage Simulation

Circuit	Estimated	Exact
A	82.12	78.60
B	83.39	81.20
C	81.80	79.00
D	80.80	79.00

- The number of faults injected was on the order of 1100.
- Exact fault coverage both for SA and ST was computed by simulation.
- Signature register length r was 5.

The expected fault coverage for SA and ST was obtained from the empirical distribution.

The empirical ß-distribution can be obtained in a manner similar to that discussed in section III. The error sequences obtained during fault simulation are categorized into disjoint sets $S_1, S_2, ... , S_{m-1}, O$. The cardinality of the union of these sets is the number of faults simulated. The estimated fault coverage is computed from the following relations for SA and ST respectively:

$$efc_{sa} = 100 \, (1 - \frac{1}{|S|} \sum_\alpha |S_\alpha| P_{sa}(m,r,\alpha)) \text{ for } 1 \leq \alpha < m/r \quad (15)$$

$$efc_{st} = 100 \, (1 - \frac{1}{|S|} \sum_\alpha |S_\alpha| P_{st}(m,k,\alpha)) \text{ for } 1 \leq \alpha < m \quad (16)$$

Figure 2 represents the normalized ß-distribution of length 64 error sequences. Figure 3 is the normalized empirical distribution obtained for circuit C under gate stuck type faults. It is clear that the empirical distribution for this circuit does not exhibit the characteristics of the ELE model. Similar behaviour was observed in other simulated circuits.

Tables 3 and 4 list the estimated and exact fault coverage values for SA and ST. For example, in circuit C (Table 4) the count value k was 34 and the ELE model would have predicted 91.2% fault coverage. In all the simulated circuits the coverage estimated by the heuristics was more accurate than that estimated assuming the ELE model. In the case of SA with r = 5 and m = 64, and assuming the ELE model, the coverage would be 96.87%. This is not significantly different from that predicted by the heuristics and should be expected because the aliasing function $P_{sa}(m,r,\alpha)$ remains fairly constant (Figure 1) over a wide range of α.

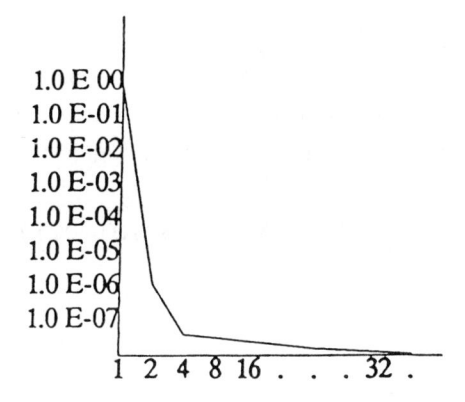

Figure 2. Normalized ß-distance distribution.

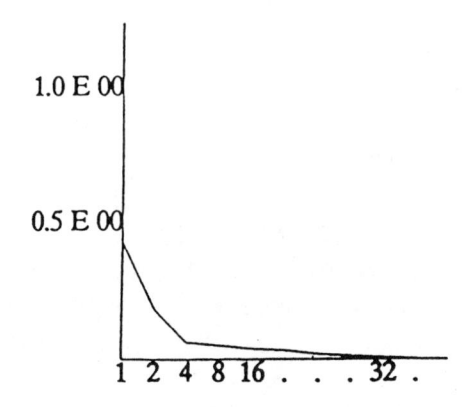

Figure 3. Normalized empirical distribution for circuit C.

V. EXPERT GATE ANALYSIS

The main thrust in the previous sections was to demonstrate that aliasing due to compression can approximately be computed using the concept of distance distribution. Computing empirical distributions requires exhaustive fault simulation. This is not feasible for large circuits. In this section we propose an algorithm called *expert gate analysis* (EGA) which approximately computes the empirical distribution without resorting to fault simulation.

The simulation model of a *logic gate* is extended to compute the minimum distance between two successive sensitizing patterns for every fault local to it. We consider single stuck type failures with respect to the gate. Given below is a model for a two-input AND gate in pseudo-PASCAL.

```
Procedure AND(L,M,G,A1,A2,B1,B2,D1,D2);
Var
  K : Integer;
Begin
  G = L AND M
  K = 2 *  Ord(L) + Ord(M);
  Case K of
  0: Begin
       If D1 > 0 then
         Begin
           If (COUNT - D1) < D2 then
             Begin
               BETA[D2] = BETA[D2] - 1;
               D2 = COUNT - D1;
               BETA[D2] = BETA[D2] + 1;
               D1 = COUNT;
             End
           Else
             D1 = COUNT
         End
       Else
         Begin
           D1 = COUNT
     End;
```

We will briefly describe the functionality of this extended gate. Let $t_1, t_2, ..., t_m$ be a sequence of two-bit patterns applied to inputs L, M of the AND gate. G is the computed output. COUNT is a global variable which maintains the value of the current position of the input pattern. If t_i is the current input then COUNT = i. K is a local variable which points to the fault sensitized by the current input t_i. For example, when K = 0, the output

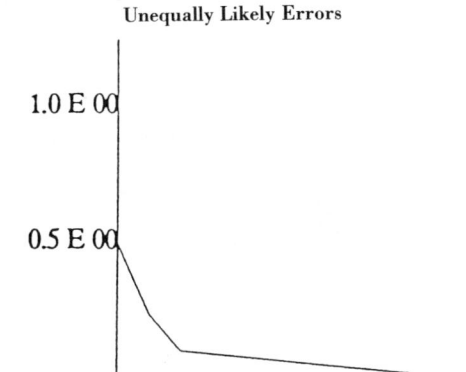

Figure 4. EGA distance distribution for circuit C.

stuck-at 1 fault of the AND gate is sensitized. In this case D1 points to the position of the last sensitized pattern. D2 contains the minimum distance between two successive sensitizing patterns that occurred in the sequence $t_1, t_2, ..., t_{i-1}$. The array BETA is global and contains the most recent distance distribution. It gets updated as the patterns are applied.

If the output of a gate is not a primary output of the circuit, then the sensitizing pattern for a fault will only be a necessary condition for its propagation. This means that the minimum distance computed by the EGA for a particular fault in a gate will be less than or equal to the exact minimum distance computed through fault simulation. Therefore the EGA will give a slightly skewed distribution. Figure 4 illustrates the distribution obtained for circuit C using EGA. We can notice that it is not significantly different from the empirical distribution (Figure 3) obtained for the same circuit by fault simulation.

We will illustrate the EGA algorithm by considering a three-input multiplexor circuit (Figure 5). Table 5 lists the various stuck type faults and their corresponding sensitizing patterns. A "y" in a column indicates that the input pattern is a test pattern for the fault in the corresponding row. Inputs that are only sensitizing patterns are indicated by an "x". Notice that faults are distinquished only up to the gate level. All equivalent faults in a gate are collapsed into a single fault. For example, EGA cannot treat a/0 (a fault local to the AND gate) and f/0 (a fault local to the OR gate) as equivalent. The analysis is on a gate basis. Table 6 lists the distance distributions (by EGA and through fault simulation) for the multiplexor circuit.

TABLE 5

MUX Example

	a	0	0	0	0	1	1	1	1
	b	0	0	1	1	0	0	1	1
	c	0	1	0	1	0	1	0	1
a/1				y	y				
d/1						y	x		
a/0								y	y
e/1			y				y		
c/1		y				y			
e/0					y				x
f/0								y	y
g/0			y				y		
f/1		y		y					
h/0			y		y	y	y	y	y

TABLE 6

Empirical Distance Distribution

	$|S_1|$	$|S_2|$	$|S_3|$	$|S_4|$	$|O|$
Expert Gate Analysis	5	1	0	4	0
Fault Simulation	4	1	0	3	2

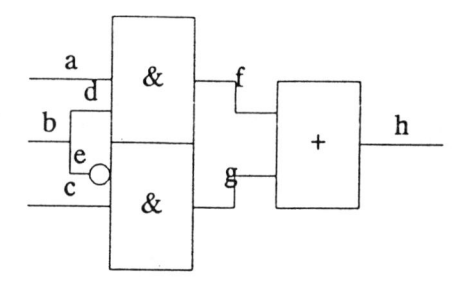

Figure 5. Multiplexor example.

Distance distribution by the EGA can be computed in a single iteration of the input patterns. For example, the computing precedence for the multiplexor circuit is as follows:

AND(a,d,f,...);
AND(e',c,g,...);
OR(f,g,h,...);

The above sequence of computations should be iterated over the number of input patterns to enumerate an approximate distance distribution. In fault simulation this iteration should be done for every simulated fault and one fault at a time. The complexity in fault simulation is therefore of the order: number of faults × number of gates × number of inputs; whereas in the EGA it is: number of gates × number of inputs.

VI. CONCLUSIONS

The distance distribution heuristic proves useful in studying the aliasing problem in compression testing. A similar approach would be to use the weight distribution of error patterns. Signature based methods seem to be less sensitive to error model. In syndrome based compression, a possible degradation in performance may take place due to a sparse distance distribution. Experimental and theoretical results demonstrate this. It is felt that other counting based methods (transition count, spectral coefficients) will exhibit similar characteristics.

ACKNOWLEDGEMENT

The author would like to thank David S. Graves for his help in the preparation of this chapter.

REFERENCES

[1] Hassan, S. Z., and E. J. McCluskey (1984), "Increased fault coverage through multiple signatures", *Proc. 14th Int. Symp. Fault-Tolerant Comput.*, pp. 354-359.

[2] Bhavsar, D. K., and B. Krishmamurthy (1984), "Can we eliminate fault escape in self testing by polynomial division", *Proc. IEEE Test Conf.*, pp. 134-139.

[3] Frohwerk, R. A. (1977), "Signature analysis: a new digital field service method", *Hewlett Packard J.*, pp. 2-8.

[4] Williams, T. W., and K. P. Parker (1982), "Design for testability - a survey", *IEEE Trans. Comput.*, **C-31**, pp. 2-16.

[5] Zorian, Y., and V. K. Agarwal (1984), "Higher certainty of error coverage by output data modification", *Proc. IEEE Test Conf.*, pp. 140-147.

[6] Robinson, J. P., and N. R. Saxena (1987) "A unified view of test compression methods", *IEEE Trans. Comput.*, **C-36**, pp. 94-99.

[7] Saxena, N. R. (1984), "Test compression methods", *M. Sc. Thesis*, University of Iowa, Iowa City.

[8] Smith, J. E. (1980), "Measures of the effectiveness of fault signature analysis", *IEEE Trans. Comput.*, **C-29**, pp. 510-514.

THE EXHAUSTIVE TESTING OF CMOS STUCK-OPEN FAULTS[†]

J. A. Bate and D. M. Miller

Department of Computer Science
University of Manitoba
Winnipeg, MB
Canada R3T 2N2

I. INTRODUCTION

Digital circuits are subject to a wide variety of physical failures. Here we are concerned with permanent failures which can be detected by observing the erroneous functional behaviour of the circuit under test. In particular, the effect of a failure on timing or operating parameters is not explicitly considered.

A fault model is a logical abstraction which attempts to describe a class of physical failures in functional terms. Fault models are used both to construct fault detection procedures and to assess their performance. The degree to which the fault model accurately represents the effect of physical failures is crucial to the success of the fault detection process.

[†] The research reported in this chapter was supported in part by grant A4851 from the Natural Sciences and Engineering Research Council of Canada.

The single stuck-at fault model (1) assumes that a failure in a circuit is equivalent to one of its lines being permanently fixed at logic 0 or logic 1. This model was originally developed for discrete component circuits where it has a close resemblance to the actual physical failures. For integrated circuits, the single stuck-at model has little correspondence to the actual physical failures. However, the single stuck-at fault model is still widely used. Often, the performance of a fault detection procedure not particularly oriented to single stuck-at faults is measured in terms of the percentage of single stuck-at faults it covers.

A multiple stuck-at fault (1) assumes the physical failure corresponds to some number of lines in the circuit being fixed at logic 0 or logic 1. The density of LSI and VLSI circuits suggests that multiple faults are quite probable. The difficulty is the number of potential multiple stuck-at faults. A combinational circuit with t lines has 2t possible single stuck-at faults. However, the same circuit has $3^t - 2t - 1$ multiple stuck-at faults. This large number prohibits the explicit consideration of multiple stuck-at faults for large circuits.

A variety of additional fault models have been developed to account for the physical failures not adequately covered by the stuck-at model. The bridging fault model (2) is motivated by physical shorts. Current circuit density makes this failure mode highly probable. In the bridging model, it is assumed that the shorting of two lines results in them being replaced by their logical AND or OR depending on the technology in question. Analysis of bridging faults is complicated by the fact that such a fault can introduce a feedback path into an otherwise combinational circuit. The faulty circuit can thus behave in a sequential manner thereby greatly complicating the fault detection process.

Fault models specific to programmable logic arrays (PLAs) have also been introduced (3). Fortunately, the highly regular structure of a PLA facilitates its testing and good PLA oriented fault detection procedures are known (4).

CMOS (complementary metal oxide semiconductor) has become a major VLSI technology primarily due to its low static power dissipation and the reasonable circuit density that can be achieved (5). Consequently, there is considerable interest in the failure modes of CMOS circuits as well as procedures for their detection. These topics are the subject of this chapter.

In section II, CMOS circuit technology is reviewed and its physical failure modes are examined. Section III looks at the applicability of the traditional test set (1) method to CMOS circuits and details certain approaches suggested in the literature. Section IV examines exhaustive testing methods in the CMOS environment. It is shown that "simple" exhaustive testing is insufficient even in the combinational case. A novel approach to the exhaustive testing of combinational CMOS is presented in section V. The chapter concludes with a discussion and several suggestions for further developments in this area.

II. CMOS CIRCUITS

The discussion of CMOS technology presented below is greatly simplified. In particular, the switch model of CMOS behaviour is emphasised since this leads to the clearest understanding of functional fault mechanisms in a CMOS circuit. The interested reader should consult a book on microelectronics (5) for a full explanation of the behaviour of CMOS circuits.

A. CMOS Technology

The CMOS circuits considered are composed of P channel and N channel enhancement mode field effect transistors. In considering the functional behaviour of a circuit, the transistors can be viewed as simple switches. We adopt the symbols shown in Figure 1.

When a logic 1 (VDD) is applied to the gate of an N channel transistor, a conducting path is created between the source and drain of the transistor. When a logic 0 (GND) is applied to the gate of an N channel transistor no conducting path is formed. For a P channel transistor, the conducting path is created when logic 0 is applied to the gate. This inverted logical behaviour

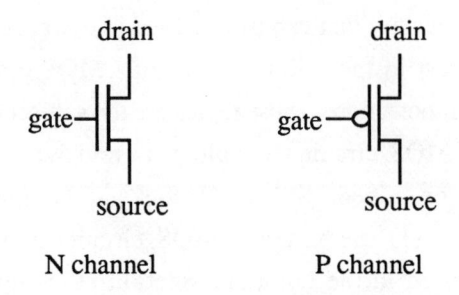

Fig. 1. CMOS enhancement mode transistor symbols.

Table 1. Switch behaviour of CMOS transistors.

GATE INPUT	N channel	P channel
logic 0 (GND)	not conducting	conducting
logic 1 (VDD)	conducting	not conducting

is symbolized by the circle on the gate of the P channel transistor. Table 1 summarizes this switching behaviour.

A conducting N channel transistor passes good logic 0s but poor logic 1s. Conversely, a conducting P channel transistor passes poor logic 0s but good logic 1s. For these reasons, a CMOS cell is constructed according to the block diagram in Figure 2. The output of the cell is 0 if the logic values of the cell inputs create a path from the output to GND through conducting N channel transistors. Likewise, the output of the cell is 1 if the logic values of the cell inputs create a path from the output to VDD through conducting P channel transistors. Under normal operation, no set of input values should create a path through both the load circuit (P channel transistors) and the driver circuit (N channel transistors).

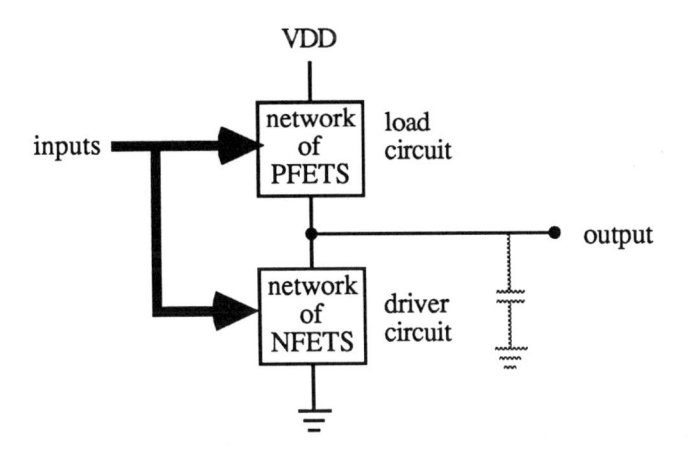

```
                    VDD
                     |
           ┌─────────┴──────┐
           │ network │ load │
 inputs ──▶│   of    │circuit│
           │  PFETS  │      │
           └─────────┴──────┘
                 │         ├──────────● output
           ┌─────┴───┬──────┐
           │ network │      │
       ──▶ │   of    │driver│
           │  NFETS  │circuit│
           └─────────┴──────┘
                 │
                ─┴─
```

Fig. 2. Block diagram of a CMOS cell.

The shadowed capacitor in Figure 2 represents the output load capacitance which is a result of the capacitance of the transistor gates or chip pins to which the output of the cell is connected together with the capacitance in the conducting paths forming the output connections.

Figure 3 shows a CMOS inverter. When the input is at 0, the P channel transistor is conducting and the output goes to 1 (VDD). When the input is at 1, the N channel transistor is conducting and the output goes to 0 (GND). Figure 4 shows a CMOS 2-input NAND cell. Either input at 0 pulls the output to 1. The output is 0 only if both inputs are 1.

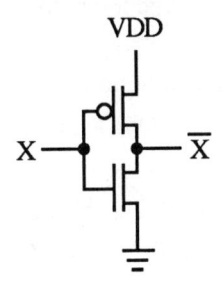

Fig. 3. CMOS inverter.

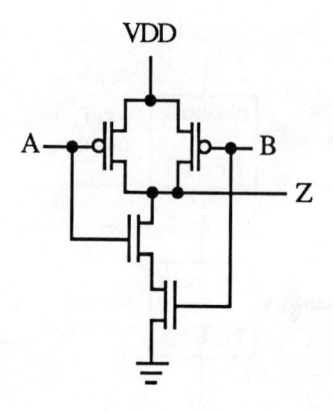

Fig. 4. A CMOS 2-input NAND gate.

Fig. 5. A 4 variable function with $\bar{f} = \bar{a}\,b\,c + \bar{c}\,\bar{d} + \bar{b}\,\bar{d}$

An arbitrary combinational function can be realized by the structure depicted in Figure 2 (see above). One approach is to find an expression for the complement of the function to be realised. This expression is used to construct the driver circuit. In doing this, a logic AND corresponds to a series transistor connection while a logic OR corresponds to a parallel transistor connection. The load circuit is then constructed as the 'dual' of the driver circuit, i.e. series connections map to parallel connections and paraalel connections map to serial. For example, applying this approach to the function depicted in Figure 5 yields the circuit shown in Figure 6. This realization could of course be simplified by factoring d' which would remove one transistor from each of the load and the driver circuits.

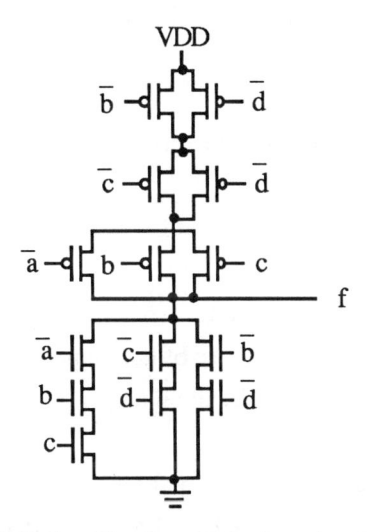

Fig. 6. CMOS realization of the function in Figure 5.

The structure in Figure 2 is termed fully complemented since for every input condition there is a conducting path through either the driver or the load circuit. Several other variants of CMOS exist e.g. pseudo-NMOS, dynamic, clocked, cascade voltage switch logic, domino logic and pass transistor logic (5). Here we restrict our attention to fully complemented static cells. Note that the discussion of failure modes below is for this particular case and the other classes of CMOS circuits should be considered separately.

B. CMOS Failure Modes

A CMOS circuit is formed by superimposing several layers of conducting material (metal and diffusion), insulating material and transistor forming material. The physical failures in a CMOS circuit which lead to permanent functional faults are broken conductors, faulty transistors and shorted conductors. Note that the latter can happen on the same layer i.e. within the same conducting material, as well as between layers due to failures in the insulating material between the layers. One study (6) of MOS failure modes involved the analysis of failures in 43 faulty 4-bit microprocessor chips. The following results were obtained:

- shorts between metal conductors   39 %
- open metal conductors   14 %
- shorts between diffusion conductors   14 %
- open diffusion conductors   6 %
- shorts between metal and substrate   2 %
- inobservable (physical cause of failure not found)   10 %
- catastrophic (complete circuit failure)   15 %

While these results are specific to a particular chip design and a particular process, they indicate that shorted and broken conductors should be closely considered when considering failures in MOS circuits.

Conductor shorts in a CMOS circuit will generally result in an erroneous path from VDD to GND and thus result in a significant increase in the static current dissipation (7). These failures should thus be observable by monitoring this current (8). Broken conductors present more of a challenge.

Consider the NAND cell in Figure 7. Suppose conductor A is broken. In this case, the driver circuit can not conduct and the output can not be pulled down to 0. If both inputs are 1, neither the driver nor the load circuit is conducting and the cell output is floating. The load capacitance will hold the previous output value for some period of time which is a function of the amount of capacitance and the leakage currents in the circuit. The circuit failure has thus introduced memory into a combinational circuit.

Table 2 shows the failures resulting from breaks in several other conductors in the circuit in Figure 7. Note that the cell inputs and the cell output are not included. Breaks in these conductors correspond to single stuck-at faults.

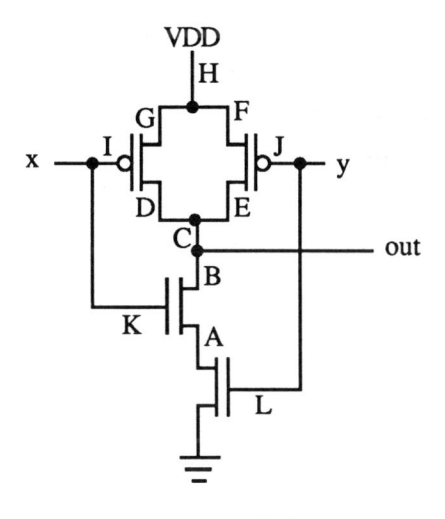

Fig. 7.

Table 2. Effect of stuck-open faults on a CMOS NAND gate.

FAULTY CONDUCTOR

x	y	out	A	B	C	D	E	F	G	H	I	J	K	L
0	0	1	1	1	M	1	1	1	1	M	1	1	1	1
0	1	1	1	1	M	M	1	1	M	M	M	1	1	1
1	0	1	1	1	M	1	M	M	1	M	1	M	1	1
1	1	0	M	M	0	0	0	0	0	0	0	0	M	M

note: M denotes capacitive memory resulting from a floating output

A stuck-open fault (9) represents any failure in a CMOS cell which results in one or more transistors which can not conduct. This can be the result of a broken conductor as illustrated above or can arise from a failure in a transistor such as a faulty channel or the gate shorted to the substrate. Note that a single broken conductor can result in a multiple stuck-open fault. For example, conductor C broken in the NAND gate in Figure 7 effectively means that neither P transistor can conduct.

The converse of a stuck-open fault is a stuck-closed or stuck-on (8) fault. This is equivalent and is in fact typically caused by a short between the source and drain. As noted above, this results in an observable increase in the static power dissipation.

Recently, Renovell and Cambon (10) have empirically examined floating gate faults in MOS circuits and have demonstrated that the stuck-at, bridging, stuck-open and stuck-on models are not sufficient for all potential physical failures in an MOS circuit.

III. DETECTION OF STUCK-OPEN FAULTS BY TRADITIONAL MEANS

The traditional approach to detecting stuck-at faults in a combinational circuit involves a test set (1) which is simply a subset of the possible input assignments to the circuit under test. The elements of the test set are termed test vectors or test patterns. The test set is selected so that for every fault in the targeted fault class, the good and faulty circuits yield different results for at least one test vector. The test set approach involves two distinct phases; the identification of the test vectors, and their subsequent application including the comparison of the observed and expected results. There are several algorithms (11) for the identification of a test set for all single stuck-at faults in a combinational circuit. Multiple stuck-at and bridging fault test set generation algorithms have also been proposed. It has been estimated that the complexity of finding a single stuck-at fault test set for a combinational circuit with t gates is on the order of t^3. Thus determining a test set for a large circuit can be extremely time consuming. However, this is a one time operation associated with the design of the circuit.

Identifying a test set for a sequential circuit is considerably more difficult than for the combinational case (1). This has led to a number of design for testability techniques (12) which allow a sequential circuit to be reconfigured so that it can be tested by combinational means. Design for testability will be considered in more detail in section IV below.

Applying the tests is also time-consuming and involves the storage of a large amount of data. A computer-based test station is normally used to store, apply and analyse the results of the test set. In the LSI and VLSI environments, the problem is compounded by the complexity of the circuitry on a single chip. The ratio of circuitry to observable points, chip pins, is quite high. Despite these drawbacks the test set remains the backbone of the test industry.

When testing a combinational circuit for stuck-at faults and certain types of bridging faults, the order in which the test vectors are applied to the circuit is irrelevant. This is because these faults leave the circuit combinational and thus the response to any one particular test vector is independent of all others. For stuck-open faults, and bridging faults which introduce feedback loops, this is not the case.

As we have seen, a stuck-open fault introduces memory into a CMOS combinational circuit. When a test vector is applied which puts the circuit output into a high impedance (disconnected) state, the output value observed is that resulting from the last driven circuit response. This assumes the retained value has not leaked off which will be true if consecutive test vectors are applied quickly enough. This will usually be the case since the goal is to test a circuit as quickly as possible.

For example, if conductor D is broken in the NAND cell in Figure 7, the outcome of the test vector 01[†] depends on whether it is preceeded by 11 or either of 00 or 01. In the former case, applying 01 will yield 0 and the fault is detected, while in the latter case it will yield 1 and the fault is not detected.

Clearly, for stuck-open faults, the order in which the test vectors are applied is important. A number of researchers (13-17) have considered the extension of the test set approach to stuck-open faults in combinational circuits. Basically, two test vectors are required for each targeted fault. The

---

[†] Test vectors are ordered with x on the left.

first 'sets up' the circuit, while the second will detect the presence of the fault. For example, to test the NAND cell in Figure 7 for conductor E broken (which results in the right P transistor being stuck-open), the vector 11 is applied to force the output to 0, and then the test vector 10 is applied. The latter will yield 1 in the fault-free case but will retain the 0 in the faulty case. In general, to test for a P (N) channel stuck-open fault, the circuit output should be 'set up' to 0 (1).

While two vectors appear to be required for each stuck-open fault considered, the sequence can be shortened by ordering the tests so that each test sets up the network for the next test in the sequence . For example, analysis of Table 2 shows that a 2-input NAND cell is completely tested for single stuck-open faults by the sequence {11,10,11,01}. {11,10} tests one P transistor, {10,11} tests the N transistor pair, and {11,01} tests the second P transistor. In fact, this sequence detects all multiple stuck-open faults in the NAND gate. In general, it is not true that a test set constructed for single stuck-open faults will detect all multiple stuck-open faults as well,

Several researchers have examined the identification of test sets for single stuck-open faults.. Typically, a method designed to find a test set for single stuck-at faults is extended to the stuck-open case. For example, in (15), the D-algorithm (18) which is probably the best known test generation procedure for stuck-at faults, is extended to the stuck-open case. Since the construction of a test set for stuck-at faults which has no ordering requirement is quite complex, constructing a compressed test vector sequence for a general CMOS circuit can be prohibitive.

Reddy and Reddy (19) have studied the effect of stuck-open faults on CMOS memory elements. It is shown that typical memory cell designs contain undetectable stuck-open faults and that certain of these undetected faults change static memory elements into dynamic elements. This is not surprising since the capacitive effects of CMOS which give rise to the stuck-open fault phenomenon are exactly those used to advantage in forming dynamic memory. Alternate memory cells avoiding these problems are proposed in (19).

The failure mechanisms which must be considered are topology dependent. Galiay *et al* (6) have suggested a set of design rules for NMOS circuits wherein all physical failures are properly modeled as stuck-at faults. This approach carries a significant penalty in chip area and is not directly applicable to CMOS. However, it emphasizes the effect of layout on the potential failures in a circuit.

Zasao (20) has shown how alternative layouts can reduce the probability of a stuck-open fault occuring. While this is likely wise design practice, the fact that the faults are not totally eliminated requires that they still be considered in testing.

Moritz and Thorsen (8) have considered the effect of layout on the testability of stuck-open faults. They give examples where the layout can invalidate a two vector test. Both glitches and large internal capacitance can cancel the effect of the setup vector resulting in the circuit output being preloaded with the incorrect value and the stuck-open fault going undetected by the second vector.

IV. THE EXHAUSTIVE TESTING OF STUCK-OPEN FAULTS

The traditional test set approach has a number of drawbacks particularly in the VLSI environment. For large circuits, the identification of the test set can be extremely time-consuming. This is a one-time design cost which can usually be accepted. However, as indicated above, stuck-open faults make this task even more complex requiring much more attention to the physical details of the circuit than was necessary for the stuck-at model. Also of serious concern is the fact that the test set and its associated valid responses must be stored, systematically applied and verified.

Several schemes have been proposed for increasing the testability of VLSI circuits. For example, level sensitive scan design (LSSD) (21) replaces the simple latches in a sequential circuit by more complex two stage latches. These latches permit the breaking of the feedback paths in a sequential circuit so that the circuit can be treated as combinational at test time. In addition during testing, the latches are configured into a shift register so that test vectors can be shifted in and out with low pin overhead. This approach, and those like it (12), solve many problems in sequential circuit testing. However the test set must still be determined and the test process can be quite lengthy due to the serial shifting of the test vectors and their responses.

Recently, considerable attention has been paid to built-in self-test (12). Figure 8 is a block diagram of this approach. The input generator exercises the network under test in an exhaustive or possibly pseudo-exhaustive fashion. The response data generated is compressed according to some scheme resulting in a few bits of data called the signature. This signature is compared to a built-in reference value and a single pass / fail signal is produced.

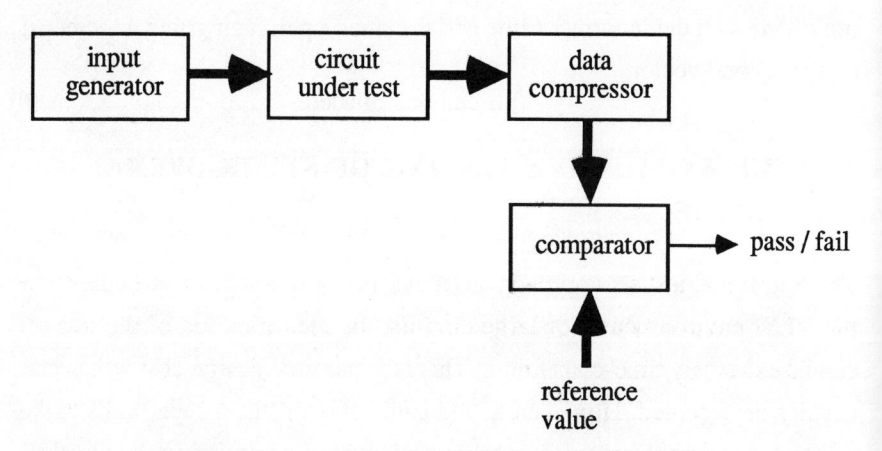

Fig. 8. Block diagram of built-in self-test.

In many schemes, e.g. built-in logic block observation (BILBO) (12), a sequential circuit is designed to be reconfigured so that the input generator and data compressor are formed from the latches used in the normal operation of the circuit. This approach is taken in order to cut down the on-chip hardware required to implement the test.

This approach has a number of advantages:
  i) no test set need be generated, applied or verified;
  ii) the hardware overhead is not usually too great;
  iii) the pins required are minimal: one to put the circuit into self-test mode and one to supply the pass / fail response.

For exhaustive testing, the input generator could be a simple counter. However, a linear feedback shift register (LFSR) (22) is usually more economical. This is the approach taken in the BILBO scheme. A properly selected LFSR serves as a random sequence generator over the set of input conditions to the combinational circuit. As noted earlier, for the traditional stuck-at fault model, the order in which the input conditions is applied is not important. For circuits with a large number of inputs, it may be appropriate to test the circuit by applying a randomly selected subset of the input conditions. Here, we restrict our attention to the exhaustive case but the problems identified apply to the case of random testing as well.

Numerous data compressors have been suggested. An LFSR can be used. For multiple-output circuits this can be replaced with a multiple input shift register (MISR) (12). In syndrome testing (23), a counter[†] is used to accumulate the number of ones in the response data. In Rademacher-Walsh coefficient testing (24), the correlation between the circuit output and certain exclusive-OR functions is measured. These latter two methods are discussed at some length later in this chapter.

The primary disadvantage to the built-in self-test approach lies in the fact that the response data is compressed. Compression, of course, leads to a loss of information and the response of a faulty circuit can be mapped onto the same signature as the correct circuit response, a phenomenon referred to as aliasing. In this case, the fault would go undetected. The faults missed depend on the particular compressor chosen.

[†] An LFSR can be used as a counter in this application since it is only the fact that we reach the correct final state and not its actual value that is important. This can result in a considerable saving in chip area.

Consider applying exhaustive built-in self-test to the detection of stuck-open faults in CMOS combinational circuits. The masking problems caused by the compression of the test response data still exist. In addition, a second form of fault masking caused by the memory effect of stuck-open faults is introduced.

Table 3. Exhaustive test of 2-input NAND cell.

x y	good cell	faulty cell*
0 0	1	1
0 1	1	1
1 0	1	1 [†]
1 1	0	0

notes: * fault for conductor E broken in Figure 7
† 1 retained from previous input

Table 4. Alternate test of 2-input NAND cell.

x y	good cell	faulty cell*
1 1	0	0
1 0	1	0 [†]
0 1	1	1
0 0	1	1

notes: * fault for conductor E broken in Figure 7
† 0 retained from previous input

Consider the case of the 2-input NAND cell in Figure 7 with conductor E broken. Applying the input conditions in the order depicted in Table 3 masks the fault, since the output sequence generated by the faulty cell is identical to that generated by the fault-free cell. Note that a different output sequence is generated if the input conditions are applied in the reverse order, as shown in Table 4.

This simple example demonstrates that an exhaustive test, even in the unlikely case that all the response data is available, will not necessarily detect all stuck-open faults in a CMOS combinational circuit. The problem is the one noted earlier. Detection of a stuck-open fault requires one test vector to set up the output, followed by a second to check for the fault in question. A randomly chosen exhaustive sequence will not necessarily have this property for all possible stuck-open faults in the network. For example, the sequence in Table 4 does not test for the a stuck-open in the P transistor whose gate is connected to input x nor does it provide a test for any stuck-open faults in the N-channel transistors.

It is of course possible to construct a test sequence for a particular circuit taking topological considerations into account when determining which potential stuck-open faults to consider. The important requirement for built-in self-test is that the sequence must be easily generated by hardware embedded in the chip. This hardware should be as simple as possible in order to reduce the area overhead associated with the test. Identification of an appropriate sequence for a particular circuit and its associated generator is a complex problem. In the next section we thus consider a universal test sequence.

V. THE EULERIAN CYCLE TEST GENERATOR

The input test sequence associated with an exhaustive test can be viewed as a tour of the squares of a Karnaugh map. For the input generator to have a simple hardware implementation, the tour must follow some regular pattern or must be totally random as in the case of an LFSR input generator. In the random case, fault simulation is required to assess the coverage with respect to a particular circuit. We here concern ourselves with structured sequences.

Figure 9 shows a readily generated input sequence displayed as a tour of the Karnaugh map from Figure 5. The block of two zeros indicated corresponds to the series of three N channel transistors at the left of the implementation in Figure 6. A stuck-open fault in this series results in these two squares acting as memory. The sequence in Figure 9 would not detect such a fault since this block is entered from a 0 on the map (the bottom left corner) and the function would appear to behave correctly.

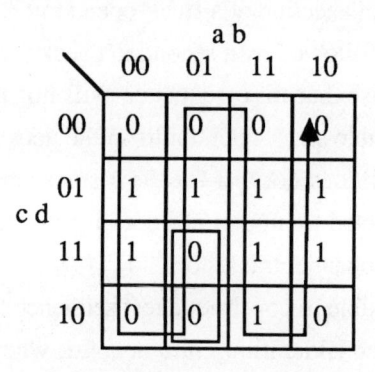

	a b			
c d	00	01	11	10
00	0	0	0	0
01	1	1	1	1
11	1	0	1	1
10	0	0	1	0

Fig. 9. An example of an easily generated input sequence.

From this example, it is clear that a single structured tour once through the map is, in general, not sufficient. Figure 10 illustrates that applying such a sequence in the forward direction and then in the reverse direction is also not sufficient since the indicated block of four zeros in Figure 10 is entered from a 0 in both directions.

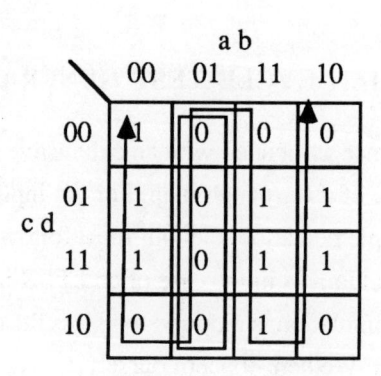

	a b			
c d	00	01	11	10
00	1	0	0	0
01	1	0	1	1
11	1	0	1	1
10	0	0	1	0

Fig. 10. An example of a forward / backward tour.

A. Eulerian Cycles

To construct a universal sequence, i.e., a sequence that will work for all circuits, we must construct a tour which enters each map square once from each adjacent square (two map squares are adjacent if their associated input assignments differ in one position). This is more readily visualized using graph theory.

An n-variable Karnaugh map corresponds to an n-dimensional hypercube. The transition from one square to an adjacent one can be represented as a directed edge in the hypercube. Representing all adjacent transitions in this way yields $n2^n$ edges. We refer to the resulting graph as an n-dimensional directed hypercube. The case for n=3 is depicted in Figure 11. Note that since all possible transitions are depicted there are two edges between each vertex pair, one in each direction.

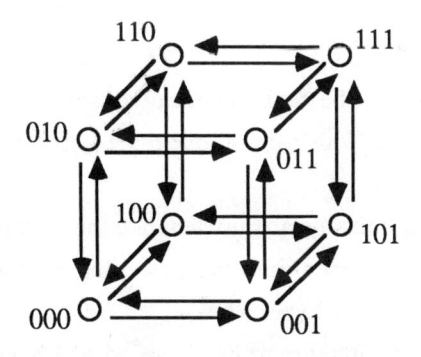

Fig. 11. The 'directed' hypercube for n=3.

The input sequence we require corresponds to a Eulerian cycle[†] in the n-dimensional directed hypercube. Since the in-degree is equal to the out-degree for each vertex in the hypercube a Eulerian cycle must exist. The classical method for finding a Eulerian cycle is as follows:

---

† A Eulerian cycle is one which begins at a vertex, passes along each edge exactly once, and returns to the initial vertex.

i) begin at any vertex and follow a random edge (never using the same edge twice) until the path can no longer be extended at which point a partial cycle must have been found;

ii) if all edges have now been visited the cycle is complete;

iii) if not, find a vertex in the partial cycle with an unused edge, construct another cycle from this vertex, and 'splice' it into the original cycle;

iv) repeat this procedure until the Eulerian cycle is completed.

This approach is clearly not practical for the problem at hand. For the case of n=16, the desired sequence has 1,048,577 vectors. The above method could be applied with the resulting sequence stored on chip in ROM, but this would require 2 megabytes. Conventional methods could be used to implement the sequence as a finite state machine, but this would involve the manipulation of some very random 20-input functions.

To obtain a *practical* method, the following requirements must be met:

i) an algorithm must be devised which will generate a test sequence (Eulerian cycle) with a definite and predictable pattern for any value of n;

ii) this algorithm must be able to generate the next vector required given only the current and the preceding vector, i.e., the last edge followed in the hypercube;

iii) a simple digital circuit must be identified which will generate the sequence.

The authors (25) have investigated this problem in detail. A test sequence (Eulerian cycle) may be constructed for any value of n using the following recursive procedure.

ALGORITHM: *Recursive generation of Eulerian cycle in a directed hypercube*

i) find the sequence for n-1 bits, and add a leading 0 to each vector;

ii) modify the sequence by replacing the **first** occurence of each nonzero vector 0X (where X denotes an n-1 bit string) by the three vectors 0X, 1X and 0X;

iii) append a second copy of the sequence for n-1 bits, with a leading 1 added to each vector;

iv) add the final vector of all zeros.

Steps (i) and (iii) generate all transitions of the lower n-1 bits when the high-order bit is 0 and 1, respectively. Step (ii) adds all transitions of the high order bit when the low order bits are non-zero. The transition of the high order bit when the low order bits are zero occurs at the boundary of the two sections. The sequence generated in this fashion has $n2^n+1$ test vectors, hence the required $n2^n$ transitions. The case of n=2 and n=3 is shown in Figure 12.

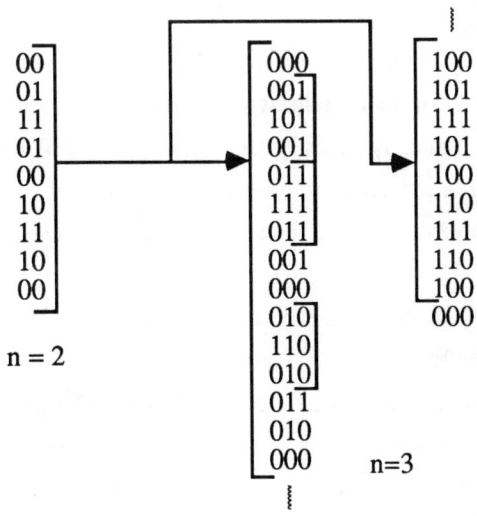

```
00       000      100
01       001      101
11       101      111
01       001      101
00       011      100
10       111      110
11       011      111
10       001      110
00       000      100
         010      000
n = 2    110
         010
         011
         010
         000     n=3
```

Fig. 12. Test sequences for n=2 and n=3.

B. The Algorithm

This procedure generates a correct sequence but it is not in a proper form for a digital circuit. However, a simple set of rules of the proper form may be derived which will generate the same sequence.

Define $B = (b_m, b_{m-1}, ..., b_0)$ to be the current vector (where $m = n-1$). Let $0 \le C \le m$ be the index of the bit which will change to generate the next vector. Let B^* and C^* denote the next values of B and C, respectively. The value of B^* may be trivially found from B and C. The value of C^* may be found by using the following rules. In these rules, x denotes any single bit (a "don't care"), and {Y} denotes any sequence of bits.

A) if $C = m$ and $B = 0\{Y\}$ then $C^* = 0$ if $Y = 0$ and $C^* = m$ if $Y \ne 0$;

B) if $C = m$ and $B = 1x\{Y\}$ then $C^* = 0$ if $Y = 0$ and $C^* = m-1$ if $Y \ne 0$;

C) if $C = m-1$ and $B = x100...0$ then $C^* = m$;

D) if $C = m-k$ ($k \ge 1$) and B has more than k leading zeroes, then $C^* = m$;

E) if none of the rules (A) - (D) apply, ignore the high-order bit of B and apply these rules recursively to the sequence on n-1 bits.

The value of C^* for n=4 is given in Table 5. This table contains two copies of the table for n=3, as shown by the dotted lines (rule E). The last column is determined by rules A (the top half) and B (the bottom half). Portions of the smaller tables are overridden by rules C and D, as indicated in the table. Rule D determines the triangular area at the top of the table, and the two isolated 3's are the result of rule C.

These rules generate the same sequence as the recursive construction given earlier. The rules were derived as follows.

The last transition in the sequence will always be from 10...00 to 00...00. When the sequence on n-1 bits is completed, the next bit to change will always be the high-order bit. Therefore, whenever $B = x10...00$ and $C = m-1$ (the sequence on n-1 bits is ending), C^* should be 0 (rule A).

Whenever bit m is 0 and a pattern on the low-order n-1 bits is about to be generated for the first time, C^* should be m. This happens whenever bit m-1 is about to change from 0 to 1 ($B = 00\{Y\}$ and $C = m-1$), or whenever bit m-1 is a 0 and a pattern on the low-order n-2 bits is about to appear for the

Table 5. C* values for n=4.

B \ C	0	1	2	3
0000	3	3	3	0
0001	1	3	3	3
0010	0	2	3	3
0011	1	0	3	3
0100	1	0	3	3
0101	1	1	1	3
0110	0	2	0	3
0111	1	0	1	3
1000	2	2	0	0
1001	1	2	2	2
1010	0	2	2	2
1011	1	0	2	2
1100	1	0	3	0
1101	1	1	1	2
1110	0	2	0	2
1111	1	0	1	2

first time. Recursively, this gives the patterns B = 000{Y} and C = m-2, B = 0000{Y} and C = m-3, etc. This gives rise to rule D.

Now consider the case when bit m is a 1 and C = m (the high-order bit is about to change from 1 to 0). This happens in two situations. First, when B = 10...0 and C = m the sequence is ending and C^* should become 0 to begin a new sequence. Second, bit m has changed from 0 to 1 and is now immediately returning to 0 (see step ii in the original algorithm). The sequence on n-1 bits must now be resumed from where it left off. This implies that rule D was used, followed by rule A, and that the following vectors have just been generated:

	B	C	
	0z0{Y}	m-k	
	0z1{Y}	m	where k≥1 and z denotes k-1 zeroes
current B, C	1z1{Y}	m	
B^*, C^*	0z1{Y}	?	

The interrupted sequence on n-1 bits must now be resumed. Therefore, the two lines where C = m may be ignored, as may the high order bit. The line prior to these should determine C^*. This line was B = 0z0{Y} and C = m-k, where z represents k-1 zeroes. If k=1, then rule i applies to the lower n-1 bits and C^* should be 0 if Y = 0 and m-1 if Y ≠ 0. If k > 1 then rule iv applies to the lower n-1 bits (in this case m is smaller by 1) and therefore C^* should be m-1. In terms of the current B and C, the rules are as follows. If B = 1z1{Y} and C = m, where z represents k-1 zeroes and k ≥ 1, the C^* should be 0 if Y = 0 and k = 1, and C^* should be m-1 otherwise. Combining this with the fact that C^* = 0 if B = 10...0 and C = m gives rule B.

The above covers all cases where C = m or C^* = m. If the top bit is not involved, then the sequence on n-1 bits should proceed normally. Thus the above rules should be applied to the smaller case recursively (rule E).

C. The Circuit

Rules A-E may be converted to non-recursive Boolean expressions which determine C^* from B and C, allowing a circuit to be constructed. The basic form of the desired circuit is shown in Figure 13.

Let c_i (0≤i≤m) be a Boolean value such that c_i=1 if C=i and c_i=0 otherwise. Let c^*_i be defined similarly. Only one c_i value may be 1 at any

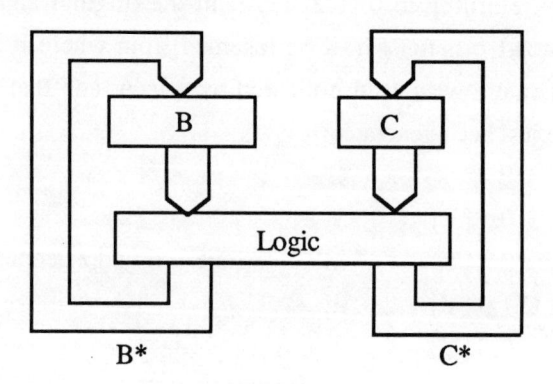

Fig. 13.

given time. The Boolean formulae that follow will use C in this form. The C register may store the c_i values directly, which would require n bits of memory. Alternatively, the value of C may be stored in encoded form, which would only require $\log_2 n$ bits of memory. However, an encoder and a decoder would then be required to convert C to and from the c_i values. In practice, storing the c_i values themselves would likely be simpler.

Rules A-E may now be converted to Boolean expressions which give c^*_i in terms of b_j and c_j ($0 \le j \le m$). To remove the recursive nature of the rules, it is only necessary to determine the cases in which the value of C^* that would normally be used by the low-order k bits is overridden by one of the rules for k+1 bits. In Figure 2, these cases were shown by outlines inside columns 0-2. There are two cases. When $b_k...b_0 = 10...0$ and $C=k$, rule B applied to these bits would give $C^* = 0$. However, this is always overridden by rule C applied to the bits $b_{k+1}b_k...b_0$, unless there is no higher bit. Thus part of rule B is never applied, and a special term is added to the expression for c^*_0 to allow for the 0 that *is* generated by this rule when there is no higher bit. The other case concerns the triangular area at the top of the table where $C = n-k$ and there are k leading zeros in B. The values in this triangular area, when considering the low-order k bits, are overridden whenever the next higher bit is a 0. Keeping these conditions in mind, the following expressions may be derived.

Let $Z_{-1} = 1$ and $Z_k = b_k' Z_{k-1}$ ($k \ge 0$). Thus Z_k indicates whether or not the lowest k+1 bits (bits 0-k) are all zeroes. Let $T_{-1} = 0$ and $T_k = b_k'(c_k + T_{k-1})$. Thus T_k indicates whether or not the current values of B and C are in the triangular area in the top of the first k+1 columns (columns 0-k).

Rules A and D can be combined to give

(1)
$$c^*_k = 1 \text{ if } b_{k+1} T_k (Z_k c_k)' = 1 \quad (0 < k \le m) \text{ and}$$

(2)
$$c^*_0 = 1 \text{ if } b_{k+1} Z_k c_k = 1 \quad (0 \le k \le m).$$

The b_{k+1} in (1) and (2) allow for the fact that the values in the upper triangle are overridden when the next higher bit is a 0, as mentioned earlier.

Rule C gives
$$c^*_k=1 \text{ if } c_{k-1}b_{k-1}Z_{k-2}=1 \quad (0<k\leq m).$$
(3)

Rule B gives
$$c^*_{k-1}=1 \text{ if } c_k b_k Z_{k-2}'=1 \quad (1<k\leq m+1) \text{ and}$$
(4)
$$c^*_0=1 \text{ if } c_k b_k b_{k-1} Z_{k-2}=1 \quad (0\leq k\leq m).$$
(5)

The b_{k-1} in (5) allows for the part of rule B that is always overridden, as mentioned earlier. The only case when this does not happen is when the entire vector is considered. Thus the special case
$$c^*_0=1 \text{ if } c_m b_m Z_{m-1}=1$$
(6)
is also necessary. In order for the above expressions to apply to all of the bits, it is necessary to define the extreme cases $Z_{-2}=Z_{-1}=b_{m+1}=1$ and $c_{m+1}=b_{-1}=0$. Combining (1), (3), and (4) gives the final result below.

$$c^*_k = b_{k+1} T_k \overline{(Z_k c_k)} + c_{k-1} b_{k-1} Z_{k-2} + c_{k+1} b_{k+1} \overline{Z_{k-1}} \quad \{1\leq k\leq m\}$$

Combining (2), (5), and (6) gives the following expression.

$$c^*_0 = b_m c_m Z_{m-1} + S(b_{k+1} c_k Z_k + b_k b_{k-1} c_k Z_{k-2})$$

However, since only one of the c_i values may be 1 at any time, it would be simpler to calculate c^*_0 as the NOR function of the other c^*_i values. This would give the simpler, but slower, expression

$$c^*_0 = \overline{(Sc^*_k)}$$

The expressions for c^*_1, c^*_2, and c^*_m may be simplified due to the presence of constants (1 or 0) in some terms in these cases. The circuit diagram for a typical bit of the resulting test sequence generator is shown in Figure 14. The circuit diagram for bit 0 is simple and will not be shown.

VI. CONCLUDING REMARKS

Having presented a new input sequence and a corresponding generator for exhaustive built-in self-test of CMOS circuits, we now comment on its relation to the data compressors used at the output of the circuit. First, it is important to note that even when the chosen input sequence properly sets up and propogates a stuck-open fault to the circuit output, it may still go undetected due to aliasing problems in the output data compressor.

The performance of an LFSR as the output compressor is difficult to assess due to its essentially random behaviour, which is in fact its primary strength. Fault simulation would likely be required to assess the stuck-open fault coverage afforded by a Eulerian sequence generator and an LFSR compressor for any particular circuit.

Syndrome testing is considerably more amenable to analysis. In syndrome testing as defined in (23), all possible input conditions are applied and the number of ones in the output stream is counted. If the observed number of ones differs from the expected value, a fault has been detected.

For a CMOS circuit constructed as in Figure 2, any single or multiple stuck-open fault restricted to the load circuit (the P channel transistors) which is propogated to the output must reduce the syndrome since expected 1s are replaced by 0s. Conversely, any single or multiple stuck-open fault restricted to the driver circuit (the N channel transistors) which is propogated to the output must increase the syndrome since expected 0s are replaced by 1s. Hence all single and unidirectional multiple stuck-open faults in a circuit of the form shown in Figure 2 are syndrome-testable if the input generator properly induces the fault. If the fault is mixed, i.e., involves both P channel and N channel transistors, the syndrome may stay the same since the number of added 1s and 0s can cancel, in which case the fault goes undetected.

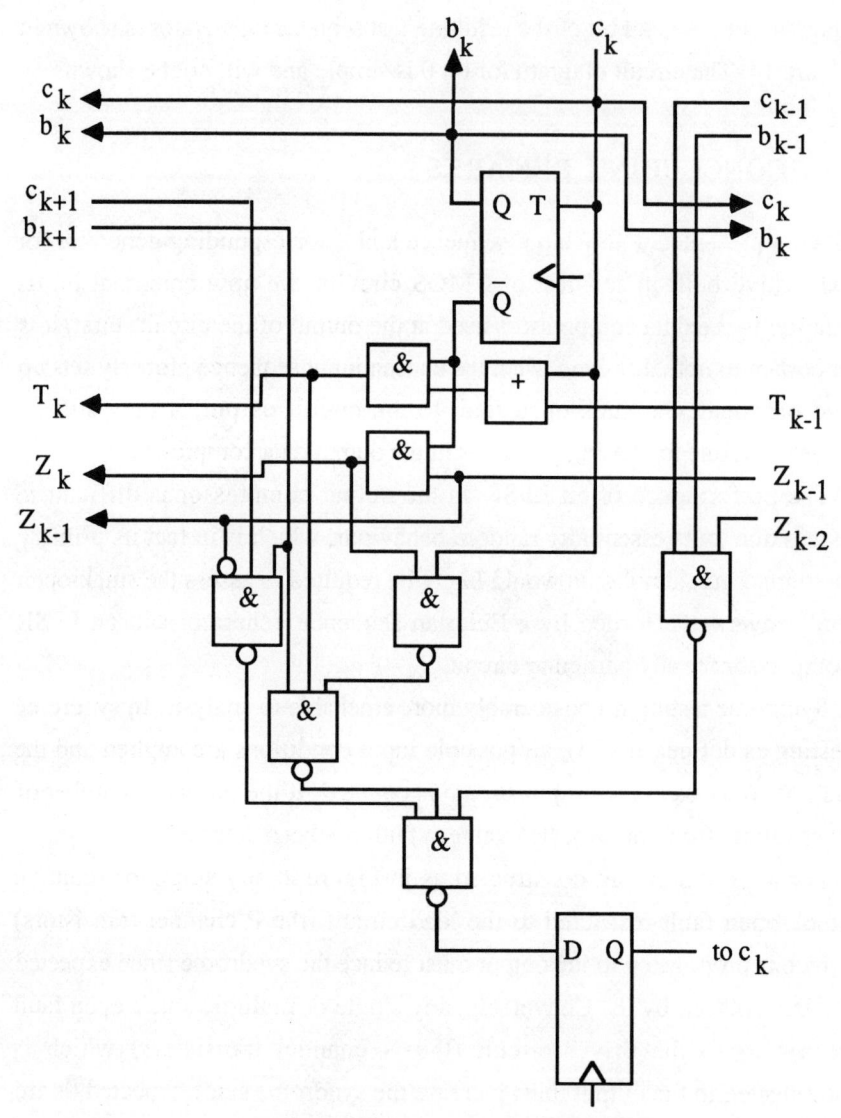

Fig. 14. Bit k of Eulerian sequence generator.

If our Eulerian cycle generator is used, the above statements hold since we are simply touring the possible input conditions multiple times. The advantage to the Eulerian cycle approach is that we are assured that all stuck-open faults will be induced and propagated to the output. Hence, all single stuck-open faults and all unidirectional stuck-open faults are tested. The faults that are missed by masking in the mixed case will of course be different than in the traditional syndrome test approach.

In general, a CMOS combinational circuit will be realized as a composition of cells, each implemented as in Figure 2. This makes the analysis much more complex. First, applying the Eulerian sequence to the primary inputs to the circuit does not ensure a Eulerian sequence is being applied to a particular internal cell. Hence, it is possible that a stuck-open fault in that cell is not induced. There is the secondary problem that even if a stuck-open fault is induced it may not be properly propogated through the intervening cells to the network output.

Given the "highly" exhaustive nature of the Eulerian cycle approach, it seems unlikely that a stuck-open fault which is not induced by applying such a test sequence will appear in normal operating conditions. There is of course the possible problem that the presence of an undetectable fault could mask a fault which on its own would be detectable. This problem is well known for the stuck-at fault model (1). Further study is required to determine if it exists for the stuck-open case.

A circuit line has been termed a unate line if all paths from that line to the network output have the same inversion parity. This removes cancelation effects. Hence, if a stuck-open fault is syndrome-testable with respect to the cell output, the cell output is a unate line, and the input sequence induces the fault,then the fault will be syndrome-testable at the circuit output.

For non-unate lines, detailed analysis is required. Work to date leads us to the following conjecture:

Conjecture 1. If a stuck-at-0 (stuck-at-1) fault on a cell output is syndrome-testable at the network output for the Eulerian cycle input sequence, a stuck-open fault which reduces (increases) the syndrome at the cell output is also syndrome testable.

The spectral coefficient testing techniques discussed in (24) will likely be of use in handling syndrome-untestable stuck-open faults. This area is being pursued. Considerably more research is required.

Several interesting questions arise from the Eulerian cycle generator presented in this chapter. Consider the following test sequence for n=2.

00,01,11,10,00,10,11,01

This sequence is made up of two copies of the standard 2-bit Gray code, one with the bits in the normal order, and one with the bits reversed. It is natural to ask whether this type of construction could be used for larger values of n. A unit-distance code, such as the Gray code, is a cyclic sequence of 2^n n-bit binary vectors in which each distinct vector appears once, and each differs from its successor in exactly one bit. This corresponds to a Hamiltonian cycle in the directed hypercube discussed earlier.

An interesting question is: can n unit-distance codes (Hamiltonian cycles) be chained together to form a test sequence of the required type (an Eulerian cycle)?

For n=2, there is a unique way in which this can be done, as shown previously. For n=3, unfortunately, it is not possible. This is easily verified either by hand, or with a simple computer program.

For n=4, there are 112 Hamiltonian cycles with the bits in a canonical order (2688 if order is ignored). These can be combined in 552 different ways to form an Eulerian cycle. Of these, there are 24 ways in which the same Hamiltonian cycle can be used 4 times, with 4 different permutations of the bits, to form an Eulerian cycle. The 4 permutations used form a latin square. This gives an interesting construction, but there is no obvious pattern to the codes which can be used in this way. The standard Gray code cannot be used.

For n=5, the following code may be used 5 times, with the bits rotated one place to the left each time, to form a valid test sequence.

0,1,3,2,10,8,12,14,30,31,15,13,9,11,27,26,24,25,29,28,20,
4,5,21,23,7,6,22,18,19,17,16

There are certain to be a large number of other sequences that can be used in this way for n=5. This gives rise to the following conjecture.

Conjecture 2. It is possible, for all n≠3, to construct an Eulerian cycle in an n-dimensional hypercube by using n copies of a Hamiltonian cycle with n different permutations of the dimensions. The permutations used will form a latin square.

The following unsolved problems remain.

i) Is there an algorithm which can be used to generate a test sequence using a simpler digital circuit than that shown in Figure 3?

ii) Is the conjecture given above valid? If so, is there a simple constructive proof which could be used to design the corresponding circuit?

iii) Is a weaker form of the conjecture true, in which n different Hamiltonian cycles may be used?

An LFSR can be used to exhaustively exercise the input space of a combinational circuit. Note that the LFSR has a cycle length of 2^n-1 and that an extra gate is required to include the input vector 00...0. The discussion above has indicated that the sequence generated by the LFSR when applied to a single CMOS combinational cell may not detect all possible stuck-open faults. In fact the problem is even worse.

Consider the situation in Figure 15. Here two bits of the LFSR serve as inputs to a NAND cell. During the sequence generated by the LFSR, the input conditions to the NAND gate are applied many times. However, the bits of an LFSR are highly correlated. In particular, b_i at time t is equal to b_{i-1} at time t-1. Hence for our example in Figure 15, the test sequence 11, 10 will never be applied to the NAND cell. From the discussion above it is clear that stuck-open faults for one of the P transistors in the NAND cell will not be tested. Note that the problem remains even if the EXORs are inserted between bits (1) since the LFSR still will contain some number of adjacent bit pairs.

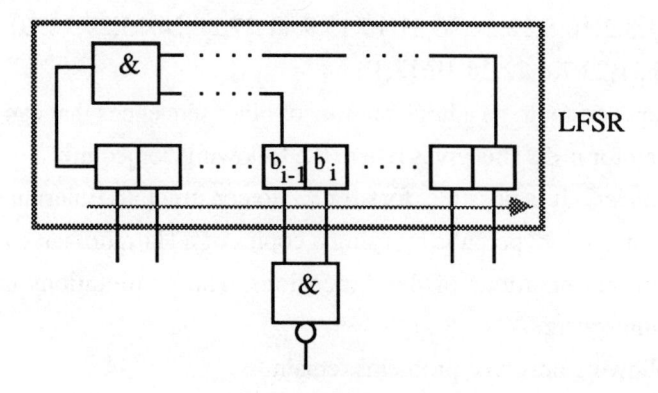

Fig. 15.

Used in the fashion depicted in Figure 15, which is the manner proposed in many articles for the use of an LFSR as the input stimulus in built-in self-test, an LFSR is a very poor pseudo-random sequence generator (26), and as we have just shown the correlation between bits means certain faults will not be tested. The Eulerian cycle approach presented in this chapter solves the problem in a structured and exhaustive fashion but at the expense of a very long test sequence of $n2^n$ for an n-input circuit. A more practical approach would be to employ a 'better' pseudo-random sequence generator in the hope that the required test pairs will be applied in reasonable time.

Random sequence generators based on cellular automata have been investigated in (27). Certain of these generators have been shown to have very good "randomness" properties. In particular the adjacent bit correlation of an LFSR is not present.

The BILBO (12) approach has been modified with the LFSR replaced by a cellular automata pseudo-random sequence generator in a scheme identified as CALBO (28). The improved randomness of the test sequence seems likely to increase the probability of detecting stuck-open faults. A detailed analysis of the extent to which this is true is ongoing.

REFERENCES

1. Breuer, M. A,. and A. D. Friedman (1975), <u>Diagnosis and Reliable Design of Disgital Systems</u>, Computer Science Press, Potomac, MD.

2. Karpovsky, M. G., and S. Y. H. Su (1980), "Detection and location of input and feedback bridging faults among input and output lines", *IEEE Trans. Comput.*, <u>C-29</u>, pp. 523-527.

3. Smith, J. E. (1979), "Detection of faults in programmable logic arrays", *IEEE Trans. Comput.*, <u>C-28</u>, pp. 845-853.

4. Fujiwara H., K. Kinoshita, and H. Ozaki (1980), "Universal test sets for programmable logic arrays", *Proc. 10th Int. Symp. Fault-Tolerant Computing*, pp. 137-142.

5. N. Weste, and K. Eshraghian (1985), <u>Priciples of CMOS Design: A Systems Perspective</u>, Addison Wesley, Reading, Mass.

6. Galiay, J., Y. Crouzet, and M. Vergniault (1980), "Physical versus logical fault models for MOS LSI circuits: impact on their testability", *IEEE Trans. Comput.*, <u>C-29</u>, pp. 527-531.

7. Malaiya, Y. K. (1982), "A new fault model and testing technique for CMOS devices", *IEEE Test Conf.*, pp. 25-34.

8. Moritz, P. S., and L. M. Thorsen (1985), "CMOS circuit testability", *IEEE Custom Integrated Circuits Conf.*, pp. 311-314.

9. Wadsack, R. L. (1978), "Fault Modelling and Logic Simulation of CMOS and MOS Integrated Circuits", *Bell System Technical Journal*, <u>57</u>, pp. 1449-1473.

10. Renovell, M, and G. Cambon (1986), "Topology dependence of floating gate faults in MOS integrated circuits", *Electronics Letters*, <u>22</u>, pp. 152-157.

11. Abraham, J. A., and V. K. Agarwal (1986), "Test generation for digital systems", in <u>Fault-Tolerant Computing: Theory and Techniques</u>, ed. D. K. Pradham, Prentice-Hall, Englewood Cliffs, NJ.

12. Williams, T. W., and K. P. Parker (1982), "Design for testability -- a survey", *IEEE Trans. Comput.*, <u>C-31</u>, pp. 2-15.

13. El-ziq, Y. M., and R. J. Cloutier (1981), "Functional-level test generation for stuck-open faults in CMOS VLSI", *IEEE Test Conf.*, pp. 536-546.

14. Chiang, K.-W., and Z. G. Vranesic (1982), "Test generation for MOS complex gate networks", *Proc. 19th Design Automation Conf.*, pp. 149-157.

15. Jain, S. K., and V. K. Agarwal (1983), "Test generation for MOS circuits using the D algorithm", *Proc. 20th Design Automation Conf.*, pp. 64-70.

16. Reddy, K. M., S. M. Reddy, and V. K. Agarwal (1984), "Robust test for stuck-open faults in CMOS combinational logic circuits", *Proc. 14th Int. Symp. Fault Tolerant Comput.*, pp. 44-49.

17. Jain, S. K., and V. K. Agarwal (1985), "Modeling and test generation algorithm for MOS circuits", *IEEE Trans. Comput.*, C-34, pp. 426-433.

18. Roth, J. P. (1980), Computer Logic, Testing and Verification, Computer Science Press, Potomac, MD.

19. Reddy, M. K., and S. M. Reddy (1985), "On FET stuck-open fault detectable CMOS memory elements", *IEEE Test Conf.*, pp. 424-429.

20. Zasao, J. J. (1985), "MOS stuck fault testing of CMOS VLSI", *IEEE COMPCON Spring*, pp. 388-391.

21. Eichelberger, E. B. and T. W. Williams (1978), "A logic design structure for LSI testability", *J. Design Automat. Fault Tolerant Computing*, 2, pp. 165-178.

22. Golomb, S. W. (1967), Shift Register Sequences, Holden-Day, San Francisco, CA.

23. Savir, J. (1980), "Syndrome-testable design of combinational circuits", *IEEE Trans. Comput.*, C-29, pp. 442-451, correction: pp. 1012-1013.

24. Hurst, S. L., D. M. Miller, and J. C. Muzio (1985), Spectral Techniques in Digital Logic, Academic Press, London.

25. Bate, J. A. and D. M. Miller (1985), "Fault Detection in CMOS Circuits and an Algorithm for Generating Eulerian Cycles in Directed Hypercubes", *Congressus Numerantium*, 47, pp. 107-117.

26. Hortensius, P. D., R. D. McLeod, and H. C. Card (unpublished), "VLSI pseudorandom number generation for parallel computation", *Dept. Elect. Eng., U. of Manitoba, Winnipeg*.

27. Wolfram, S. (1986), "Random sequence generation using cellular automata", *Advances in Applied Math.*, *7*, pp. 123-169.

28. Podaima, B., P. D. Hortensius, R. D. Schneider, H. C. Card, and R. D. McLeod (1986), "CALBO - cellular automata logic block observation", *Proc. Canadian VLSI Conf.*, pp. 171-176.

TESTABILITY OF COMBINATIONAL NETWORKS OF CMOS CELLS[*]

J. A. Brzozowski

*Department of Computer Science
University of Waterloo
Waterloo, Ontario
Canada N2L 3G1*

ABSTRACT

The following types of faults are considered in CMOS cells: input and node stuck-at-0 and stuck-at-1 faults, and transistor stuck-on and stuck-open faults. A fault is called clean if there is a sequence of input vectors resulting in one output value in the good network and the complementary value in the faulty network, without the presence of any connections between V_{DD} and ground. Testing a conventional CMOS cell is difficult because many of its faults are not clean, and therefore their detection requires inconvenient electrical measurements. A clean fault is called static if it can be detected by a single input vector, and dynamic if the dynamic memory of an isolated node must be used in order to detect it. It is shown that input stuck faults are static, transistor stuck-open faults are dynamic, and no other faults are clean. Next, a modified cell

---

[*] This research was supported by the Natural Sciences and Engineering Research Council of Canada under grant No. A0871.

design is proposed: two control transistors are added, each with a control input. It is proved that all the faults studied are clean in combinational networks of modified CMOS cells, where two control inputs are used for each level of logic. Throughout this chapter it is assumed that only a single fault may occur at any time. Additional topics discussed include a one-transistor scheme that also leads to improved testability, hazards that may invalidate dynamic tests, and extensions to sequential circuits.

I. INTRODUCTION

In recent years CMOS technology, in the "pure" fully-complementary form and several of its variants, has been emerging as the dominant VLSI technology (38). Consequently, problems related to the detection of faults in CMOS circuits deserve particular attention. In this chapter we study the detection of logic faults in combinational static CMOS circuits. An attempt is made to present a comprehensive, unified treatment of the known results concerning faults in CMOS circuits, along with new results on the use of modified gate designs for improving testability.

The "stuck-at" fault model has been widely used in the "classical" theory of logic faults in gate circuits, both because of its simplicity and its relative success in testing (1,21,30). However, CMOS circuits present new problems in fault modeling and detection, as was first observed by Wadsack (37) in 1978. He pointed out that transistor "stuck-open" faults can transform combinational circuits into "dynamic" sequential circuits, and may not be detectable by the classical tests. For this reason, and also because of the bi-directional conductivity of MOS transistors, it is more appropriate to study CMOS faults using a switch-level, rather than a gate-level model. This is, indeed, the approach used here.

CMOS Cells

Wadsack also noted that "stuck-on", or shorted, transistors may lead to a state in which both the P-network and the N-network are conducting. This causes the output to assume an intermediate voltage level, between those representing logic 0 and logic 1. Wadsack considered such faults "analog in nature" and did not treat them as logic faults. We also adopt this approach initially. However, it is shown later that, by using additional control transistors, it is possible to design CMOS circuits in which these analog faults become logic faults.

In summary, most of this chapter is concerned with faults of the following types:

(a) *Input line stuck-at-0 and stuck-at-1 faults*.

These faults are easiest to detect in a natural way in CMOS circuits, for each such fault can be detected by a single input vector which drives the output of the good circuit to one logic value and that of the faulty circuit to the complementary logic value. We call such faults *static* in analogy with static CMOS design: in both cases the output signal does not decay with time.

(b) *Transistor stuck-open faults*.

These faults are not static, but can be detected by applying a sequence of two test vectors. The first vector drives both the good and faulty logic gates to the same logic value. The second vector drives the good gate to the complementary logic value, while the faulty gate's output node is isolated and retains the first value. Because the dynamic memory of the capacitance associated with the output node is used here, we call such faults *dynamic*. Static and dynamic faults are considered *clean*: their presence in a circuit is detectable as an incorrect logic signal at the gate output.

(c) *Transistor stuck-on faults*.

We classify these faults as *unclean* because any attempt at their detection must lead to a connection between 0 and 1, which is likely to result in an intermediate voltage level.

(d) *Node stuck-at faults*.

These faults are also classified as unclean.

As mentioned before, it is shown later that faults of type (c) and (d) can be made clean by using modified gates with additional control transistors. Throughout this chapter it is assumed that at most one fault may be present in any circuit. This so-called "single-fault assumption" simplifies the analysis, and usually produces satisfactory fault detection coverage.

The chapter is structured as follows. In Section II we introduce the basic terminology and properties of CMOS logic gates, which we prefer to call cells, in order to avoid confusion between the gate terminal of a single CMOS transistor and a CMOS logic gate. In Section III we describe faults in conventional CMOS cells, and we extend these observations to combinational networks of such cells in Section IV. In Section V we introduce a modified cell design using two extra control transistors, and show that all the faults of the type (a) through (d) are clean in networks of modified cells. In Section VI we present similar results using a scheme with only one extra control transistor. In Section VII we summarize briefly some results about hazards which may invalidate certain tests. In Section VIII we discuss the extension of the modified-cell techniques to sequential circuits.

As much as possible, the results mentioned above are stated and proved for general CMOS cells, with the hope that this work will become a first step towards a more general theory of testability of CMOS networks.

Finally, Section IX contains a discussion of the literature. We felt that such a discussion would be more meaningful to the reader at the end, after the terminology has been defined and the basic concepts have been explained.

II. CMOS CELLS

CMOS technology is based on N-channel and P-channel enhancement mode transistors, which we will call simply *N-transistors* and *P-transistors*. Commonly used transistor symbols are shown in Fig. 1 (a) and (c); in that figure we also show in (b) and (d) a simplified notation in which a transistor is represented as an edge in a graph. As a first approximation, an N-transistor may be considered to be a switch which is closed when its *gate* input a is 1, and open when a is 0. In Fig. 1 (b) the label a^N indicates that the labeled edge corresponds to an N-transistor whose gate signal is a. A P-transistor operates in a dual fashion, being closed when $a = 0$ and open when $a = 1$. This is denoted by the label a^P as shown in Fig. 1 (d).

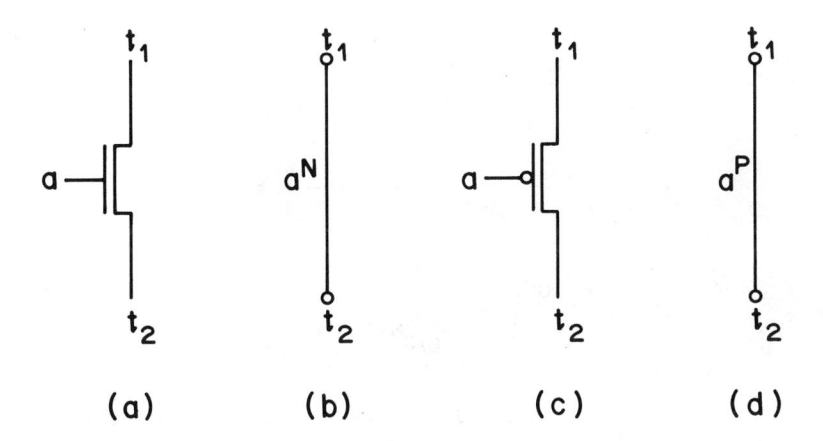

Fig. 1. (a) N-transistor; (b) graph for (a); (c) P-transistor; (d) graph for (c).

The switch model is only an approximate representation because a closed N-transistor does not transmit 1-signals well in the sense that, if $a = 1$ and $t_1 = 1$, the signal at t_2 is not a "strong" 1-signal. On the other hand, an N-transistor does transmit 0-signals well.

Similarly, a P-transistor transmits 1-signals well, but 0-signals poorly. Further discussion of these device-level phenomena is beyond the scope of this work, and we refer the reader to the literature (38). However, the above-mentioned properties of transistors lead to the basic CMOS design principle of using N-transistors to transmit 0's and P-transistors to transmit 1's. This principle is illustrated by the network of Fig. 2. Here, we follow the notation of (13). The dark nodes labeled 1 and 0 are *input nodes* connected to the strong signals 1 (representing the drain voltage V_{DD}) and 0 (representing the source voltage V_{SS} which is usually ground).

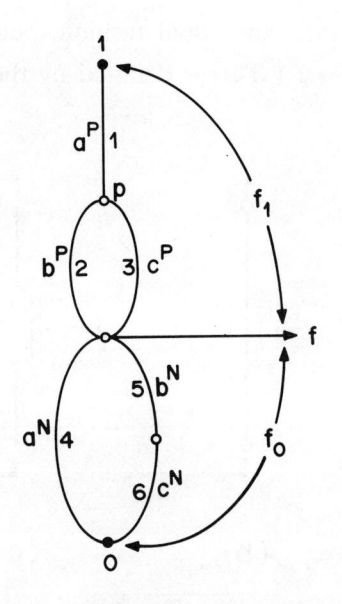

Fig. 2. A CMOS cell Γ_1.

Internal nodes are denoted by white circles, and an outgoing arrow on an internal node denotes an *output node*. The output node f is connected to 0 (by a path of closed transistors) when $a = 1$ or $b = c = 1$ or both. Let $f_0 = 1$ iff f is connected to 0; here

$f_0 = a + bc$, where bc denotes the *AND* of b and c and $+$ denotes the (inclusive) *OR*. Similarly, let $f_1 = 1$ iff f is connected to 1; here $f_1 = a'(b' + c')$, where x' is the complement of x. We refer to f_0 and f_1 as *transmission* functions. Note that $f_1 = f_0' = f$; thus the output node is either connected to 0 by a path consisting of closed N-transistors (an *N-path*), or it is connected to 1 by a path of closed P-transistors (a *P-path*).

In general, a "fully complementary" static CMOS logic gate, which we will call a *CMOS cell*, has the structure shown in Fig. 3. The *N-part* is any two-terminal network (or graph) consisting solely of N-transistors, and the *P-part* is a complementary network of P-transistors, in the sense that $f_1 = f_0'$. Each transistor in the cell is controlled by an input x_i from the *input vector* $X = x_1, \cdots, x_n, n \geq 1$. In particular examples we will use a, b, etc. for input symbols in order to avoid subscripts.

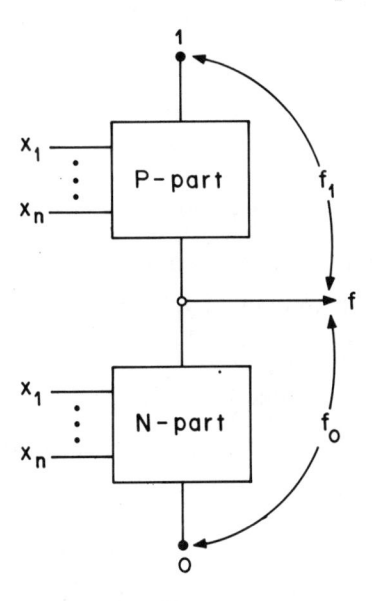

Fig. 3. General CMOS cell.

We will need certain concepts from Boolean algebra (12). Note that for $a, b \in \{0,1\}$, we have $a < b$ iff $a = 0$ and $b = 1$. The partial order $\leq$ on $\{0,1\}$ is extended to $\{0,1\}^n$, the Cartesian product of n copies of $\{0,1\}$, in the natural way:

$$A \leq B \text{ iff } a_i \leq b_i \text{ for } i = 1, \cdots, n,$$

where $A = a_1, \cdots, a_n$ and $B = b_1, \cdots, b_n$ are elements of $\{0,1\}^n$. A Boolean function $f: \{0,1\}^n \to \{0,1\}$ is *positive* if there exists a sum-of-products expression for f that does not contain any complemented variables. It can be verified (12,23) that f is positive iff for all $A, B \in \{0,1\}^n$,

$$A \leq B \text{ implies } f(A) \leq f(B).$$

Similarly, f is *negative* iff f' is positive; this is also equivalent to the existence of a sum-of-products expression for f which uses only complemented variables. A function is negative iff

$$A \leq B \text{ implies } f(A) \geq f(B).$$

Returning now to CMOS cells, we note that the transmission function f_0 is always positive and f_1 is always negative. Consequently, the output $f = f_1 = f_0'$ of any CMOS cell is always a negative function[*] of the inputs $x_1, \cdots, x_n$. Conversely, in principle, every negative function can be realized by a single CMOS cell. In practice, the number of transistors in any single cell is limited by the technology.

We will find it convenient to use the notation A_{i0} for any vector in $\{0,1\}^n$ with the i-th component equal to 0, and A_{i1} for the same vector as A_{i0}, but with the i-th component equal to 1. It follows that, for any negative function $f: \{0,1\}^n \to \{0,1\}$, we always have

---

[*] It is assumed that complemented inputs are not available; a complemented input requires an inverter, which is another CMOS cell.

$f(A_{i0}) \geq f(A_{i1})$, since $A_{i0} \leq A_{i1}$. We say that a Boolean function f *depends on input* x_i iff there exists a vector A_{i0} such that $f(A_{i0}) \neq f(A_{i1})$. If f is negative and depends on x_i as above, then $f(A_{i0}) = 1$ and $f(A_{i1}) = 0$. Clearly, if f does not depend on an input, then that input is redundant. The all-0 (all-1) vector $0, \cdots, 0$ $(1, \cdots, 1)$ will be denoted by $0...0$ $(1...1)$. Assuming that a negative function f is not identically 0 or 1, we always have $f(0...0) = 1$ and $f(1...1) = 0$.

Let f be the output function of a CMOS cell Γ, and let P be a transistor controlled by x_i in the P-part of Γ. We say that P is *tie-essential* iff there exists an input vector A, called *tie-critical for* P, such that

(i) $f_1(A) = 1$, and

(ii) $f_1^*(A) = 0$, if P is open,

where f_1^* denotes the transmission function of the P-part after the edge corresponding to transistor P has been removed. In other words, P is tie-essential iff there is an input vector A such that f is connected to 1 by at least one path of conducting transistors, and all such paths go through P. Then, if P is removed (or permanently open), $f_1(A)$ becomes 0. It follows that A must have $a_i = 0$, for P is controlled by x_i, i.e. a tie-critical input has the form A_{i0}. Note that for A_{i1} there is no path from f to 1. Note also that, if P is not tie-essential for f, then for any input A either there is no path from f to 1, or there is a path, but that path also exists with P removed. In either case, there is no change in f_1 if transistor P is removed entirely. Thus, if P is not tie-essential for f then P is a redundant transistor. Similar concepts apply also to transistors in the N-part.

The reader should note that, if a cell has two or more transistors that are not tie-essential, it is always possible to remove one of them without affecting the network behavior. However, it is not necessarily possible to remove both of them, because the removal of one transistor may make the second transistor tie-essential.

To illustrate these concepts consider the cell of Fig. 4(a). Transistor 1 is tie-essential, for $a = 0$ implies $f_1 = 1$, and f_1 becomes 0 if transistor 1 is cut. However, neither transistor 2 nor 3 is tie-essential because, if one of them is cut, the other still provides a path. In Fig. 4(b) none of the transistors 1, 2, or 3 is tie-essential.

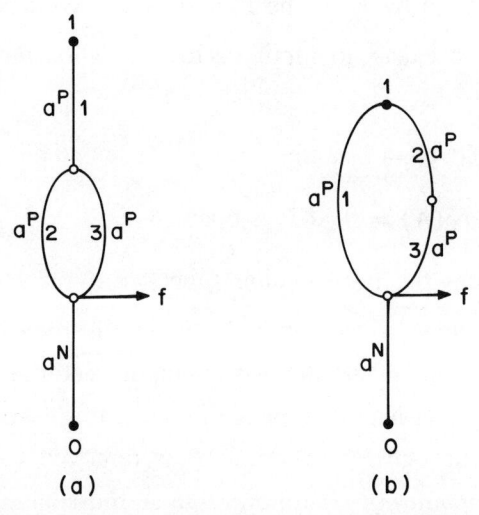

Fig. 4. Illustrating tie-essential and cut-essential transistors.

In a similar fashion we define a transistor P to be *cut-essential* iff there exists an input vector A, called *cut-critical for P*, such that

(i) $f_1(A) = 0$, and

(ii) $f_1^*(A) = 1$, if P is shorted,

where f_1^* denotes f_1 of the P-part with the gate input of P permanently set to 0.

One verifies that none of the transistors 1, 2, or 3 in Fig. 4(a) is cut-essential. In Fig. 4(b) transistor 1 is cut-essential, but neither 2 nor 3 is cut-essential. Similar concepts apply to N-transistors. Again one verifies that, if a transistor is not cut-essential, then it is redundant, and can be permanently shorted without affecting the transmission function. Note again that shorting one of two transistors that are not cut-essential may make the other cut-essential.

Finally, we define a certain concept of redundancy for nodes. We say that a node p in the P-part of a CMOS cell is *1-independent* iff there exists an input vector A, called *critical for p*, such that

(i) there is a P-path from f to p, and

(ii) there is no P-path from f to 1.

We claim that, if p is not 1-independent, then p and 1 can be connected together without changing the network behavior: in this sense p is then a redundant node. To show this consider any input vector A. If f is not connected to p when $X = A$, then f is not affected if p is permanently connected to 1. On the other hand, if f is connected to p then it is also connected to 1, since we have assumed that p is not 1-independent. Again, nothing changes if p is permanently connected to 1. In summary, a node that is not 1-independent can be eliminated.

In a similar way we define the concept of *0-independence* for N-nodes. These notions will be used later.

III. FAULTS IN CMOS CELLS

The cell of Fig. 2 will be used to illustrate the basic ideas. Let f, f_0, and f_1 be the functions in the good cell Γ and f^*, f_0^*, and f_1^* the corresponding functions in the faulty cell Γ^*. We remind the reader that the single-fault assumption holds throughout this paper.

Suppose input a is stuck-at-0. When the input vector (a, b, c) is 100 (short for (1,0,0)) we find: $f(100) = 0$ and $f^*(100) = f(000) = 1$. Thus this fault is detected by the single input vector $X = 100$. As mentioned before, such a fault is called *static*. This is summarized as follows:

	a	b	c	f	f^*
a stuck-at-0	1	0	0	0	1

Next suppose that transistor 2 is stuck-open. We first show that this fault is not static. To distinguish the good cell from the faulty one, we need an input vector giving a path from f to 1 through transistor 2, but no path from f^* to 1 in the faulty cell. Thus we must set $b = 0$ to "exercise" the faulty transistor, $c = 1$ to avoid masking the fault by a parallel path, and $a = 0$ to complete the path from f to 1. Now $f(001) = 1$, and $f^*(001)$ is connected neither to 0 nor to 1, i.e. it is isolated. Thus there is no input vector which drives f to one logic value and f^* to the opposite logic value; consequently the fault is not static.

It is possible to detect the stuck-open fault above by another method. The capacitance associated with the output node can store charge; thus the output node "remembers" its previous value when isolated. This memory is not permanent because of "leakage" current and is therefore called *dynamic* memory. However, suppose the test vector 101 is applied to drive the output node to 0 in both the good and the faulty cells, and then the vector 001 follows quickly enough. While f is driven to 1, f^* is isolated and retains the value 0. We will denote by **0** and **1** the values 0 and 1 remembered by an isolated node. We summarize these observations as follows:

	a	b	c	f	f*
Transistor 2 stuck-open:	1	0	1	0	0
	0	0	1	1	**0**

As mentioned before, faults detectable in this fashion by two vectors are called *dynamic*.

Static and dynamic faults are *clean*, for their detection does not involve any connection between 1 and 0 (i.e. between V_{DD} and ground). Consider now transistor 2 stuck-on. We need $b = 1$ to exercise the faulty transistor, $c = 1$ to avoid masking the fault, and $a = 0$ to complete the path from f to 1. Then $f_1(011) = 0$ and $f_1^*(011) = 1$. But, at the same time, $f_0(011) = f_0^*(011) = 1$. Consequently there is a path (1,2,5,6) in the faulty cell between 1 and 0. Assuming that the resistances of the conducting and stuck-on transistors are all approximately the same, we find that the output voltage in the faulty circuit is roughly $V_{DD}/2$. This means that the output is neither logic 0 nor logic 1, but has some intermediate value which we will denote by ×. Although some authors, e.g. (4), consider certain faults of this type detectable, we consider them *unclean*, i.e. not detectable by logic means alone. Faults that are not detectable because they involve redundant transistors will also be considered unclean. Note that such faults do not affect the cell behavior; for example, a shorted P-transistor in series with another P-transistor controlled by the same input has no effect.

For a second example of an unclean fault consider node p stuck-at-1 in Fig. 2. For any input vector where f is connected to 1, the fault can only add another path from f to 1, and hence cannot be detected. Also, if f is not connected to node p, the fault is undetected. Thus we need $a = 1$ and $b = 0$ or $c = 0$; say we use 101. This yields $f = 0$, and $f^* = \times$ because there is a path (2,4) from 0 to node p which is assumed to be stuck-at-1. Again the fault is unclean.

For a third example, suppose node p is stuck-at-0. To exercise the fault we must have f connected to p. If f is connected to 0 then the fault can only add another path to 0 and remains undetected. Hence f must be connected to both 1 and p, which is assumed to be stuck-at-0. Thus there is a connection between 0 and 1 and the fault is unclean. Suppose we try to use the vector 010; then the situation is as shown in Fig. 5. The output node f is connected by a P-transistor to 0, if the short to 0 is an ideal, zero-resistance short. Using a switch-level model like that of Bryant (6), one concludes that $f^* = 0$, but it is a "poor" 0 because 3 is a P-transistor. However, suppose the short is not ideal, but has a resistance comparable to that of transistor 1. Then the voltage at p, and also at f, will be some intermediate value. We will adopt the "worst-case" approach and consider this stuck-at-0 fault unclean.

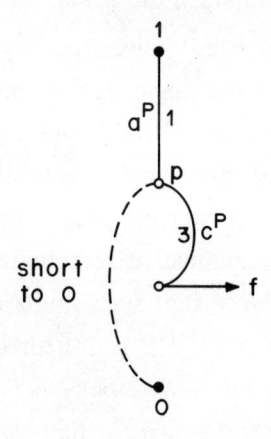

Fig. 5. Illustrating a stuck-at-0 node fault.

Finally, consider the output node f stuck-at-0. If the fault is caused by an ideal short from f to 0, then it is detectable. However, if the short is not perfect, f may take on an intermediate value. In any case, it is clear that to detect f stuck-at-0 we must drive f to 1, creating a connection between 1 and 0 in f^*. In our approach this constitutes an unclean fault.

The observations made above for the example of Fig. 2 are generalized to arbitrary CMOS cells as follows (8).

Theorem 1: Let Γ be an arbitrary CMOS cell with inputs $x_1, \cdots, x_n$ and output f. Then

(a) A stuck-at-0 or stuck-at-1 fault in input x_i is clean iff f depends on x_i. If such a fault is clean then it is static.

(b) A stuck-open fault in a transistor T is clean iff T is tie-essential. If such a fault is clean then it is dynamic, but not static.

(c) Transistor stuck-on faults and internal node or output node stuck-at-0 and stuck-at-1 faults are unclean.

Proof:

(a) This follows from the definitions of input dependence and static faults.

(b) Suppose transistor P with gate input x_i is stuck-open and is tie-essential. Let A_{i0} be a tie-critical vector for P. Then we have:

	X	f	f^*
P stuck-open	A_{i1}	0	0
	A_{i0}	1	**0**

Thus the fault is dynamic and therefore clean. Note also that the two test vectors differ only in the one variable x_i. We will make use of this observation in Section VII.

Conversely, suppose the fault is clean. Then there must exist an input A_{i0} such that $f_1(A_{i0}) = 1$ and $f_1^*(A_{i0}) = 0$. But this means that P is tie-essential. Note also that $f_0(A_{i0}) = f_0^*(A_{i0}) = 0$; thus the output node in the faulty cell is isolated, proving that the fault is not static.

A similar argument holds if an N-transistor is stuck-open.

(c) The arguments used in the examples of unclean faults above are easily generalized. □

IV. FAULTS IN COMBINATIONAL NETWORKS OF CELLS

In this section the ideas of Section III are extended to arbitrary feedback-free networks of CMOS cells, i.e. to the usual combinational networks constructed using CMOS cells as gates. An example of such a network is shown in Fig. 6. The functions realized by the four cells are:

Γ_1: $e = a'$

Γ_2: $f = (a + b)'$

Γ_3: $g = (f + ab)' = a \oplus b$ ($\oplus$ is exclusive or)

Γ_4: $h = (g + ce)' = (a \oplus b + a'c)'$

Note that the inputs to cells Γ_1 and Γ_2 are external inputs only, whereas Γ_3 depends on a, b, and the output f of Γ_2. Also, Γ_4 depends on the external inputs, on e, and on g.

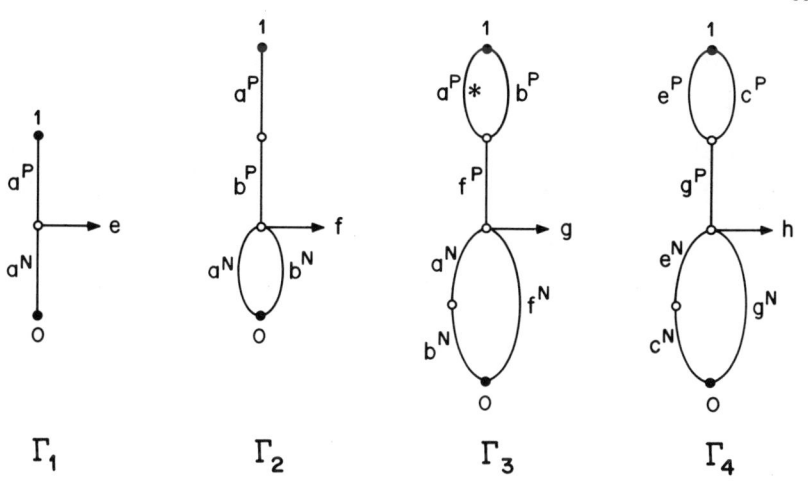

Fig. 6. Cascade of CMOS cells.

For convenience, we consider the cells in any feedback-free network to be linearly ordered as follows. A *cascade* of CMOS cells is a sequence $\Gamma = \Gamma_1, \cdots, \Gamma_k$ of cells having cell outputs $f_1, \cdots, f_k$, where the cell inputs to Γ_i are either external inputs from $X = x_1, \cdots, x_n$ or they are some outputs of the first $i-1$ cells from $\{f_1, \cdots, f_{i-1}\}$. For example, the network of Fig. 6 can be interpreted as the cascade $(\Gamma_1,\Gamma_2,\Gamma_3,\Gamma_4)$ or $(\Gamma_2,\Gamma_1,\Gamma_3,\Gamma_4)$ or $(\Gamma_2,\Gamma_3,\Gamma_1,\Gamma_4)$. The cascade is assumed to have a set Z of *observable outputs*; this is any subset of $f_1, \cdots, f_k$ that contains f_k.

Any Boolean function can be realized by a cascade of CMOS cells. For example, one can use the standard "two-level" NAND or NOR realization, with inverters for input variables forming an additional level. Also, there is an algorithm for finding a cascade using the minimum possible number of cells for a given function (31). However, such realizations do not necessarily lead to minimal transistor circuits (13). For a further discussion of cascades see (13,31).

Now consider faults in cascades. In this section we study somewhat idealized cells in the following sense. Suppose input A is to drive an output f_j to 0, and the input B is to drive f_j to 1, but leave the faulty output f_j^* isolated and retaining the value 0. If A and B differ in more than one component, the input may temporarily assume a value C while changing from A to B because the inputs may not change all at once. It may happen that $f_j(C) = 1$; this would cause $f_j^*(B)$ to have the value 1, and the fault would go undetected. The theorem below ignores such hazards, and assumes that the change from A to B happens instantaneously. We will deal with hazards in Section VII.

The inputs to cell Γ_j in a cascade are the external inputs and the outputs of the first $j-1$ cells. Thus f_j has the form $f_j(x_1, \cdots, x_n, f_1, \cdots, f_{j-1})$. However, the $f_1, \cdots, f_{j-1}$ are uniquely determined by $X = x_1, \cdots, x_n$, since we are dealing with combinational circuits. Thus there is Boolean function g_j such that

$$f_j(x_1, \cdots, x_n, f_1, \cdots, f_{j-1}) = g_j(x_1, \cdots, x_n).$$

This is true for any $j = 1, \cdots, k$. Now one easily verifies that a stuck-at-0 or stuck-at-1 fault in any cascade input x_i is detectable and static iff there is an observable cascade output f_j such that g_j depends on x_i.

Next, we consider transistor stuck-open faults in a cascade Γ. Let P be a P-transistor in Γ_j with gate input y_i, which is one of $x_1, \cdots, x_n, f_1, \cdots, f_{j-1}$. Let $(f_j)_0$ and $(f_j)_1$ denote the transmission functions of Γ_j, and let g_j be defined from f_j as above. Suppose that a stuck-open fault in P is to be detected by output f_m of cell Γ_m. Consider the following transformation. Cut the line f_j and, wherever f_j was an input to another cell, replace it by a new independent input z_j. Now f_m can be represented as a function $h_m(x_1, \cdots, x_n, z_j)$, where normally $z_j = g_j(x_1, \cdots, x_n)$. We say that transistor P is *tie-essential for* Γ iff there exists an input A such that

(i) $y_i = 0$,

(ii) $(g_j)_1(A) = 1$,

(iii) $(g_j)_1^*(A) = 0$ if P is open, and

(iv) $h_m(A,0) \neq h_m(A,1)$,

for some observable cascade output f_m. The first three conditions guarantee that the faulty transistor is exercised when $X = A$, and the fourth condition makes sure that the faulty behavior is observable. As before, we will call A *tie-critical* for P.

Clearly, if P is tie-essential for Γ as above, a stuck-open fault in P can be detected by first driving f_j to 0 and then applying a tie-critical input A. By definition this change is detected by some f_m. On the other hand, one verifies that a transistor P is redundant and can be removed if it is not tie-essential for Γ. The same type of reasoning applies to N-transistors. Altogether, we have:

Theorem 2: Let Γ be an arbitrary cascade of CMOS cells. Then

(a) A stuck-at-0 or stuck-at-1 fault in input x_i is clean iff there is some observable cascade output f_j such that g_j depends on x_i. If such a fault is clean then it is static.

(b) A stuck-open fault in transistor T is clean iff T is tie-essential for Γ. If such a fault is clean then it is dynamic, but not static.

(c) Transistor stuck-on faults, and internal node or output node stuck-at-0 and stuck-at-1 faults are unclean.

Proof: The arguments for (a) and (b) have been sketched above, and (c) is easily verified. □

In the cascade of Fig. 6 let P be the transistor marked by * in Γ_3. Then the sequence (110)(010) detects the fault with $h = 0$ and $h^* = 1$.

V. CELLS WITH TWO CONTROL TRANSISTORS

We now describe a modified CMOS cell to which have been added two control transistors. We then show that the fault properties of the modified cell are quite different. The cell $\hat{\Gamma}$ of Fig. 7 consists of the usual P-part and N-part of some cell Γ, but it also has two additional control transistors each with its own control input α and β. When $\alpha = 0$ and $\beta = 1$ the cell $\hat{\Gamma}$ behaves exactly like Γ.

The fault properties of $\hat{\Gamma}$ are summarized in the following theorem. Note that in the proof we use two adjacent vectors i.e. two vectors that differ only in one variable for all dynamic faults. This is done to avoid hazards.

CMOS Cells 335

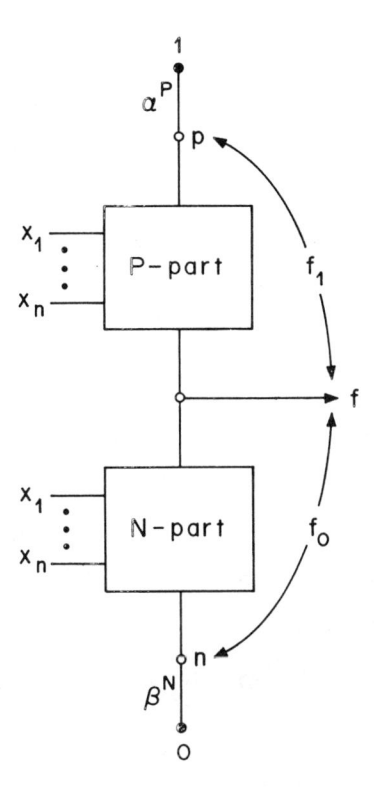

Fig. 7. A cell with two control transistors.

Theorem 3: In the cell $\hat{\Gamma}$:

(a) A fault in input x_i is clean iff f depends on x_i. If such a fault is clean then it is static.

(b) A stuck-open fault in a transistor T in the P-part or N-part is clean iff T is tie-essential. If such a fault is clean, then it is dynamic.

(c) A stuck-on fault in a transistor T in the P-part or N-part is clean iff T is cut-essential. If such a fault is clean, then it is dynamic.

(d) A stuck-at-0 or stuck-at-1 fault in the output node f is dynamic.

(e) A stuck-at-0 or stuck-at-1 fault in an internal node q in the P-part (N-part) is clean iff node q is 1-independent (0-independent). If such a fault is clean, then it is dynamic.

(f) All the faults associated with control inputs, control transistors, and nodes p and n are dynamic.

Proof:

(a) This is easily verified as before.

(b) Set $\alpha = 0$, $\beta = 1$ and proceed as in Section III.

(c) Suppose transistor P controlled by x_i is stuck-on. If P is not cut-essential, then it is redundant and can be shorted without affecting f. Suppose then that P is cut-essential, and let A_{i1} be a cut-critical input for P. We then have:

α	β	X	f	f^*
0	1	A_{i1}	0	$\times$
0	0	A_{i1}	**0**	1

N-transistors stuck-on are treated similarly.

(d) For f^* stuck-at-0 use

α	β	X	f	f^*
0	1	0...0	1	$\times$
1	1	0...0	**1**	0

f^* stuck-at-1 is treated similarly.

(e) Suppose a node q in the P-part is stuck-at-1. If q is not 1-independent, then it can be removed (i.e. identified with the node 1) and the fault cannot be detected. If q is 1-independent, let A_q be a critical input for q. Then we have:

α	β	X	f	f^*
0	1	A_q	0	×
0	0	A_q	0	1

Note that the connection of f^* to 1 (i.e. to node q stuck-at-1) is through P-transistors. This results in a "strong" 1 at f^*.

If q is stuck-at-0, we can use:

α	β	X	f	f^*
0	0	B_q	1	×
0	0	A_q	1	0

where A_q is critical for q, B_q is any vector adjacent to A_q such that $B_q < A_q$, and $f_1(B_q) = 1$ but $f_1(A_q) = 0$. Such vectors A_q and B_q can always be found because of the monotonicity of f_1. Note that here f^* is connected to 0 (i.e. to the node q stuck-at-0) through P-transistors only. This results in a "weak" 0 at f^*. The model of (13) does not accept such a signal as 0, because the CMOS design rules are violated here. However, most switch-level models such as that of (6) accept such signals as logic 0's. Thus the detection of this fault requires this more liberal interpretation.

(f) (i) Suppose α is stuck-at-0. Set $\alpha = 1$, $\beta = 1$. Let A_{i1} be such that $f_0(A_{i1}) = 1$ and $f_0(A_{i0}) = 0$. (Such an input always exists because of the monotonicity of f_0.) Then A_{i1} followed by A_{i0} results in $f = 0$, $f^* = 1$. For α stuck-at-1 use $\alpha = 0$, $\beta = 1$ and A_{i1} followed by A_{i0}; this gives $f = 1$, $f^* = 0$. Faults in β are treated similarly.

(ii) Since a stuck-open fault in the control P-transistor (N-transistor) is equivalent to α stuck-at-1 (β stuck-at-0), the claim follows by (i).

(iii) Suppose the control P-transistor is stuck-on. This is equivalent to α stuck-at-0 and the claim follows by (i). Similar comments apply to the control N-transistor.

(iv) Suppose that node p between the P-part and 1 is stuck-at-1. Use two adjacent vectors A and B such that $f_1(A) = 0$ and $f_1(B) = 1$. Then

α	β	X	f	f^*
1	1	A	0	$\times$
1	1	B	**0**	1

For p stuck-at-0 use:

α	β	X	f	f^*
0	0	0...0	1	$\times$
1	0	0...0	**1**	0

Here again, f^* has a weak 0.

Faults in the node n are treated similarly. □

Next, we extend the results obtained for modified cells to cascades of such cells. We first illustrate the ideas with an example. A two-level circuit realizing the majority function

$$h = (e+f+g)' = ((a+b)' + (b+c)' + (c+a))'$$
$$= (a+b)(b+c)(c+a)$$

is shown in Fig. 8, where we have augmented each cell using the

CMOS Cells

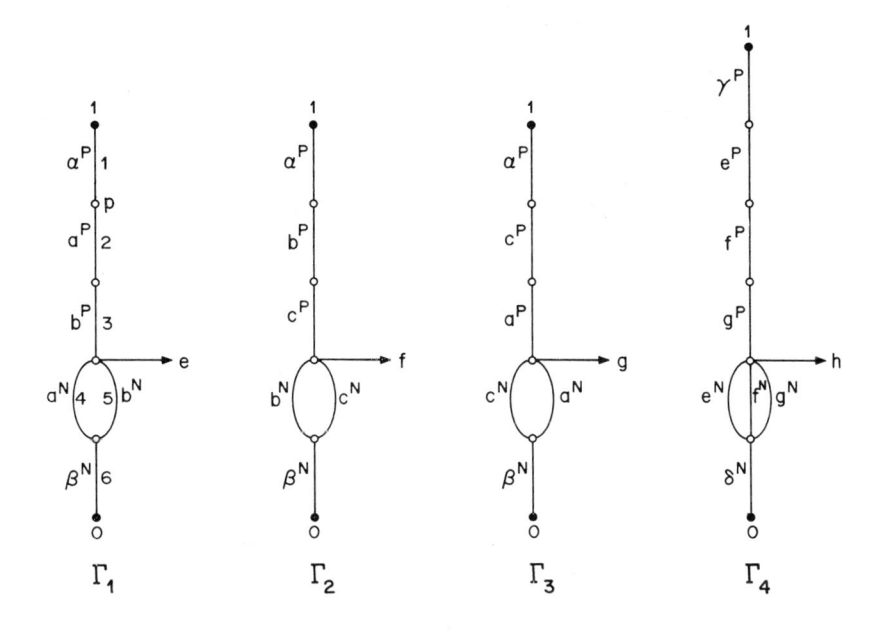

Fig. 8. Two-level circuit.

scheme of Fig. 7. Note that the three cells e, f, and g of the first level share the control inputs α and β.

We show below some examples of tests for stuck-on transistors in the first cell, where we assume that the control inputs γ and δ are set at their "passive" values $\gamma = 0$ and $\delta = 1$:

Transistor	α	β	a	b	c	e	e^*	f	g	h	h^*
1	1	1	1	0	1	0	0	0	0	1	1
	1	1	0	0	1	**0**	1	0	0	1	0
2	0	1	1	0	1	0	×	0	0	1	×
	0	0	1	0	1	**0**	1	0	**0**	1	0
4	0	1	0	0	1	1	×	0	0	0	×
	1	1	0	0	1	**1**	0	0	0	0	1

The cases of the remaining transistors are similarly treated. Transistors in cells 2 and 3 are handled by symmetrical arguments. For the level-2 cell Γ_4 set $\alpha = 0$, $\beta = 1$ and use γ and δ to obtain clean faults. Thus all transistor stuck-on faults are cleanly detectable.

Node stuck-at faults are also easily handled. Two examples are given below for node p of the first cell.

	α	β	a	b	c	e	e^*	f	g	h	h^*
p stuck-at-1	1	1	1	0	1	0	0	0	0	1	1
	1	0	1	0	1	**0**	**0**	**0**	0	1	1
	1	0	0	0	1	**0**	1	**0**	0	1	0

Note that it is assumed that the capacitance of output node e (connected to gate terminals of other transistors) is much larger than the capacitance of internal node p (not connected to any gate terminals). Thus node e will retain a valid logic 0, even after it is connected to isolated node p.

	α	β	a	b	c	e	e^*	f	g	h	h^*
p stuck-at-0	0	1	0	0	1	1	×	0	0	0	×
	1	1	0	0	1	**1**	0	0	0	0	1

In general, we have the following result:

Theorem 4: Let N be any cascade of modified cells using two control transistors, where all the cells in level i share control signals α_i and β_i. Then all the faults in N are clean under the assumption that N has no redundant inputs, transistors or nodes.

Proof: The extensions of the definitions and ideas described so far are relatively straightforward and are left to the reader. □

VI. CELLS WITH ONE CONTROL TRANSISTOR

In Fig. 9 we show a modified cell $\tilde{\Gamma}$ with an additional, control P-transistor inserted in series between the P-part and the N-part. The gate input of P is the control input α. Note that one additional node g is created; we assume that this node is observable, i.e. that g is an additional output of the cell - a control output used only in testing. When $\alpha = 0$, transistor P is ON and $\tilde{\Gamma}$ behaves like the original cell Γ of Fig. 3. When $\alpha = 1$ the P-part and the N-part are separated.

Theorem 5: In the cell $\tilde{\Gamma}$ parts (a) through (e) of Theorem 3 also hold. Furthermore, all faults associated with the control input α, the control transistor, and the control output g are dynamic.

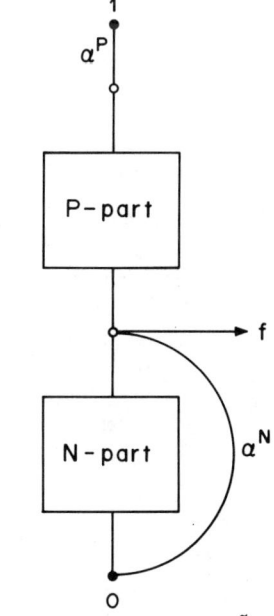

Fig. 9. Modified cell $\tilde{\Gamma}$.

Proof: The arguments are briefly sketched below.

(a) and (b) are easily verified.

(c) Suppose that P, controlled by x_i, is stuck-on, and that P is cut-essential with a cut-critical input A_{i1}. Then we find:

α	X	g	g^*
0	A_{i1}	0	$\times$
1	A_{i1}	**0**	1

Note that, in the first step, g is driven to 0 through a path of N- and P-transistors, requiring the more liberal switch-level model.

Next, let an N-transistor N controlled by x_i be stuck-on, and suppose B_{i0} is a cut-critical input for N. Then:

α	X	f	f^*
0	B_{i0}	1	$\times$
1	B_{i0}	**1**	0

(d) One easily verifies the following:

	α	X	f	f^*
f stuck-at-0	0	0...0	1	$\times$
	1	0...0	**1**	0
f stuck-at-1	1	A	0	$\times$
	1	B	**0**	1

where A and B are two adjacent input vectors such that $f_0(A) = 1$ and $f_0(B) = 0$.

(e) Choose A_p, B_p so that A_p is critical for node p, B_p is adjacent to A_p, and $f_1(B_p) = 1$. Then:

	α	X	g	g^*
p stuck-at-0	1	B_p	1	×
	1	A_p	1	0
p stuck-at-1	0	A_p	0	×
	1	A_p	**0**	**1**

Again, these tests require the more liberal model.

(f) For faults in the control input α use adjacent A and B such that $f_0(A) = 0$ and $f_0(B) = 1$.

	α	X	f	f^*
α stuck-at-1	0	B	0	**0**
	0	A	1	**0**
α stuck-at-0	1	B	0	0
	1	A	**0**	**1**

Next, the control transistor stuck-open (stuck-on) is equivalent to α stuck-at-1 (stuck-at-0). Finally:

	α	X	g	g^*
g stuck-at-0	1	A	1	×
	1	B	1	0
g stuck-at-1	0	1...1	0	×
	1	1...1	**0**	**1**

This concludes the proof. The same type of result can also be obtained by using an N-transistor instead of a P-transistor for the control transistor. □

Although the scheme of Fig. 9 uses one less transistor and one less control input than the scheme of Fig. 7, it requires an additional observable output. At this time we do not have any good way of taking care of such additional outputs; consequently, the scheme will not be pursued any further.

VII. HAZARDS

Consider again the circuit of Fig. 2, and suppose that transistor 2 is to be tested for a stuck-open fault. In principle, one could use the test vectors 100 followed by 001. The first vector gives $f = f^* = 0$, and the second vector gives $f = 1$ and f^* isolated, remembering the value 0. Note, however, that inputs a and c must both change in going from the first to the second vector. In a real circuit it is possible for one input to change before the other because of stray delays or time-skews in input variable changes. If input a changes first, the input vector goes through the transient state 000 which causes f^* to become 1. Consequently, when the final vector 001 is applied, we have $f = 1$ and $f^* = \mathbf{1}$, and the fault goes undetected. This can be viewed as a type of output hazard: In the faulty cell, the output was to change from 0 to "isolated", but it went temporarily to 1 during that transition. As Reddy et al. point out, such faults cannot be assumed to be always undetectable, because a chip may pass testing, but may fail when it becomes part of a larger circuit where its new inputs are delayed by different amounts (35). These difficulties were first noted in (24,34); a detailed study of such hazards was presented in (35). A method for generating hazard-free tests is described in (33).

The following result can be easily seen from the discussion in Sections III, V and VI.

Theorem 6: There exist hazard-free tests for all transistor stuck-open faults detectable in a standard CMOS cell Γ, and for all the dynamic faults in the modified cells $\hat{\Gamma}$ and $\tilde{\Gamma}$.

Unfortunately, it is not possible to claim this result for all combinational networks of CMOS cells, as is shown by the example of Fig. 10 taken from (35). To test the transistor marked with * for a stuck-open fault, we need $\bar{a} = 0$ to exercise the fault, $\bar{c} = 0$ to complete the path from f to 1, and $\bar{b} = 1$, $\bar{d} = 1$ to prevent masking the fault. Thus the second test vector can only be $abcd = 1010$. There are only three vectors that can be used as the first test vector, i.e. that result in $f = 0$, namely: 0101, 0110, and 1001. It is easily verified that a transition from each of these vectors to 1010 may involve a hazard.

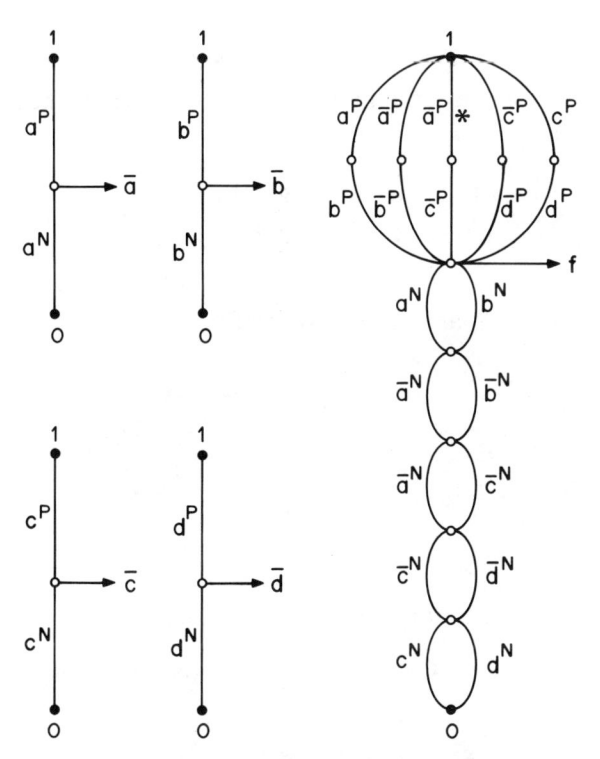

Fig. 10.

Reddy et al. (35) next show that it is possible to avoid hazards by modifications of the cells using additional control transistors. They first note that any Boolean function can be realized by a two-level circuit in which the first level contains some inverters providing complemented inputs, and the second level consists of a single cell. In particular, one can always write the complement f' of any function f as a sum of products of variables and complemented variables and thus obtain this "inverter-cell" form. The modification used to avoid hazards is precisely the two-transistor scheme of Fig. 7. The following result is proved by Reddy et al. (35). Let Γ be a two-level network in the inverter-cell form, where the second-level cell Γ has been modified to $\hat{\Gamma}$ by the addition of two control transistors. Let f be the output of Γ and let g be the function of the external inputs associated with f as in Section IV.

Theorem 7: In the inverter-cell network Γ all stuck-open faults in tie-essential transistors and in the control transistors have hazard-free test sequences.

Proof: Let transistor P in the original P-part of Γ be tie-essential for Γ. Let A be a tie-critical input for P, and let B be an input such that $g(B) = 0$. Then the following test sequence detects a stuck-open fault in P and is hazard-free:

X	α	β	f	f^*
B	1	1	0	**0**
A	1	0	**0**	**0**
A	0	0	1	**0**

The rest of the verification is left to the reader. □

A second scheme used by Reddy et al. (35) is illustrated in Fig. 11.

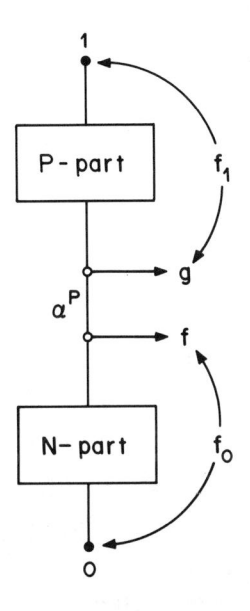

Fig. 11. Modified cell $\dot{\Gamma}$.

When $\alpha = 0$, we have the original cell Γ. The following test detects a stuck-open fault in a P-transistor in the original P-part with a tie-critical input A for P:

X	α	f	f^*
A	1	0	0
A	0	1	**0**

These ideas can be extended by the addition of two more transistors and another control input to provide hazard-free detection of all stuck-open faults in inverter-cell networks and other forms. For further details see (35).

VIII. FAULTS IN LATCHES

Sequential circuits create additional difficulties in fault detection. Such problems were considered in (10,32). We illustrate the problem with the NOR latch of Fig. 12.

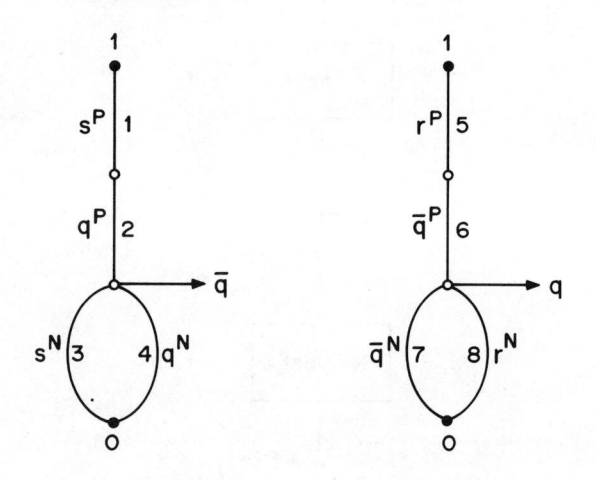

Fig. 12. CMOS NOR latch.

Suppose transistor 4 is stuck-open and consider the left NOR cell. The only input vector that differentiates in any way between the good cell and the bad one is $s = 0$, $q = 1$, for then $\bar{q} = 0$ and $\bar{q}^*$ is isolated. Thus, if a dynamic test exists, the second vector must be $s = 0$, $r = 0$, for this is the only vector consistent with $s = 0$ and $q = 1$. Now, on the one hand, the first vector must have $s = 1$ in order to achieve $q = 1$ when s becomes 0. But, on the other hand, $\bar{q}^*$ must be forced to 1 by the first vector in order to detect the fault. This can only happen if $s = 0$ and $q = 0$. Therefore the two requirements are conflicting, showing that the fault is not detectable. It is shown in (10) that a scheme like that of Fig. 11 can be used to make this fault detectable. A different scheme has also been developed by Reddy and Reddy. For more details see

(10,32). Finally, the scheme of Fig. 11 can be combined with the scheme with two control transistors to result in latches in which all the faults are clean (10).

IX. PREVIOUS LITERATURE

In this section we describe previous contributions, but only to the extent that they relate to the present work. As has been said before, the first paper discussing faults in CMOS was that of Wadsack (37). He pointed out the dynamic nature of stuck-open faults, and mentioned stuck-on faults. His approach to modeling the good and faulty behaviors of a CMOS cell involved the construction of an equivalent gate circuit, with a latch added to represent the dynamic memory of an isolated output node. At that time no other method existed for simulating nonclassical logic faults (37). However, the switch-level models developed later are more natural.

Galiay et al. (22) studied failures in NMOS, but some of their observations apply also to CMOS. They reported that an examination of a set of failed microprocessor chips showed that a great majority of failures consisted of shorts and opens at the metallization level and at the diffusion level; this confirmed that the stuck-at fault model was inadequate. They also observed that bridging faults resulted in a modification of the function of a gate. They concluded that a switch-level, rather than a gate-level representation was needed. Furthermore, they described the basic conditions for detecting open and short faults in NMOS circuits, and suggested techniques for the generation of test sequences. Finally, they recommended a set of layout rules to minimize the chances of failures and to improve testability.

Elziq (18) pointed out some disadvantages of Wadsack's gate-latch model. For combinational circuits composed of primitive gates, he proposed examining test sequences that have been generated to detect classical stuck-at faults in order to determine which stuck-open faults also happen to be detected by these sequences; this involves comparing pairs of consecutive test vectors. He then suggested that tests for the remaining undetected open faults be generated. This work was extended by Elziq and Cloutier (19) to arbitrary CMOS cells. It was noted that all stuck-open faults in transistors "connected only in series" (19) are equivalent. A method for test pattern generation to detect all stuck-open faults was described; this involved alternately testing P- and N-transistors. Stuck-open faults in CMOS flip-flops were also considered. It was pointed out that some stuck-open faults in transistors controlled by fed-back signals are not logically testable (as discussed in Section VII). "Time delay" tests were suggested for the detection of such faults. Finally, the difficulties associated with testing CMOS transmission gates were pointed out.

Chandramouli (14) showed that, in networks without reconvergent fan-out, test vectors that detect input stuck-at-0 and stuck-at-1 faults can be arranged in a suitable sequence that will detect all single stuck-open faults. Similar results appear also in (15).

McCluskey and Bozorgui-Nesbat (29) proposed designing networks as collections of subnetworks which are small enough to permit exhaustive testing. To permit detection of stuck-open faults in CMOS they use an additional test transistor for each CMOS cell, a test line, and a charge/discharge line. The output line is charged or discharged after each test vector. If there is no stuck-open fault, the output returns to its correct value; otherwise it acquires the new value. Flaherty and Walther (20) propose modifying NMOS and CMOS cells by including switchable load resistors controllable by separate test inputs. These switchable resistances change the isolated state into a logical value, thus making the fault detectable. Similarly, such switches can be used to break the connection between V_{DD} and ground present because of a stuck-on fault. The use of extra control transistors by Reddy et al. (33-35) has already been discussed in Sections VII and VIII.

A switch-level model for MOS circuits has been developed by Bryant (6). Fault simulation at the switch level was considered by Bose et al. (5): line stuck-at faults, and transistor stuck-open faults were modeled. Stuck-on faults were not included; the authors noted that a connection between V_{DD} and ground must be present if one attempted to detect such faults (i.e. they are unclean in our terminology.) A fault simulator based on the switch-level model of (6) is described in (7); the faults modeled include both node and transistor faults.

Malaiya and Su (27) discussed fault detection by leakage current measurements. Banerjee and Abraham (3) discussed the effects of physical failures leading to intermediate logic levels, timing errors, and changes in the logic function performed. See also (36) for a discussion of logical models of physical failures.

Chiang and Vranesic (16,17) model a CMOS network as a collection of graphs, where each cell is represented by a P-graph and an N-graph. The edges of each graph are labeled by variables, in a manner similar to that of Section II. For cascade networks they substitute graphs of previous cells for edges in later cells in order to facilitate test generation. They also develop some heuristics for deriving test sequences.

Chiang and Vranesic consider a stuck-on N-transistor to be testable if its shorted resistance is much lower than its normal ON resistance. They state that, "if a test can detect an open fault on transistor k' in the p-part it should be able to detect the short fault in transistor k in the n-part", where k and k' are controlled by the same input. This is not quite correct, as can be seen from the CMOS network for the function $f = (a + b)(c + ab)$, where there are two N-transistors and two P-transistors controlled by a.

However, the following statement is true: For each N-transistor N there exists a P-transistor P and an input A such that

$$f_1(A) = 1 \text{ and } f_1^*(A) = 0, \text{ if } P \text{ is stuck-open, while}$$

$$f_0(A) = 0 \text{ and } f_0^*(A) = 1, \text{ if } N \text{ is stuck-on}.$$

Mandl (28) stresses the need to develop techniques particularly tailored to CMOS. He points out the difficulties met in testing circuits using transfer gates (i.e. transmission gates) - a problem that we have not addressed in this paper. Boolean methods do not suffice here, for one must distinguish between strong and weak signals. Baschiera and Courtois (4) also survey the challenges of CMOS testing. They describe some methods for testing stuck-on faults which taking into account resistance values.

Jain and Agrawal (24,25) use an equivalent gate circuit and a memory element to model CMOS cells. They assume that either the N-network prevails in the case of stuck-on faults, or such faults are not testable. They then apply the D-algorithm to generate tests for both stuck faults and transistor faults.

P. Agrawal (2) uses a graph representation and the D-algorithm for deriving tests which detect stuck-at faults and stuck-open faults without hazards.

The detection of multiple faults in networks of CMOS cells has been recently studied by Jha (26).

Our own work on the one-transistor scheme of Section VI was briefly described in (11) and fully discussed in (8) and (9). The two-transistor scheme for clean faults was introduced in (9).

ACKNOWLEDGMENT

The author wishes to thank B. F. Cockburn and C-J. Seger of the University of Waterloo for their useful comments about this work.

REFERENCES

(1) Abraham, J. A. and Agarwal, V. K., "Test Generation for Digital Systems", in *Fault-Tolerant Computing*, Vol. I, D. K. Pradhan, ed., Prentice Hall, Inc., Englewood Cliffs, N.J., 1986, 1-90.

(2) Agrawal, P., "Test Generation at Switch Level", Proc. *IEEE Int. Conference on Computer-Aided Design*, November 1984, 128-130.

(3) Banerjee, P. and Abraham, J. A., "Generating Tests for Physical Failures in MOS Logic Circuits", *Proc. 1983 International Test Conference*, October 1983, 554-559.

(4) Baschiera, D. and Courtois, B., "Testing CMOS: A Challenge", *VLSI Design, V*, No. 10, October 1984, 58-62.

(5) Bose, A. K. et al., "A Fault Simulator for MOS LSI Circuits", *Proc. 19th Design Automation Conference*, June 1982, 400-409.

(6) Bryant, R. E., "A Switch-Level Model and Simulator for MOS Digital Systems", *IEEE Trans. Computers, C-33*, No. 2, February 1984, 160-177.

(7) Bryant, R. E. and Schuster, M. D., "Performance Evaluation of FMOSSIM, a Concurrent Switch-Level Fault Simulator", *Proc. 22nd Design Automation Conference*, June 1985, 715-719.

(8) Brzozowski, J. A., "Testability of CMOS Cells", Dept. of Computer Science, University of Waterloo, Waterloo, Ont., Canada, Tech. Report No. CS-85-31, September 1985.

(9) Brzozowski, J. A., "Testability of CMOS Cells", Proc. Technical Workshop: *New Directions for IC Testing*, Victoria, B.C., March 1986, 8-1 - 8-14. Also poster paper, *9th Annual IEEE Workshop on Design for Testability*, Vail, Co., April 1986.

(10) Brzozowski, J. A., "On the Testability of Static CMOS Latches", *Technical Digest 1986 Canadian Conference on VLSI*, October 1986, 153-158.

(11) Brzozowski, J. A. and Sayed, M., "Design of Testable CMOS Cells", *Technical Digest 1985 Canadian Conference on VLSI*, November 1985, 225-228.

(12) Brzozowski, J. A. and Yoeli, M., *Digital Networks*, Prentice-Hall, Inc., Englewood Cliffs, N.J., 1976.

(13) Brzozowski, J. A. and Yoeli, M., "Combinational Static CMOS Networks", *pp. 271-282 in Proc. 1986 Aegean Workshop on Computing: VLSI Algorithms and Architectures*, Lecture Notes in Computer Science, Vol. 227, Springer-Verlag, Berlin, 1986. Also Dept. of Computer Science, University of Waterloo, Waterloo, Ont., Canada, Tech. Report No. CS-85-42, December 1985.

(14) Chandramouli, R., "On Testing Stuck-Open Faults in CMOS Combinational Circuits", *Proc. 19th Annual Allerton Conf. on Communication, Control, and Computing*, September 1981, 707-712.

(15) Chandramouli, R., "On Testing Stuck-Open Faults", *Proc. 1983 International Test Conference*, October 1983, 258-265.

(16) Chiang, K. W. and Vranesic, Z. G., "Test Generation for MOS Complex Gate Networks", *Proc. 12th International Symposium Fault-Tolerant Computing*, June 1982, 149-157.

(17) Chiang, K. W. and Vranesic, Z. G., "On Fault Detection in CMOS Logic Networks", *Proc. 20th Design Automation Conference*, June 1983, 50-56.

(18) Elziq, Y. M., "Automatic Test Generation for Stuck-Open Faults in CMOS VLSI", *Proc. 18th Design Automation Conference*, June 1981, 347-354.

(19) Elziq, Y. M. and Cloutier, R. J., "Functional-Level Test Generation for Stuck-Open Faults in CMOS VLSI", *Digest 1981 International Test Conference*, October 1981, 536-546.

(20) Flaherty, R. J. and Walther, R. G., "Three-State Device and Circuit Testing", *IBM Technical Disclosure Bulletin*, 25, April 1983, 6286-6290.

(21) Fujiwara, H., *Logic Testing and Design for Testability*, MIT Press, 1985.

(22) Galiay, J., Crouzet, Y. and Vergniault, M., "Physical Versus Logical Fault Models MOS LSI Circuits: Impact on Their Testability", *IEEE Trans. Computers*, C-29, June 1980, 527-531.

(23) Gilbert, E. N., "Lattice Theoretic Properties of Frontal Switching Functions", *J. Mathematics and Physics, 33*, 1954, 57-67.

(24) Jain, S. K. and Agrawal, B. D., "Test Generation for MOS Circuits using D-algorithm", *Proc. 20th Design Automation Conference*, June 1983, 64-70.

(25) Jain, S. K. and Agrawal, V. D., "Modeling and Test Generation Algorithms for MOS Circuits", *IEEE Trans. Computers*, C-34, No. 5, May 1985, 426-433.

(26) Jha, N., "Detecting Multiple Faults in CMOS Circuits", *Proc. 1986 International Test Conference*, September 1986, 514-519.

(27) Malaiya, Y. K. and Su, S. Y. H., "A New Fault Model and Testing Technique for CMOS Devices", *Digest 1982 International Test Conference*, October 1982, 25-34.

(28) Mandl, K. D., "CMOS VLSI Challenges to Test," *Proc. 1984 International Test Conference*, October 1984, 642-648.

(29) McCluskey, E. J. and Bozorgui-Nesbat, S., "Design for Autonomous Test", *IEEE Trans. Computers, C-30*, No. 11, November 1981, 866-875.

(30) Miczo, A., *Digital Logic Testing and Simulation*, Harper & Row, 1986.

(31) Muroga, S., *VLSI System Design*, John Wiley, 1982.

(32) Reddy, M. K. and Reddy, S. M., "On FET Stuck-Open-Fault-Detectable CMOS Memory Elements", *Proc. 15th International Symposium Fault-Tolerant Computing*, June 1985, 424-429.

(33) Reddy, S. M., Reddy, M. K. and Agrawal, V. D., "Robust Tests for Stuck-Open Faults in CMOS Combinational Logic Circuits", *Proc. 14th International Conference Fault-Tolerant Computing*, June 1984, 44-49.

(34) Reddy, S. M., Reddy, M. K. and Kuhl, J. G., "Testable Design for Stuck-at-Open Faults in CMOS Circuits," Presented at *Sixth Annual IEEE Workshop on Design for Testability*, Vail, Co., April 1983.

(35) Reddy, S. M., Reddy, M. K. and Kuhl, J. G., "On Testable Design for CMOS Logic Circuits", *Proc. 1983 International Test Conference*, October 1983, 435-441.

(36) Timoc, C. et al., "Logic Models of Physical Failures", *Proc. 1983 International Test Conference*, October 1983, 546-553.

(37) Wadsack, R. L., "Fault Modeling and Logic Simulation of CMOS and MOS Integrated Circuits," *Bell System Tech. J.*, May-July 1978, 1449-1474.

(38) Weste, N. and Eshraghian, K., *Principles of CMOS VLSI Design*, Addison-Wesley, 1985.

CONCURRENT CHECKING TECHNIQUES - A DFT ALTERNATIVE

I.L. Sayers,   G. Russell,   D.J. Kinniment

Department of Electrical and Electronic Engineering
University of Newcastle upon Tyne   NE1 7RU
England

I. INTRODUCTION

One of the major issues in the design of VLSI circuits is that of curtailing the exponential growth in test generation costs (1) with the increase in circuit complexity. In an attempt to alleviate this problem three approaches have been adopted.

    (a) The development of more sophisticated test generation schemes

    (b) The adoption of Design for Testability Techniques.

    (c) The adoption of Built-in Test Methods.

The development of more sophisticated test generation schemes has been carried out at both gate and higher levels of abstraction. It has been reported that automatic test generation programs have been developed, for example PODEM-X (2), which are capable of generating tests in unpartitioned combinational logic functions comprising 50,000 gates. The D-Algorithm has also been modified so that it can be used on higher level functional blocks, such as adders, counters, multiplexors etc. or on behavioural descriptions of functions written in a special purpose high level language. An improvement in the efficiency of the test generation process has been sought through the use of Special Purpose Engines (3) (4) or Hardware Accelerators, which under certain circumstances can run test generation algorithms much faster than a general purpose computer. More recently, the use of Knowledge Based Systems (5) to assist designers in the generation of test programs has also been reported. In general, however, this approach to reducing test generation costs only attacks the symptoms of the problem, namely complexity, and not the cause and hence can only provide an interim solution to the problem.

The root cause of the test generation problem, as the circuit complexity increases, is the inability to control or observe signal values in the circuit. It is this facet of the problem to which the Design for Testability (DFT) (6) approach has been directed. The DFT techniques ease the testing problem either by making internal nodes more accessible or by reconfiguring the circuit for the purposes of testing. Although these techniques are highly desirable, they ultimately incur some penalties with respect to either their effect upon the circuit performance or upon the designer. Most DFT techniques involve the use of additional hardware thus the physical size of the circuit will increase with a corresponding reduction in the yield; the additional hardware will also introduce further delays into the circuit and hence reduce the overall performance of the circuit. Furthermore, if a particular DFT technique is to be adopted by the design community at large it must not be too constraining as to inhibit the ingenuity of the designer and also there must be adequate CAD support software to check a design for compliance with a given set of DFT rules (7). In the light of these penalties it must be demonstrated that a particular DFT technique produces a marked reduction in test generation costs measured in terms of reduced test generation time and personal effort.

Although DFT techniques reduce the amount of effort in generating test vectors for a circuit, a vast amount of test data (input test vectors and output responses) has to be handled during the testing phase. Furthermore, the most popular DFT techniques, for example those using the basic Scan Path principle, cannot test the circuit at speed and also require long test times. In an attempt to overcome these limitations in the DFT techniques, another approach evolved called Built-in-Test, which incorporated hardware on the circuit to generate the test vectors in the first instance and also a mechanism to compress the test vector responses into either a 'Signature' (8) or a 'Syndrome' (9) which can be checked to determine if a given circuit is fault free.

The main function of the current testing strategies is the detection of faulty devices either before or immediately after packaging. The present trend, however, of scaling down device sizes has not only increased the complexity of the testing problem but also has led to an increase in the occurrence of transient errors, a fault mechanism which is not covered by the current range of DFT and Built-in-Test methods. Furthermore, with the advent of Wafer Scale Integration (WSI) current test strategies are again inadequate, since the basic approach to combat complexity in testing, is the "divide and conquer" principle, the test patterns applied are thus local to a function block subsequently leaving global interconnection paths untested; global testing strategies are fundamental to the testing of WSI circuits.

To overcome the limitations of current design for testability techniques and also to improve the reliability of complex VLSI circuits, designers have been resorting to the use of either hardware or information redundant techniques to perform Concurrent Circuit Testing (CCT). It is considered, however, that the area penalty incurred by hardware redundancy is excessive hence favouring the use of information redundant methods. Although many error detection codes exist for use in information redundant schemes, few, however are suitable for incorporation into VLSI designs; since they are either applicable only to a limited range of functions and hence can not be applied readily to a VLSI design without the use of code conversion blocks, incurring a high area penalty, or the codes require decoders to extract the checkbits from the codewords and hence introduce unwanted delays in the circuits. It has been demonstrated, however, that the Mod 3 Low Cost Residue Code can be incorporated cost effectively into VLSI designs providing a facility not only for the detection of transient errors but also as a mechanism for testing the integrity of the global interconnections between functions which is essential in testing WSI circuits; furthermore the checking hardware can in general, be implemented independently from the hardware manipulating the information bits.

II. NON-CLASSICAL FAILURE MODES IN VLSI CIRCUITS

In addition to the problems of testing VLSI circuits, brought about by the increase in complexity and the reduction in accessibility to internal nodes, the improvements and changes in device technology have introduced failure modes which do not manifest themselves as simple stuck-at-faults (classical faults) for the purposes of testing. In the absence of a detailed understanding of how to model these non-classical fault mechanisms (10), circuits continue to be tested using the simple stuck-at fault model without any guarantee of its validity for the current dominant manufacturing technologies, namely NMOS and CMOS. Although current test strategies give a high fault coverage for the known types of faults, the question which subsequently arises, what is the effect of the 'one's that get away' on system reliability and also on the cost of detecting these faults when the device is in operation. It is estimated that the cost of detecting a fault at chip level and when it is in its operational environment differ by a factor of 10^3.

The limitations in the use of the stuck-at-fault model with current technologies and some of the physical failure mechanisms which affect the logical behaviour of a circuit but cannot be modelled with a stuck-at-fault model will be discussed, briefly.

A. Deficiencies in the Stuck-at Fault Model (11)

The limitations in the stuck-at-fault model when used with present day technologies result from using a gate level representation of the curcuit for the purposes of test pattern generation. First, the gate level representation is not topologically equivalent to the layout of the circuit on silicon, as shown in Fig. 1, with the layout of a complex gate structure. Second, fault conditions can occur which do not manifest themselves as an output stuck-

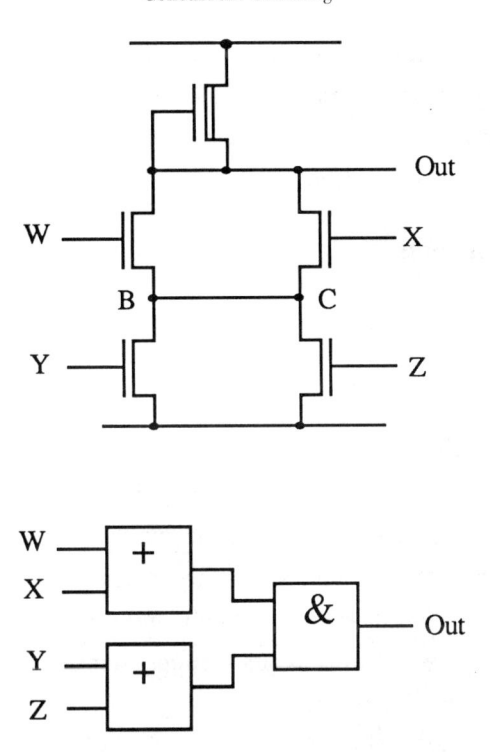

Fig. 1. NMOS Complex gate and its logical representation.

at-fault but, for example, alter the Boolean function realised by the gate structure, which would occur if the intraconnection, B-C, in Fig. 1. is open circuited; in general, this type of fault is only detected by testing the structure of the gate, rather than testing for the gate output stuck-at-1/0.

B. Delay Faults (12)

Delay faults do not affect the logic behaviour of a circuit but alter the propagation characteristics of gates in the circuit. Delay faults may be caused either by the degradation in the transconductance of a device, so limiting the rate at which capacitances can be charged or discharged, or it may be due to partially open inter-connect lines increasing the circuit resistance. The main reason that delay faults are not detected during normal testing is because testing is carried out, normally, at sub-operational clock rates due to the limitations in the tester. However, if the tester clock rate can be increased above the normal clock rate of the circuit, the delay fault will manifest itself as a stuck-at-fault.

C. Intermittent and Temporary Faults.

Electrical noise can cause problems in electronic circuits, in general, and hence it may be regarded as a potential source of intermittent faults in VLSI circuits. It has been shown, however, that with decreases in device sizes, errors due to electrical noise problems (13) also decrease; the fundamental limits for fabricating devices will be reached before voltages/currents are small enough to be adversely affected by electrical noise. Furthermore, the noise margins associated with the switching levels in a logic function reduce the effects of electrical noise on the circuit, and if scaling of device size is performed in such a way that the voltage supply remains constant, electrical noise effects will be insignificant.

A major source, however, of intermittent faults in integrated circuits is internal electromagnetic interference (11) which may be injected from the power supplies or caused by capacitive coupling between conductors. Large injected current 'spikes' are produced in synchronous circuits when a large number of devices switch on, simultaneously, when the circuit is activated by the clock; these current 'spikes' can cause numerous errors unless appropriate steps are taken, during the design phase, to minimise their effect. The capacitive coupling between conductors can also produce intermittent faults by causing a change in voltage in one conductor to be reflected into an adjacent conductor. The effect of internal electromagnetic interference on normal circuit operation is, to some extent, increased as devices are scaled down, since their switching times are much faster, hence the effect of a transient caused by the interference can propagate further into a circuit, where it may set/reset a latch and produce an error in the system.

When devices are scaled down the amount of charge, Q_{crit}, which differentiates a logic 1 from a logic 0 is also reduced, rendering the device more susceptible to soft errors caused by alpha-particle strikes(14). This error source was first observed in high density dynamic RAMs and is considered to be a potential error source in complex VLSI circuits, in general, as device sizes decrease. The alpha-particles which cause the soft errors originate from the decay of Uranium and Thorium inside the device package (13,14,15). The decay of these elements also produces beta-particles which have a smaller charge and also a smaller mass and consequently have a longer mean free path; subsequently, these particles may become more troublesome as device sizes decrease. Whenever an alpha or beta-particle strikes, a number of atoms in the material are ionised along its propagation path, leaving free electron-hole pairs. As the particle slows down the Energy Shedding Rate (ESR) of the particle increases, consequently intense radiation is produced at the end of the particles range or depth of penetration. Hence severe problems can be caused in devices whose junction depths are greater than or equal to the penetration depth of the particle. When the electron-hole pairs are created by the ionising radiation, two current components are produced, a drift current and a diffusion current. The drift current is due to the nearby electric fields of the isolating junctions in a device and is relatively fast. The diffusion current, which is slow, results from the carrier concentration gradient in the device and is particularly troublesome in dynamic logic circuits. The particle strikes may affect more than one node in the circuit if the angle of incidence is low; this is generally the case when the source of radiation is the packaging material. A necessary condition, however, for errors to result from ionisation radiation is that the injected transient must propagate to a latch, consequently as device sizes decrease and propagation delays are reduced, induced errors due to ionisation radiation are likely to increase. Furthermore, if voltage scaling is also used, then the transistor saturation current will be reduced, although the ionisation current will not, thus increasing the probability that a particle strike will result in an error being produced.

From the brief description of non-classical failure modes it is seen that the stuck-at-fault model is incapable of simulating their effect. Intermittent faults have been modelled using a stuck-at-fault model in conjunction with a random number generator, and the D-Algorithm has been modified to generate tests to detect delay faults. However, these solutions increase the burden on our presently overstressed testing strategies, requiring alternative testing techniques to be developed.

III Concurrent Checking Techniques

As surveys (16) have shown intermittent faults are emerging as the dominant failure mode in VLSI circuits. Current testing strategies using Exhaustive, Structural or Functional tests are incapable of detecting this type of fault, since these testing techniques, whether applied internally or implemented as a Built-in Test Scheme, were designed to detect permanent faults resulting from defects in the manufacturing process or due to wear out from normal circuit operation, and as such are applied too infrequently to detect a fault which can occur at random and exist only for a short time period. The characteristics of an intermittent fault requires a test strategy which continuously monitors the operation of a circuit and compares it to some 'reference'. Whenever a mismatch occurs the circuit can flag the occurence of a fault, the action taken thereafter depends upon the system, it may, for example,

  (a) Simply log the occurrence of the fault.
  (b) Implicitly mask the effect of the error.
  (c) Explicitly correct the error which has occurred.

The 'reference' used in concurrent circuit testing can be implemented using either hardware or information redundancy.

A. Hardware Redundancy

The simplest form of hardware redundancy is duplication, where two identical versions of a circuit are run concurrently with the same input signals being applied to each, the outputs are subsequently compared with each other and any mismatch is flagged as an error, with little indication of which version of the circuit failed or when the failure occurred since the comparison is only carried out at the circuit outputs. Thus, this type of concurrent testing only permits fault detection; furthermore the possibility of the circuit recovering from an intermittent fault, by invoking a 'retry' procedure, is small, since the occurrence of the intermittent fault is not necessarily flagged immediately as comparisons are only made at the outputs, consequently the information required to successfully implement a 'retry' procedure may have been discarded.

With duplication, it is possible to have an identical fault in both circuits due to a fault in a mask layer or a design fault in a common cell. This situation, however, can be avoided if Complementary Logic Redundancy (17) is used. In this technique instead of the original circuit being duplicated, the inputs/outputs of each combinational element in the original circuit and complemented version are of opposite polarity. The storage elements although identical in both circuits are not only driven and activated by control signals, but also store data, of opposite polarity. This approach has the inherent advantage of avoiding mask faults which may cause identical errors to occur in each version of the duplicated circuit, and which would go undetected. Furthermore, comparisons of both versions of the circuit are not limited to the outputs, but intermediate control and data results are also monitored, it has been shown that this approach enhances the detection and isolation of intermittent faults and increases the success rate of 'retry' procedures.

Duplication can only detect the presence of errors, however, many situations exist where it is required to mask the effect of a fault on the system whenever an error is detected. Triple Modular Redundancy (TMR) (18) schemes have the capability of masking the effect of a fault, although they do not explicitly detect the presence of the fault. The masking effect is

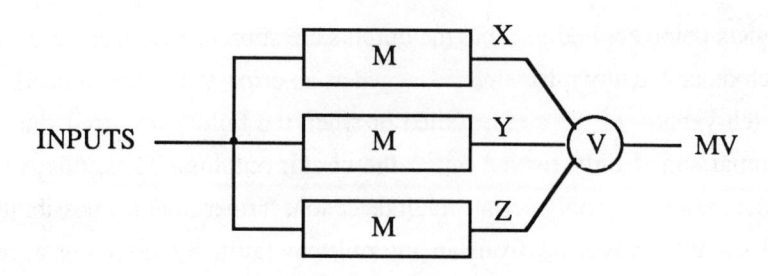

Fig. 2. A basic TMR system

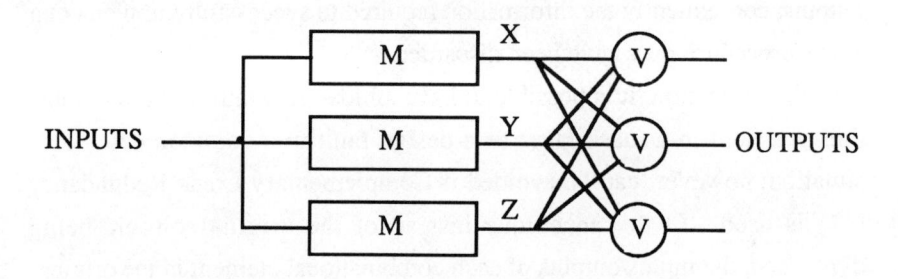

Fig. 3. A TMR system with three voters.

achieved through the use of a voting circuit, which accepts inputs from each version of the circuit, as shown in Fig. 2, the output from this circuit is a majority vote on the inputs. Consequently, if one circuit fails momentarily and produces an incorrect output, its effect would be masked, explicitly by the other two circuits out voting it. The reliability of this system is dependent upon the reliability of the voting circuit; in some instances the voting circuit is also triplicated as shown in Fig. 3. The basic concept of TMR can be extended to N-modular Redundancy (NMR) which can tolerate n-failures in the system where $n = (N-1)/2$, N in general is an odd number.

B. Information Redundancy

Information Redundant schemes use extra bits, called checkbits, appended to the data bits to detect faults which result in errors occurring in the data bits, without explicitly applying a test pattern or sequence to detect a fault.

There are numerous techniques, called coding schemes, which can be used to implement information redundancy in a system. Some of these coding schemes not only permit error detection but also error correction. The best known coding techniques are the Hamming type codes (19) and these will be discussed briefly.

The simplest of all codes is the Parity Code in which a single checkbit is appended to the data bits. The Parity Code bit is defined such that the total number of one-bits in a codeword (i.e. data bits plus checkbits) is even (even parity) or odd (odd parity). In this way any single bit error can be detected, since any change in the data bits or the parity bit itself will result in a mismatch between the parity indicated by the checkbit and the parity derived from the data bits. The error detection capability is quantified by the Hamming distance 'd' which defines the number of bits by which two codewords can differ. For a simple parity code $d = 2$, hence this code can only detect a single bit in error but not correct it. If, however, the number of checkbits is increased a more powerful code can be produced which also corrects errors. For example, if the Hamming distance is increased to three, i.e. $d = 3$ the resulting parity check code will have the capability of single error correction. The number of checkbits, c, required to perform single error correction in a data word comprising m-bits, is given by the relationship.

$$2^c \geq m + c + 1$$

Thus if $m = 4$, then $c \geq 3$, resulting in a codeword of 7-bits. In a conventional Hamming code, the checkbits are located in bits positions in the codeword which are powers of two. Hence if the data bits are denoted by B1 - B4 and the checkbits are denoted by C1 - C3, the complete codeword would appear as

C1 C2 B1 C3 B2 B3 B4

where C1 is the least significant bit position.

The relationship which defines the values to be assigned to the checkbits in a given codeword is derived from the Parity Check Matrix shown below

$$P = \begin{matrix} C1 & C2 & B1 & C3 & B2 & B3 & B4 \\ 1 & 0 & 1 & 0 & 1 & 0 & 1 \\ 0 & 1 & 1 & 0 & 0 & 1 & 1 \\ 0 & 0 & 0 & 1 & 1 & 1 & 1 \end{matrix}$$

The checkbits are formed by summing (modulo-2) the information bits which have a 1 in the same row of the parity matrix as the checkbits, hence

$C1 = B1 \oplus B2 \oplus B4$
$C2 = B1 \oplus B3 \oplus B4$
$C3 = B2 \oplus B3 \oplus B4$

Thus if the data bits are 0101 then C1, C2, and C3 = 010 respectively and the corresponding Hamming Codeword would be 0100101. If this word is transmitted in a digital system and a fault occurs in which bit 5 changes from a 1 to a 0, then when the word is received and the checkbits recalculated from the information bits a mismatch would be detected. In this instance $C_1' = C_2' = C_3' = 1$. The 'address' of the error is then obtained by Exclusive-ORing the received and generated checking bits; i.e.

$E1 = C1 \oplus C1' = 0 \oplus 1 = 1$
$E2 = C2 \oplus C2' = 1 \oplus 1 = 0$
$E3 = C3 \oplus C3' = 0 \oplus 1 = 1$

The address of the error bit is 101 (i.e. 5) and bit position five would be inverted to correct the transmitted data word. It has been shown that coding techniques can be used not only to detect but also to correct errors in digital systems, the hardware necessary to implement these functions will be less than that required for full duplication.

Although many coding schemes exist they are not generally applicable to all types of circuit. For example, parity checking codes are used in data transmission and also for error detection/correction in memory systems, however they cannot be readily applied to arithmetic circuits since arithmetic functions do not preserve parity. It has been demonstrated, however, that a class of separable codes, discussed in the next section, called Residue

Codes can be applied to a wide range of functions which includes arithmetic, logic and storage functions. Residue Codes also have the important characteristic that the checkbits of the result of an operation can be derived independently from the checkbits of the operands.

IV. PROPERTIES OF RESIDUE CODES

Residue codes have the desirable property that for arithmetic operations such as addition, multiplication etc, the checkbits of the result of an operation can be derived, independently, from the checkbits of the operands; a similar result can also be shown to exist for logical operations. However, before demonstrating these properties of residue codes several informal definitions will be presented.

Definitions:

(a) N is a separable code if the information and checkbits can be obtained without decoding the codeword; that is N has the form $N = \{X : X = IC\}$, where I = Information digits and C = check symbol.

(b) N is defined as a residue code with a check base 'b' if,

    (1) N is a separable code

    (2) $N = \{X : X = C\}$ given that C is the binary representation of the residue, modulo b, of I considered as an integer number.

(c) N is said to be a Low Cost Residue (LCR) Code if N is a residue code with a check base 'b' such that $b = 2^p - 1$, $p \geq 2$, where p is the number of checkbits.

(d) In the separable residue codes, the number n will be represented as a pair $(n, |n|_b)$. Where $|n|_b$ is the check symbol of the number n and is given by the least non-negative integer congruent to n modulo b.

The applicability of these codes to checking arithmetic functions will be demonstrated (20) with respect to addition and multiplication. Let $\{n_i\}$ be a set of numbers with checkbits $C_i = |n_i|_b$ and let $|\Sigma n_i|_b$ and $|\Pi n_i|_b$ be the residue of the sum and product of the numbers respectively. It is also assumed that $n_i = a_i b + r_i$ where $0 \leq r_i < b$.

Hence $|\Sigma n_i|_b = |\Sigma(a_i b + r_i)|_b$
$= |\Sigma a_i b|_b + |\Sigma r_i|_b$

However, since km mod m = 0 (k is an integer),
$|\Sigma n_i|_b = |\Sigma r_i|_b = |\Sigma |n_i|_b|_b$

i.e. the sum of the checkbits modulo b is equal to the modulus of the sum of the numbers n_i.

Similarly
$|\Pi n_i|_b = |\Pi(a_i b + r_i)|_b$
$= |\Pi a_i b|_b + |\Pi r_i|b$
$= |\Pi r_i|_b = |\Pi |n_i|_b|_b$

i.e. the product of the checkbits modulo b is equal to the modulus of the products of the numbers n_i. Similar results can also be derived for subtraction and complementation.

A. Using Residue Codes

As an example of these results consider the four-bit numbers A = 1010 and B = 0111, the residues of each of the numbers are, assuming b = 3 R_A = 01 and R_B = 01.

Thus for addition
A + B = 10001 and R_{A+B} = 10
$R_A + R_B$ = 10

and for multiplication
A x B = 1000110 and R_{AxB} = 01
R_A x R_B = 01

Although residue codes can be readily applied to arithmetic operations, logical operations are much more difficult since residue codes are not closed under bitwise logical operations, for example AND, OR and Exclusive-OR. However by using the simple relationships (21) shown below which exists between arithmetic and logical operations, residue codes can be used to check logical operations with relative ease.

Let N_1 and N_2 be two n-bit binary numbers, then
$$N_1 + N_2 = (N_1 \cdot N_2) + (N_1 \vee N_2)$$
and
$$N_1 \oplus N_2 = (N_1 + N_2) - 2(N_1 \cdot N_2)$$
where the bitwise logical operations for AND, OR and Exclusive-OR are represented by $\cdot$, $\vee$, $\oplus$ respectively. Using these relationships, expressions can be derived for the output residue of the OR and Exclusive-OR operations, as shown below.
$$|N_1 \vee N_2|_b = |\ |N_1|_b + |N_2|_b - |N_1 \cdot N_2|_b |_b$$
and
$$|N_1 \oplus N_2|_b = |\ |N_1|_b + |N_2|_b - 2|N_1 \cdot N_2|_b |_b$$

In deriving the above relationships an arbitrary value has been assumed for the base; however with respect to minimising the size of the checking hardware, which increases with the base size, a base value of three is used in the implementation of the system. This results in a simplification of the relationships used to derive the residues for the logical operations as shown below.

Let $N_3 = |N_1 \cdot N_2|_3$ then
$$N_1 \vee N_2 = |N_1 + N_2 + N_3|_3$$
$$N_1 \oplus N_2 = |N_1 + N_2 + N_3|_3$$

The mod 3 checkers can be easily constructed using full and half adders.

It can be readily demonstrated that provided the base is odd, all single bit errors occurring in either the information bits or the checkbits will be detected.

Consider that the binary representation of an information word represents the number N, if the ith bit in the word is changed due to some error the number N' represented by the bits will be equal to $N \pm 2^i$. Since the base, b, is odd then the checkbits C given by N mod b will not be equal to $(N \pm 2^i)$ mod b, permitting the error to be detected. Similarly, if the error occurs in the checkbits then $C' = (C \pm 2^i)$ mod b, and if b is odd $C' \neq C$, and the error is detected.

V. IMPLEMENTATION OF A MOD 3 LCR CHECKING SCHEMA (22)(23)

The additional hardware required to implement Mod 3 checking comprises,

(a) A Mod 3 Residue Generator which produces the appropriate residue for a given dataword. The checkbits are generated by performing modulo ($2^p - 1$) addition on the information bits using a tree of binary adders with end-around-carry as shown in Fig. 4. The information word is divided into n groups of p-bits, and a modulo ($2^p - 1$) addition on the n p-bit groups results in the checkbits.

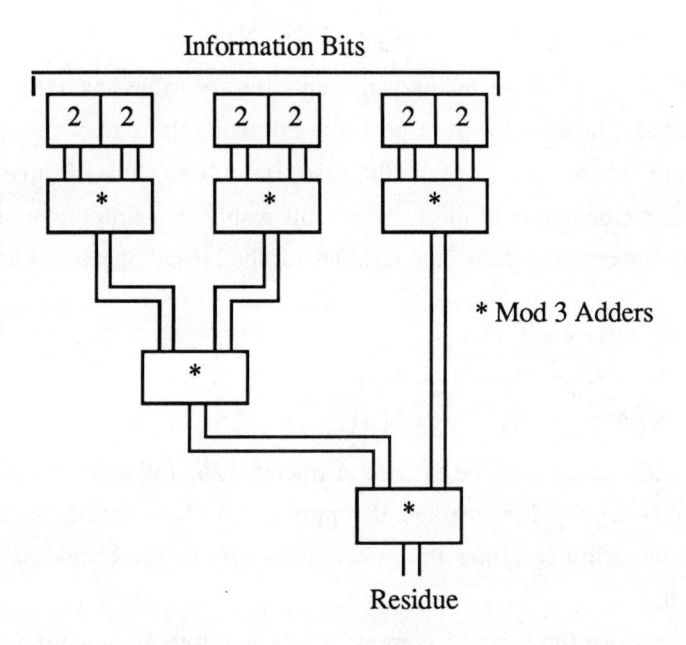

Fig. 4. Residue generation tree.

(b) A Mod 3 Comparator which compares the actual and generated residue for a given operation. The comparator is implemented as a two rail checker. Since the comparator essentially decides if a function block is faulty, it must be designed to be totally self checking. The logic diagram of a totally self

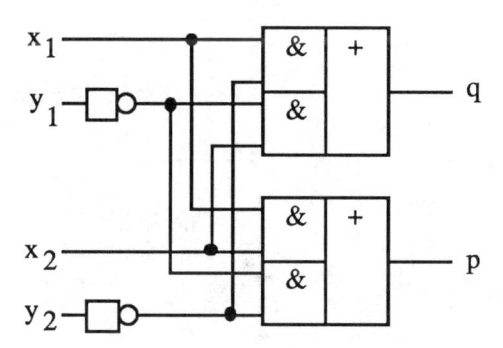

Fig. 5. Two rail checker.

checking two-rail checker is shown in Figure 5. The checker has two sets of inputs (x_1,x_2) and (y_1,y_2) and two outputs p and q which are complementary provided each pair of inputs is also complementary i.e. x_1 and y_1, and x_2 and y_2. Thus, if $x_1x_2 = 10$ and $y_1y_2 = 10$ the outputs p and q will be equal to 0 and 1 respectively. However, if a fault occurs such that $y_1y_2 = 00$, the outputs p and q will assume identical values indicating that an error has occurred.

(c) A Two-Rail Encoder which compresses the comparator outputs from the various function blocks in the circuit, into a single two-rail code; this reduces the number of 'test-pins' required on the package.

(d) A Mod 3 Multiplier to generate the two residues when checking the output of a multiplier.

Figure 6, illustrates the general schema for applying Mod 3 checking to an arbitrary circuit. However, to show its general applicability to VLSI a range of functions were designed; for example, N x N bit multiplier, N-bit ALU, PLA structures, together with some miscellaneous functions such as rotate and shift operations.

A. N x N Multiplier: (14)

The application of Mod 3 checking to the multiplication operation is straightforward; the checkbits of the product being equal to the product of

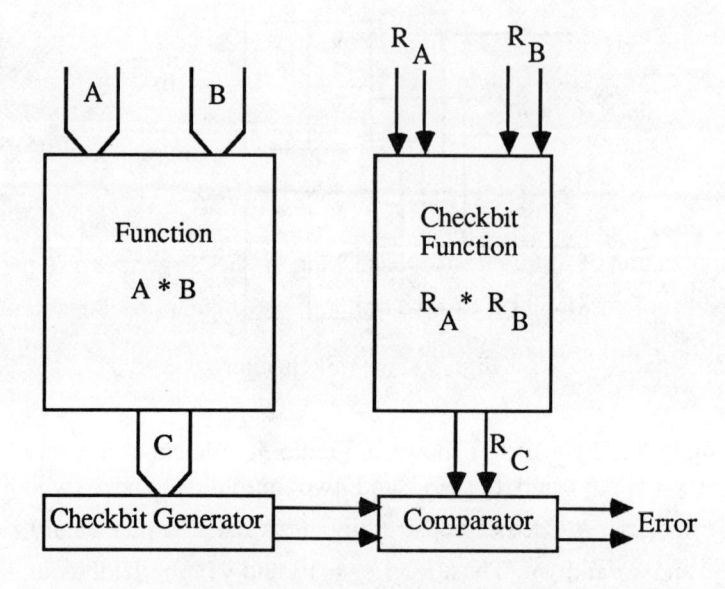

Fig. 6. General checking scheme.

the checkbits of the operands, Mod 3. The multiplier used in this instance was a N x N bit parallel multiplier which was constructed hierarchically from two atomics which comprised a 2 x 2 bit multiplier and a full adder. This multiplier was also used as a reference vehicle in order to compare the additional area required by Mod 3 checking, as the circuit complexity increased, in comparison to other methods of design for testability, for example Scan Path, BILBO and SASP (25); the results of the comparison will be discussed later.

B. Arithmetic Logic Unit:

The major problem in applying residue checking schemes to an ALU design is that it can perform both arithmetic and logical operations. Consequently additional logic has to be incorporated into the design to implement the Mod 3 Residue Generator for the AND operation which is used in deriving the residues of the OR and Exclusive-OR operators. Furthermore, since an ALU can perform many functions a residue selector circuit must also be included in the design so that the residues needed to derive the final residue for a given operation can be selected. The configuration of the selector circuit will be dependent upon the functions performed by the ALU.

C. PLA:

The application of Mod 3 Residue Checking to PLA structures also presents several problems, in that PLAs do not perform standard functions unlike a multiplier, for example, which performs a given operation, the internal architecture being irrelevant to the checking problem; consequently in arithmetic operations the checker is essentially independent of the hardware to be checked and since both systems are separate any unchecked common hardware does not exist. The objective in checking a PLA is to provide a checking scheme which is indepedent of the hardware generating the functional outputs without incurring the overheads involved with duplication. The methods of applying Mod 3 checking to PLAs are shown in Figure 7 and 8.

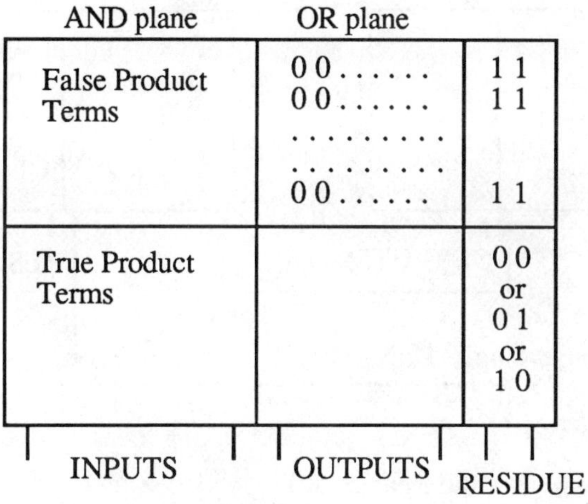

AND plane	OR plane	
False Product Terms	0 0 . . . . . . 0 0 . . . . 0 0	1 1 1 1 1 1
True Product Terms		0 0 or 0 1 or 1 0

INPUTS OUTPUTS RESIDUE

Output = 0 Valid Residue 11
Output ≠ 0 Valid residues 00 or 10 or 01

Fig. 7. Mod 3 PLA Technique - Method 1

Method 1: In this technique, shown in figure 7, two extra residue bits are appended onto the side of the 'OR' plane before the product terms are minimised. In this instance, however, the system being checked contains the checkbits and it is possible for certain faults to occur in the system and remain undetected. For example, if a product term became stuck-at-0 the checker would also fail because the output including the residues would be all zeroes, giving a valid codeword. To overcome this problem false product terms can be included in the 'AND' plane; the false product terms are those terms associated with an all zero output and normally omitted from PLA designs. The false product terms have a residue of 11 associated with them, consequently when a true product term now becomes stuck-at-0, its output would be considered illegal since a fault free all zero output would have a 11 residue. Most errors in the AND plane can be detected in this way. Errors in the 'OR' plane usually produce single bit output errors which can be detected Mod 3.

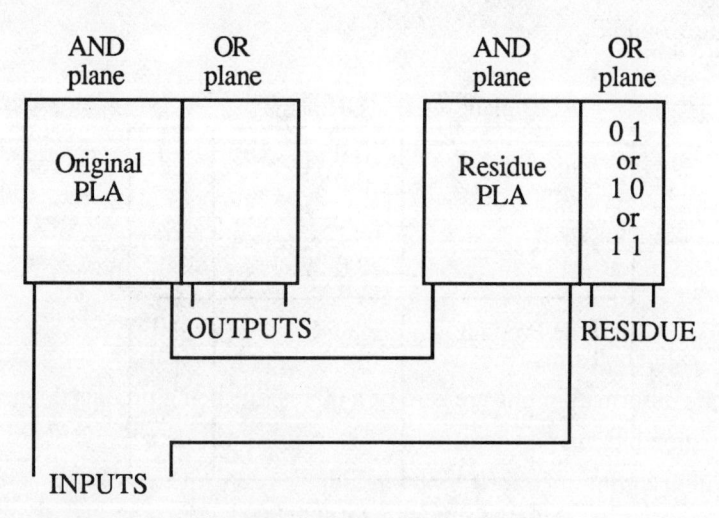

Output = 0 Valid Residue 00
Output ≠ 0 Valid Residues 11 or 01 or 10

Fig. 8. Mod 3 PLA Technique - Method 2.

Method 2: In this technique, shown in Figure 8, a Second Residue PLA is added to the original PLA. The Residue PLA has exactly the same inputs as the PLA being checked, but only has two outputs, namely the residues of the output of the original PLA. In comparison to the first method, this technique offers several advantages, namely

(a) Since an error is unlikely to occur in both PLAs simultaneously, the product terms in both the original and Residue PLAs can be identical, hence the necessity to include the false product terms and the associated area overhead is avoided.

(b) The operating characteristics of the original PLA are not affected when residue checking is applied in this instance, since the Residue PLA is essentially independent of the functional PLA.

VI. ANOMALIES IN THE USE OF MOD 3 LOW COST RESIDUE CODES

In general the application of Mod 3 checking to arithmetic operations is straightforward; however, in some situations problems arise due to the use of the two bit adders with end around carry in the residue generators, producing a spurious result. These adders produce a 11 residue when the input data is divisible by 3, except for a zero input; consequently on occasions corrective action must be taken to restore the residues to a consistent value. However, the occasions on which this corrective action must be taken are predictable, for example, a 11 residue being produced when the information bits are zero or a 00 residue being produced when the information bits are non zero.

Another problem arises when residue codes are applied to circuits performing modulo 2^n arithmetic. It is possible in performing a binary addition to generate a carry out in the (n+l)th bit position; the residue generated will be correct for (n+l) bits and must therefore be modified for the n-bits in the normal word.

Possible solutions to these anomalies are discussed below with respect to the one's and two's complement number systems and the rotate and logical shift operations (26).

A. One's Complement System:

Mod 3 checking can be applied to the one's complement system with ease since the carry out bit is usually added back into the number using an 'end-around carry'.

One's complement negation simply involves performing the NOT operation on every bit in the number. This operation can also be performed on the residue to produce the residue of the negated number. However because of the form of the residue coding implemented, corrective action must be taken whenever a number with a residue 11 is negated, unless the negation operation produces an all zero output. However, since the one's complement system has two representations of zero the all one's case could be used, hence the residue 11 would never have to be inverted. The circuit which can be used to correct the residues is shown in figure 9.

The addition of one's complement numbers can be accomplished in the normal way. The residues are summed in the Mod adder and the result is the residue of the result of the addition operation.

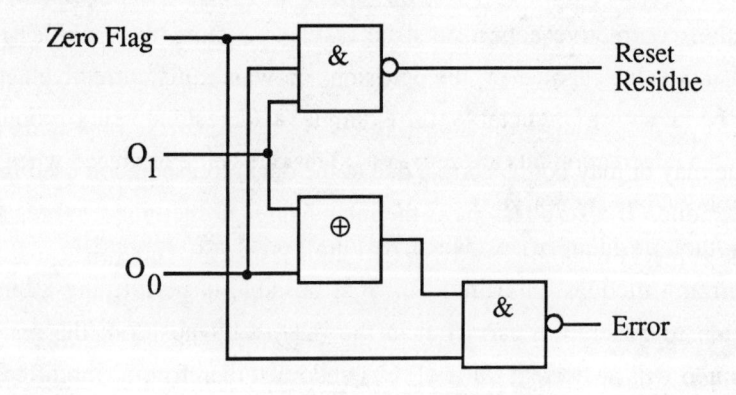

Fig. 9. Residue correction for the 1's complement number system.

Subtraction is also performed in the normal way, the subtrahend is first negated and then added to the minuend, the residues undergo the same operations.

B. Two's Complement Number System

The application of Mod 3 checking to the two's complement number system presents several problems due to the possible generation of a carry bit from the most significant bit position. This carry bit is not incorporated back into the system as in the one's complement case, hence information will be lost, the generated residue will also be incorrect since this applies to the complete word including the carry bit. Several solutions to this problem exist, the first is to regenerate the residue from the data whenever a carry is generated; however, this will infer that the checker hardware is not independent of the functional hardware, making the system unreliable. Second, a technique of using 'Pseudo Residues' (26) may be employed, whereby any carries that occur are collected and added using the same adders as the residue number system. In this way the carry information is preserved and the residue is then taken over the whole word including the 'Pseudo Residue'. However, a high area overhead is incurred by this technique. A third technique may be adopted which utilises information normally available in the adder, that is the carry bit itself and the zero flag; the final residue can be corrected by subtracting the carry bit from the predicted residue, a correct result will be produced provided the answer is not zero. This anomaly can be resolved, however, by examining the zero flag bit. If the zero flag bit is true the residue may or may not be correct due to the dual representation used for a zero residue. If a zero residue is the intended result then the final residue will be either 11 or 00 but not 10 or 01. Thus if the zero flag is true and the residue is either 10 or 01 then an error has occurred, otherwise the residue can be cleared to zero for the zero datum, providing the required corrective action, assuming that any corrective action for the carry bit is first performed; hence the zero flag bit is checked. The carry bit is also checked, since if it was incorrect an error of $\pm 2^n$ would occur in the residue and this can be detected by the Mod 3 residue code, the error would be detected when the comparison was made between the residue and the residue generated from the word.

The method used to negate two's complement numbers is to invert all bits in the number and add one on to the result. This operation can also be applied to the residue, the residue is subsequently corrected if required.

The operations of addition and subtraction are carried out in the normal way, the residue of the numbers being added/subtracted and corrective action taken if necessary.

C. Rotate and Logical Shift Operations

In either a right or left rotate all the bits are preserved, the residue will remain constant and can simply be rotated along with the numbers.

However, logical shifts are more difficult to check because of the loss of bits during the shift operation. The circuit shown in Figure 10 illustrates

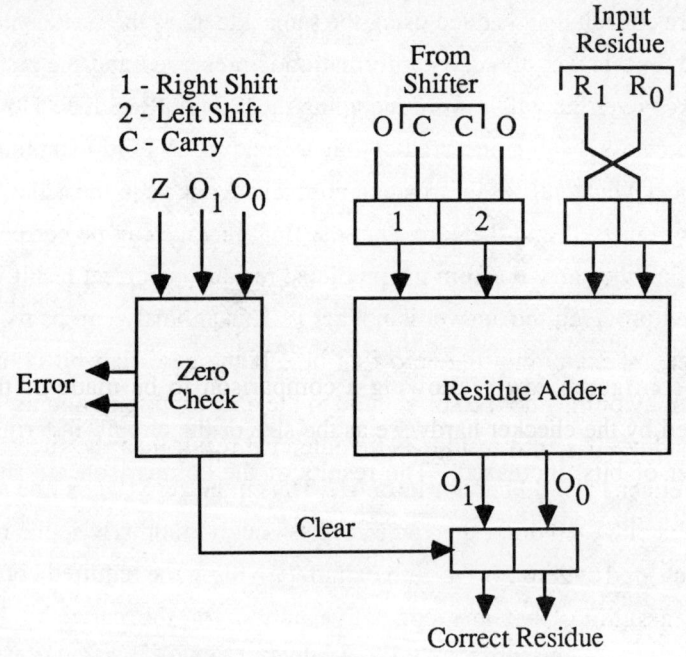

Fig. 10. Logical shift operations.

how to deal with this loss of bits by using the same corrective action adopted in the two's complement numbering system. The left shift operation involves rotating the residue to the left and subtracting the carry out of the information bits from the rotated residue. For the right shift operation the residue is rotated to the right and the carry bit added to the residue. If the shift operation produces a zero result the residue must be cleared, this situation can be detected as previously discussed.

VII. Discussion

In adopting any design for testability technique the important factors to be considered are,

(a) The cost, in terms of silicon area, required to implement the chosen test method in comparison to the area occupied by the design without the test structures.

(b) The performance penalty imposed upon the design by any parts of the test hardware.

(c) The ease with which a given technique can be applied to a design.

In an attempt to quantify the increase in silicon area resulting from the use of Mod 3 checking in a design, two test vehicles were considered in the first instance, namely a N x N bit parallel multiplier and an 8-bit datapath circuit.

The parallel multiplier which was designed comprised two basic building blocks, namely a 2 x 2 bit multiplier and a full adder. The basic architecture of the multiplier permitted multipliers having eight through to sixty-four bits to be configured easily allowing a comparison to be made of the area required by the checker hardware as the size of the circuit, in terms of the number of bits, increased. The results of the comparison are shown in Figure 11; it is seen that as the size of the circuits increase the percentage area taken up by the checker hardware overall tends to decrease, although the size of the residue prediction hardware increases as the number of bits in the multiplier increases. As a comparison with other design for testability

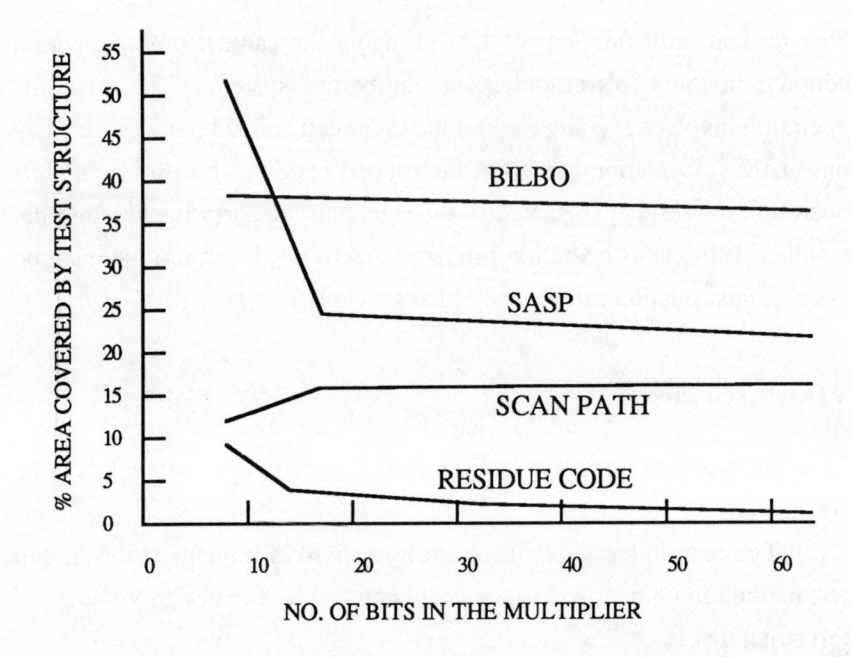

Fig. 11. Area comparison for different DFT techniques

techniques a similar exercise was carried out implementing Scan Path, BILBO and SASP techniques in the multiplier, the results of this comparison are also shown in Figure 11, indicating that the incorporation of Mod 3 checking into a design offers a distinct advantage, with respect to the area occupied, over the standard DFT implementations.

The datapath which was considered is shown in Figure12. It comprises eight 8-bit wide registers which are used for the program counter, stack pointer and general use. The ALU can perform twenty-two arithmetic and three logical operations. The datapath controller performs two basic functions, first it controls the operations of the ALU and second it controls the overall operation of the datapath through the use of a microcode PLA or ROM.

The data path was designed as a single programmable cell to be used in the control of digital signal processing chips, where it will be embedded

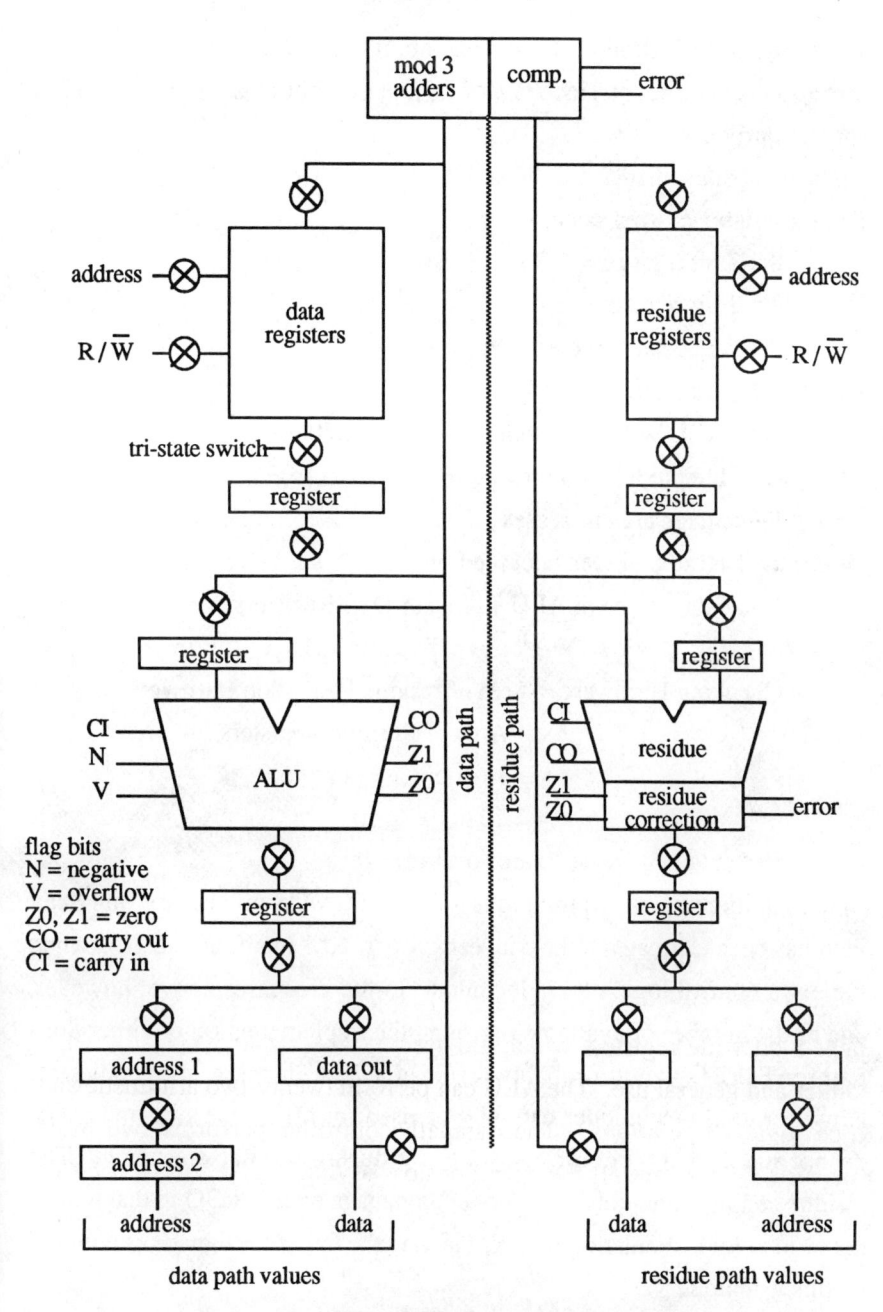

Fig 12. Block diagram of the residue coded data path
(IEE Proc. Pt. E, V. 133, No. 3., pg. 170, reprinted with permission)

within the overall design. Hence, the ability of the CPU to check its own operation is paramount, since few of its input/output pins will be accessible for test purposes.

The datapath statistics are shown below,

Data Registers: Area occupied by,

8 x 8 bit register = $1.3 \times 10^6 (\mu m)^2$

8 x 2 bit register = $6.2 \times 10^5 (\mu m)^2$

Percentage area increases = 47%

ALU: Area occupied by,

8 bit ALU with carry look ahead = $3.1 \times 10^6 (\mu m)^2$

Residue prediction hardware = $1.2 \times 10^6 (\mu m)^2$

Percentage area increases = 38%

Datapath Cell: Total area occupied by,

Datapath = Area of ALU + Area of Data Registers

= $3.1 \times 10^6 + 1.3 \times 10^6 (\mu m)^2 = 4.4 \times 10^6 (\mu m)^2$

Checking Hardware = Area of residue Prediction Hardware

+ Area of Residue Registers

+ Area of Checker

= $2.13 \times 10^6 (\mu m)^2$

Percentage increase in area overall = 48%

The additional area required to test a 32 bit CMOS processor using Scan Path has been estimated to be between 25% to 40% increasing to 70% when the extra control logic etc. is included. In the processor above, however, due to the number of available registers, the implementation of either Scan Path or BILBO would probably result in a 20% increase in overall area. However, these techniques can only be used for off-line testing and hence do not monitor the circuit continuously during normal operation. The residue coding technique has a further advantage over BILBO in that it does not require long simulation runs to generate the fault free signatures required by this technique.

With respect to the effect that Mod 3 Residue Checking has on performance, in the case of the parallel multiplier since the residue prediction is performed concurrently with the normal circuit operation, the only delay incurred is during the generation of the residue of the result which will increase with bit size. In the datapath circuit some additional delay occurs in the residue prediction hardware for the ALU since this has to wait for the generation of the carry and zero flags in order to predict the final residue.

Finally, with respect to ease of use, it has been demonstrated that Mod 3 checking can be applied to a wide range of functions; consequently, a unified error detection scheme can be applied throughout a design and if required the incorporation of this technique into a design can be made transparent to the user.

Thus, in conclusion, Mod 3 Low Cost Residue checking offers the following advantages

(a) A unified error detecting scheme can be applied throughout a design, since Mod 3 can be used on arithmetic, logical and storage functions.
(b) Concurrent circuit testing is performed permitting both permanent and intermittent faults to be detected.
(c) The extra area required for its implementation is much less than that required by hardware redundancy schemes and also by some standard DFT techniques.
(d) The technique can be applied hierarchically.
(e) The checking hardware is independent of the hardware processing the data.

A further advantage, of incorporating an information redundant scheme hierarchically in a design, is that it permits the integrity of the interconnections between modules in a system to be tested, this will be advantageous in the design of WSI circuits.

VIII. REFERENCES

1. Eichelberger, E.B., Lindbloom, E., (1983), "Trends in VLSI Testing" *VLSI '83*, Elsevier Science Publishers, 339-348.

2. Goel, P., Rosales, .B.C., (1981), *18th Design Automation Conference Proceedings*, 260-268.

3. Ambrovici, M., Menon, P.R., (1983), *Digest of Technical Papers, International Conference on Computer Aided Design*, 27-29.

4. Blank, T., (1984), *Design and Test of Computers*, 21-39.

5. Bellon, C., Robach, C., Saucier, G., (1983), *Digest of Technical Papers, International Conference on Computer Aided Design*, 32-33.

6. Williams, T.W., Parker, K.P., (1983), *Proceedings of IEEE*, Vol. 71, Number 1, 95-122.

7. Bhavasar, D.K., (1983), *20th Design Automation Conference Proceedings*, 300-307.

8. Hewlett-Packard, (1977), "A Designer's Guide to Signature Analysis" *Application Note 222*.

9. Savir, J. (1983), *IEEE Transactions on Computers*, Vol. C-29, Number 6, 442-451.

10. Mangir, T.E., and Avizienis, A., (1980), *Proceedings 1st IEEE Conference on Circuits and Systems*, 658-688.

11. Burgess, N. et al., (1985), *Proceedings IEE*, Pt G, Vol. 132, No. 3, 32-89.

12. Timoc, C., (1983), *International Test Conference*, 546-553.

13. Savaria, Y. et al., (1984), "Characterisation of soft error sources", report no. 84-11, McGill University, VLSI Design Laboratory, Canada.

14. May, T.C., (1979), *Proc. 29th IEEE Conference on Electronic Components*, 247-256.

15. Sai-Halasz, G.A., Wordeman, M.R., and Dennard, R.H., (1982), *IEEE Transactions on Electronic Devices*, Vol. ED-29, No. 4, 725-731.

16. Tendolaker, N. and Swann, R., (1982), *IBM Journal of Research and Development*, Vol. 26, 78-88.

17. Sedmak, R.M., and Liebergot, H.L. (1980), *IEEE Transactions on Computers*, Vol. C-29, No. 6, 492-500.

18. Lala, P.K., (1985), *Fault tolerant and fault testable hardware design*, Prentice Hall International.

19. Hamming, R.W., (1950), *Bell Systems Technical Journal*, No.2, 147-160.

20. Breuer, M.A., Friedman, A.D., (1977), *"Diagnosis and Reliable Design of Digital Systems"*, Pitman.

21. Monteiro, P., Rao, T.R.N., (1977), *2nd Fault Tolerant Computing Symposium*, 8-13.

22. Sayers, I.L., Kinniment, D.J., (1985), *IEE Proceedings*, Vol. 132, Part E, No. 4, 197-202.

23. Sayers, I.L., Kinniment, D.J., Chester, E.G., (1986), *IEE Proceedings*, Vol. 133, Part E, No. 3, 169-179.

24. Yung, H.C., Allen, C.R., (1984), *IEE Proceedings*, Vol 131, Part G, No. 2, 56-60.

25. Sayers, I.L., Russell, G., Kinniment, D.J., (1985), *Proceedings of ISCAS*, 1547-1550.

26. Sayers, I.L. (1986), "An Investigation of Design for testability techniques in very large scale integrated circuits", *Ph.D. Thesis*, Newcastle University, UK.

THE TESTING OF INTEGRATED CIRCUITS INCORPORATING ANALOGUE SECTIONS

A.P. Dorey, P.J. Silvester, and R.J. Ball,

*Department of Engineering
University of Lancaster,
Bailrigg,
Lancaster LA1 4YR. U.K.*

ABSTRACT

The continuing reduction in cost and increased complexity of integrated circuits has made it possible to incorporate more of an electronic system on to a single chip. Although many functions previously achieved in analogue form have been translated into their digital equivalent there often remains a requirement for an interface between the analogue real world and the digital processing power at the heart of many systems.

Examples include systems for data acquisition, control and instrumentation, speech recognition, and encoders and decoders for voice telephony. A further interesting class of circuits are those which incorporate digital 'intelligence' at the 'front end' of a measurement system to compensate for imperfections in transducers.

Thus hybrid integrated circuits which mix analogue and digital functions are becoming more widely used and are likely to become more important as the technology develops towards the 'system on a chip'.

These circuits provide new problems of testability. Digital circuit testing is highly developed with an understanding of fault mechanisms, models and effects having been evolved to provide design and test methods giving high fault coverage.

Simplified tests for analogue circuit blocks are being developed which are based on an 'open circuit' and 'short circuit' fault model. Using piecewise linear macro models of the circuit blocks, tests may be derived for the analogue part of a circuit for application using modified digital test equipment.

I. INTRODUCTION

An increasing proportion of a total electronic system can be incorporated on a single silicon chip. Despite the translation of many analogue electronic functions into their digital equivalent there still exists a need to incorporate analogue sections on many such chips. Under these circumstances the digital section of the chip may well be masked by the analogue circuits that are frequently needed at the input and output of the chip. Thus there is a requirement to develop strategies which will allow the often peripheral analogue and internal digital parts of the circuit to be tested simultaneously.

This makes it attractive to use an approach similar to a digital test strategy to identify hard faults in the analogue section of such circuits as such faults will give rise to fault effects similar to those encountered in the digital part of the circuit. It also should enable digital test equipment to be used, thereby reducing costs and allowing one test program to perform both the analogue and digital tests. This may be done by reducing the analogue input stimuli to two values corresponding to a logical '1' and '0' and setting an analogue threshold voltage level at the output to detect the presence of a faulty circuit. These input and output levels have to be set as the result of simulation on a computer model of the circuit. Such analogue circuit simulators already exist, but are aimed at giving highly accurate solutions to relatively simple circuits, typically with up to 50 nodes. Larger circuits than this can be simulated but ultimately the limiting factor is the time taken. However as a 'go' or 'no go' result is sufficient, accuracy may be traded for an increase in the speed at which the simulation runs by using simplified models of the circuit under test.

Two separate approaches have been considered to solve this modelling problems:

    i) Digital modelling;

    ii) Macro modelling.

The first of these possibilities has been examined using an existing digital simulation program (HILO) to model an analogue circuit. This has the added advantage of having a test pattern generation facility available within the simulator. The second was more conventional high level macromodelling techniques in which the 'macros' are approximate functional descriptions of analogue circuits designed to give a quicker, and hence cheaper, simulation of complex systems. The macros have been implemented in SUICIDES (Southampton University Investigative Design Engineering System) and IBM'S ASTAP (Advanced Statistical Analysis Program).

Fault modelling is the representation of typical faults that can occur in real circuits by changes to their models. Some fault mechanisms, such as electromigration leading to open circuits or track to track short circuits can be modelled particularly simply at the circuit or functional block (macro) levels. Other fault types are more difficult to model, for example an incorrect ion implantation dose causes a shift in the threshold voltage for a MOSFET device. It is convenient to divide faults into 'hard' (irreversible open or short circuits) and 'soft' or 'spec' faults (where the device, although nominally functioning as intended, does not meet the design specifications). The relationship between faults in analogue circuits and appropriate fault models have also been investigated.

```
                          B
                 OUT     0  1
  A ──┤+\            0 │ X │ 1 │
       │ ├── OUT  A    ├───┼───┤
  B ──┤-/             1 │ 0 │ X │
```

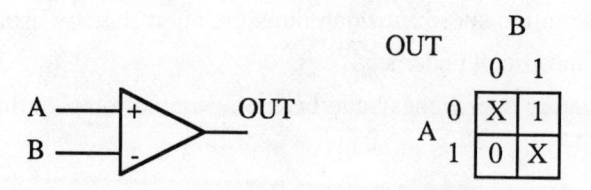

Fig. 1. Comparator and its Karnaugh map.

II. CIRCUIT MODELLING AND FAULT DIAGNOSIS

A. Digital Modelling

Analogue circuits may be modelled in an approximate way using logic gates. As a simple example, the comparator shown in figure 1 has two inputs A and B.

If a logic convention is used where any positive voltage is regarded as a one and any negative voltage is regarded as a zero then the response of the circuit may be described in very approximate terms by the Karnaugh map in figure 1.

This approach to modelling may be extended to a range of analogue circuits such as comparators, sample and hold circuits and amplifiers. For certain applied voltages a useful measure of agreement can be obtained between the behaviour of the analogue circuits and that of a digital model in their response to standard TTL (0V and 5V levels) inputs. This correspondence enables test patterns (consisting entirely of logic 0 and 1's) to be derived which can be evaluated quickly on a logic simulator, in the knowledge that the analogue circuit would behave in a similar manner under these restricted conditions. An additional benefit is gained from the fact that most logic simulators incorporate their own Automatic Test Pattern Generation facilities, thus formalising, and often accelerating, this phase of the test design process.

Faults in analogue circuits are predominantly 'hard', open or short circuits. Consideration of the likely location of such faults can reduce the set still further to, amongst others, the important set of short and open circuits between pairs of terminals of active devices.

By analogy with digital circuits these 'hard' faults may be considered to be in some way equivalent to s-a-1 and s-a-0 faults. Using a fault dictionary of likely open and short circuits an equivalent stuck-at fault dictionary can be derived.

Even from this crude representation, a very high percentage coverage of actual faults may be obtained from digital test patterns produced using these fault models.

The use of a digital simulator to evaluate the behaviour of an analogue circuit seems at first sight to be rather unusual. There are, however, several good reasons why such an approach could give benefits:

Digital simulation software has existed for some time, with consequent benefits to its availability, reliability and price.

The output produced by this type of program (test patterns) is of a form compatible with digital test equipment.

The speed of a logic simulator is many times higher than that of a circuit-level simulator (for similar circuits).

Combined analogue/digital circuitry could be simulated by the same program, giving a unified approach to what have previously been two separate tasks.

The fault simulation facilities of most of these programs give a definite figure for fault coverage, a feature absent from analogue simulators.

B. Macro Modelling

The digital modelling of analogue circuits has two major deficiencies. Firstly it trades accuracy for simulation speed to the extent that the model is only accurate for particular discrete values of voltages or currents applied in the analysis. In itself this is not too serious provided that under such conditions the circuit is operating within a permitted operating range. The second deficiency is more significant. That is that in the module represented by the digital equivalent there is no correspondence, except at the inputs and outputs to the real physical connections in the analogue circuit. Thus the identification and location of faults is limited to the circuit module under consideration.

An alternative approach which provides more detail corresponding to the real circuit and allows a more flexible choice of circuit operating conditions is to use a macromodelling approach.

Such macro models use high level circuit elements, such as controlled voltage and current sources. They allow the circuit simulation to take place at a much greater speed at some expense of accuracy but preserve a closer correspondence to the real circuit although the number of components in a circuit represented in this way is greatly reduced.

Macro modelling greatly reduces complexity by representing a circuit at the sub-system block level, thereby decreasing simulation time. The model may be devised to include only those aspects of circuit behaviour that are specifically required, further simplifying the description.

Boyle (1) demonstrated the macro modelling technnique on typical operational amplifier circuits. The models produced were run on SPICE and showed improvements in run time on the order of 6 to 10 times over the full circuit description run on the same simulator, as well as being much simpler to describe.

To devise a macromodel two related alternative processes may be used, simplification or synthesis. In simplification, portions of the circuit are replaced with simple ideal elements that perform an equivalent function in the same manner as the original. Synthesis achieves a similar reduction in complexity, but by replacing sections of the circuit with ideal elements configured to meet the original circuit performance specification, but not necessarily with a one to one correspondence with the original.

Boyle models are built up of a number of circuit elements, both real (such as resistors, capacitors) and abstract (such as controlled voltage sources) which each represent a parameter of the real circuit. In this way, if certain features are not relevant to a particular application, further simplifications can be made to the model. For example non linearities, such as limiting the output voltage swing, may be dealt with by introducing perfect diodes to clip the output to the voltage of the supply rail, reduced by the voltage drops in other circuit components. However an unfortunate consequence of this method, which is a convenient way of introducing such nonlinearities on a conventional circuit simulator, is that the simulation is slowed down considerably.

C. Fault Modelling

In addition to representing the fault free operation of a circuit, any modelling technique must be capable of representing a circuit with a particular fault present.

In digital circuit modelling, the Stuck-At fault model has been used with success since it was introduced in 1959 by Eldred(2) and was applied particularly successfully to TTL devices. This is because the 'real' faults that frequently occur in a board of TTL devices correspond closely with the fault models; for instance an open circuit TTL input behaves as though it is Stuck-At-1. Over the years this simple Stuck-At model has been considerably enhanced to cover other types of faults and to cope with other technologies. Recent work, for example by Damper (3), explores fault models that are applicable to other integrated circuit technologies, particularly MOS types, where more important additional fault effects can occur.

In the case of analogue circuits the fault effect of an open or short circuit will give rise to a new macromodel for each fault. This is analogous to the process by which individual faults within a digital circuit give rise to a particular stuck-at condition at the terminal of a gate. As in the digital case many different faults may give rise to the same stuck-at fault model, so in the case of an analogue circuit different open and short circuits may give rise to the same overall fault effect which may be described by the same macromodel which will be referred to as a "fault macro". These "fault

```
    ─▶│ A   f0 │─▶│ B   f0 │─▶│ C   f0 │─

           ─│ B   f1 │─

           ─│ B   f2 │─

           ─│ B   f3 │─
```

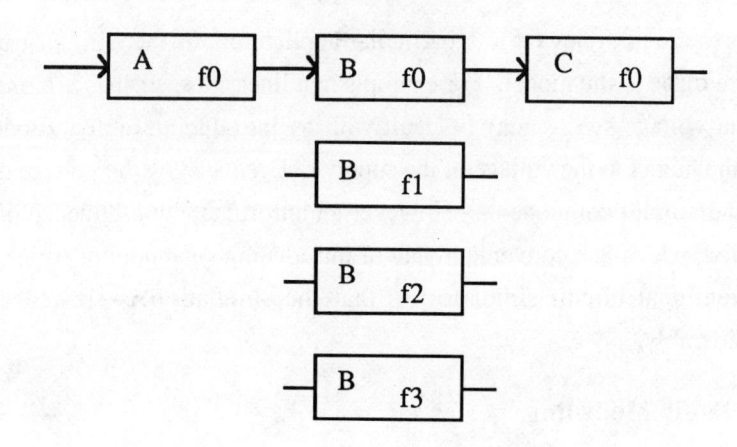

Fig. 2. Substitution of fault macros (f1 - fn) or fault free
macros (f0) to compute observable fault effects.

macros" may then be substituted for the fault free model to simulate the effect of the set of faults on the complete circuit (figure 2).

D. Fault Diagnosis

The techniques presently used in the diagnosis of faults in analogue circuits can be classified into two groups, depending on whether simulation takes place before or after testing.

In the simulation before test (SBT) approach, the most likely faults or combination of faults are included in the simulation of the circuit and the fault effects noted for various inputs. Hence all computation is performed in advance of testing the real circuit and is directed at producing a table of faults and their related fault effects in the form of a Fault Dictionary.

The alternative is to apply a standard set of test inputs to the circuit and by monitoring the results and with a knowlege of the circuit topology, work back to find the value of the component parameters. Thus by comparing measured and expected values for these parameters even 'soft' faults, in which the out of tolerance component continues to function, can be identified.

This approach, called simulation after test (SAT) obviously requires the intervention of the design engineer to hypothesise the likely faults and a great deal of computing to be done after testing, but none before. Each approach has its advantages, the SBT having all its computing done in advance, but usually only being able to diagnose catastrophic failures. The SAT method is able to cope also with component parameter drifts, but entails considerable computing effort after testing each circuit and is less suitable to production testing.

IV. TEST PATTERN GENERATION FOR ANALOGUE CIRCUITS

In testing any circuit, analogue or digital, the aim is to drive it so that the output reflects the good/faulty nature of the device. For a digital circuit, the choice of input logic levels and output threshold are straightforward and it is the test patterns that require the effort in test design. In an analogue circuit, the relatively simple internal structure means that often a much shorter test pattern can be employed. It is the voltage levels at which the data is fed in and read out that require careful consideration.

The objective of a set of tests can be either to distinguish a good device from a faulty one with a high degree of confidence, or alternatively to diagnose the cause of failure. These alternative objectives give rise to two different test objectives, fault detection and fault isolation.

To test a circuit a particular combination of input voltages will be applied and will produce a particular output voltage for a good circuit. Under these same input conditions faulty circuits may produce outputs which are different from those of a good circuit or deviate by some arbitrary amount from that value. Variation of the input conditions will change the set of faults for which the output is distinguishably different from that of a good circuit.

If a particular input set of voltages is considered the output will either
    a) deviate significantly positively from the good output,

```
           |── f1
           |── T1
     ↑     |── good circuit output
  voltage  |── T2
           |── f2
           |── f3
```

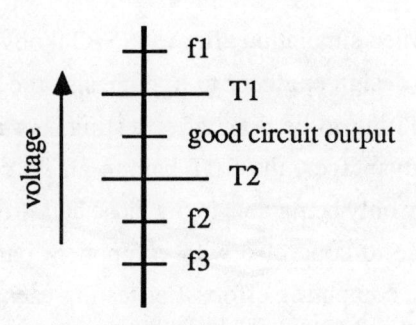

Fig. 3. Regions of operation for faulty and fault free circuits.

    b) be insignificantly different, or

    c) deviate significantly negatively.

A different input combination will change both the good output and the distribution amongst the subsets of the faulty circuits.

To detect a faulty circuit a set of input tests is required such that for at least one of the tests the output measured will deviate significantly either way from the value for a good circuit. To use this method of detection it is implicit that two thresholds should be set, above and below the good output voltage for each test as illustrated in figure 3. The test design has to be concentrated upon the minimum number of input test vectors required to detect all possible faults.

To isolate a fault a further condition is required. The set of input test vectors must provide combinations of conditions to identify a particular fault from amongst the total set. Even so some faults will remain ambiguous. For example an open circuit collector or emitter of a bipolar transistor will give rise to the same lack of current in a particular part of a circuit and consequently the same measured output.

In both cases the accuracy of the modelling of the circuit is critical. The more accurate the model the closer the thresholds of detection of a faulty circuit may be set to the response of a good circuit. Using piecewise linear macromodels a band of 0.5V either side of the good output is a sensible margin for confident detection.

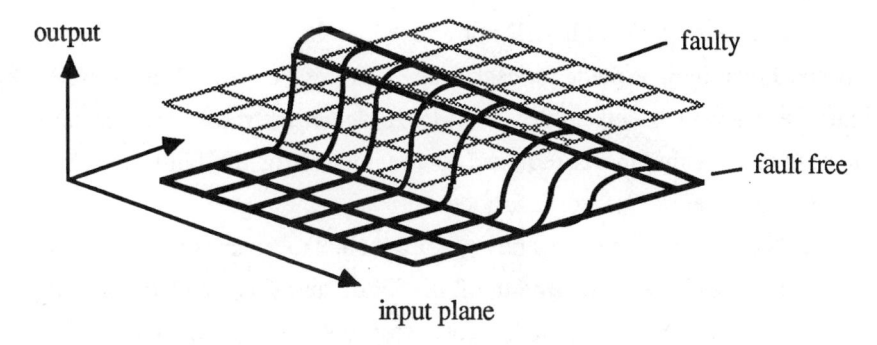

Fig. 4. A fault free and a faulty operating surface for a 2-input, 1-output circuit.

In more general terms an analogue circuit may have a number of outputs and inputs the behaviour of which may be described graphically in the form of surfaces. Consider a 2 input, 1 output device, in this case a 3 dimensional surface with the x and y axes representing the input voltages, and the z axis the output is required, e.g. figure 4. This method of description can be applied to other configurations of input and output, an m output circuit being described by m separate surfaces, an n input circuit being described by a set of (n + 1) dimensional surfaces each representing the value of a single output.

In the same way that a good circuit has an operating surface, so too does a faulty circuit. If a circuit has q faults applied to it one at a time, then there will be q fault surfaces generated for the one good circuit. In a large number of cases, these fault surfaces will be very simple, a fault resulting in a 'stuck' output for example would give a fault surface entirely in a plane. By considering the good operating surface with each fault surface in turn, an indication may be obtained of those input voltages that make the faults detectable.

A fault is detectable for those regions of input voltages where the good circuit output is distinguishable from the faulty circuit output. The fault is undetectable when the circuit is driven to those points where the good surface and fault surface are so close that the accuracy of the macromodel and measuring equipment will not allow it to be discriminated. By this means a map may be made for each fault, showing the areas of the operating surface where it is detectable. In order to detect the most faults with the fewest tests, a method is required to consider all of the fault areas both 'above' and 'below' the good surface.

In order to be able to trace a device failure back to a particular fault, or set of faults a different treatment of the fault areas is needed. Firstly, considering just the 'above' fault areas, a test voltage is selected to occur on either side of the edge of every fault area; these tests then each identify a unique set of faults.If a particular test identifies a single fault only, then that fault is already isolated by the one test. The task of isolating the other faults is performed by specifying multiple tests; for example, if a fault is present then the result of one test may be negative and that of another test positive. If a fault cannot be isolated from the 'above' areas alone, then the results of tests using the 'below' areas must be examined.

V. CONCLUSIONS

The aim is to devise ways of testing mixed analogue and digital circuits. A computer-based system is envisaged for both these purposes, the essential elements of which have been identified (by analogy with digital logic design/test systems) as:

  Circuit Simulator;
  Fault Database;
  Test Pattern Generator;
  Circuit Database.

A. Simulator

The central element is the simulator. It is this with which the effects of the various faults can be seen when the circuit is driven from the output of the test pattern generator.

In its intended application, the target analogue system is likely to form a significant part of a VLSI device. Existing circuit simulators work at the transistor, resistor, capacitor level of modelling and have typically a 50 node limit on circuit size, (equivalent to 3 or 4 average opamps). Even within this size limit, run times become excessively long as circuit complexity increases. This type of simulator is hence unsuitable for use in evaluating fault effects in large-scale circuits, where a run is necessary for each fault applied.

A fairly recent solution to this problem has been to represent convenient circuit blocks by combinations of higher level circuit elements such as controlled voltage/current sources. Circuits described in this way (macromodels) are generally much simpler to specify and an order of magnitude quicker to evaluate. Tradeoffs are invariably made here between the speed and accuracy of the macromodel but, because of our intended use of digital test equipment to test for a 0 or 1 result only, accuracy (provided the 0/1 threshold is not crossed) is not of prime importance.

To date the macromodelling that has been done on the existing circuit level simulators SPICE and SUICIDES has been confined to modelling largely linear devices (voltage rail clamping being available by use of diodes, at the expense of simulation speed). An initial review of the circuit blocks that an analogue circuit designer might reasonably require has been made, and includes several blocks that presently it has not been found possible to simulate at the macro level. The common factor of these blocks is a non-linear element, an example being the voltage-controlled switch. Even if they were modified to include such elements, existing simulators (generally using the Newton Raphson algorithm to solve the circuit equations) fail to find a solution for some circuits. More advanced circuit level simulators such as ASTAP or ASTEC3 are able to handle non linearities by means of a table-driven transfer function and will generally reach a solution in these cases.

B. Fault Database

As the complexity of the analogue circuit increases, fault effects can no longer be visualised by the designer unaided; the circuit therefore needs to be simulated operating in these fault modes. This technique is used not only in failure-analysis work (with the circuit driven by its normal system inputs) but also in checking the effectiveness of the test patterns used on the circuit.

Some form of modelling is required to implement typical failure mechanisms. Preliminary work has identified the open and short circuit fault models as being more useful in this respect than the Stuck At 0/Stuck At 1 models that are traditionally applied to logic circuits.

In application, these fault models can be applied to the periphery of circuit blocks if the macromodelling approach is used, thus replicating faults at chip connections (if working at the board level) or metallisation failures at cell connections (if working at chip level). Faults within circuit blocks can be addressed using a new technique of modelling a block with a particular fault as a fault block. This fault block is in many cases very simple to model and hence to simulate. Thus, instead of applying a fault model to a circuit, a fault block is substituted for its good block.

C. Test Pattern Generation

Proven software already exists for the automatic generation of test patterns for the proving of logic circuits. It is an aim to devise software that fulfils the equivalent function for analogue circuits (the added complication being the choice of voltages at which the logic levels are set).

Two different algorithms are needed for this task, one to obtain maximum fault coverage, the other to isolate a fault as precisely as is required.

A preliminary approach to the test pattern generation problem has been made and gives, on paper, an efficient set of test patterns for best coverage or best isolation.

D. Circuit Database

A modular approach to the design of analogue circuits is now starting to be used by semicustom IC manufacturers such as Plessey, with their 'Analogue Macrocells'. This approach lends itself well to our circuit block level of simulation, with the establishment of a set of standard cells which would each have an associated model. A preliminary attempt at defining this set of circuit blocks has been made. As an alternative, existing analogue circuits may be decomposed into collections of these blocks by means of an appropriate algorithm. This approach currently is more popular with circuit designers, who dislike being constrained to design with a limited set of blocks.

REFERENCES

[1] Boyle, G. R., B. M. Cohn, D. O Pederson, and J. E. Soloman (1974), "Macromodelling of integrated circuit operational amplifiers", *IEEE J. Solid-State Circuits*, SC-9, 6, pp. 353-364.

[2] Eldred, R. D. (1959), "Test routines based on symbolic logic", *J. Assoc. Comput. Mach.*, 6, pp. 33-36.

[3] Damper, R. I. (1984), "Fault models and failure mechanisms", *IEE Vacation School on IC Testing*, Brunel University, UK.

APPENDIX

SELECTED BIBLIOGRAPHY[†]

Compiled by P. G. Bessler and D. M. Miller

Department of Computer Science
University of Manitoba
Winnipeg, MB
Canada R3T 2N2

I. BOOKS

Bennetts, R. G. (1985), <u>Design of Testable Logic Circuits</u>, Addison-Wesley, London.

Breuer, M. A,. and A. D. Friedman (1975), <u>Diagnosis and Reliable Design of Digital Systems</u>, Computer Science Press, Potomac, MD.

Hurst, S. L., D. M. Miller, and J. C. Muzio (1985), <u>Spectral Techniques in Digital Logic</u>, Academic Press, London.

Fujiwara, H. (1985), <u>Logic Testing and Design for Testability</u>, MIT Press, Boston, MA.

Karpovsky, M. G. (editor) (1985), <u>Spectral Techniques and Fault Detection</u>, Academic Press, Orlando.

Lala, P. K. (1985), <u>Fault-Tolerant and Fault Testable Hardware Design</u>, Prentice-Hall, Englewood Cliffs, NJ.

[†] This bibiography was compiled with partial support from the Natural Sciences and Engineering Research Council of Canada under grant A4851.

McCluskey, E. J. (1986), Logic Design Principles with Emphasis on Testable Semicustom Circuits, Prentice-Hall, Englewood Cliffs, NJ.

Miczo, A. (1986), Digital Logic Testing and Simulation, Harper & Row, New York, NY.

Pradham, D. K. (editor) (1986), Fault-Tolerant Computing: Theory and Techniques, Prentice-Hall, Englewood Cliffs, NY.

Roth, J. P. (1980), Computer Logic, Testing and Verification, Computer Science Press, Potomac, MD.

Siewiorek, D. P., and R. S. Swarz (1982), The Theory and Practice of Reliable System Design, Digital Press, Boston, MA.

Williams, T. W. (editor) (1986), VLSI Testing, North-Holland, Amsterdam.

II. JOURNAL AND CONFERENCE PAPERS

Abadir, M. S., and M. A. Breuer (1985), "Constructing optimal test schedules for VLSI circuits having built-in test hardware", *Proc. 15th Int. Fault-Tolerant Comput. Conf.*, pp. 165-170.

Abadir, M. S., and M. A. Breuer (1985), "A knowledge based system for designing testable VLSI chips", *IEEE Design and Test of Comput.*, 2, pp. 56-68.

Abadir, M. S., and H .K. Reghbati (1985), "Functional specification and testing of logic circuits", *Int. J. Comput. Math.*

Abadir, M. S., and H. K. Reghbati (1986), "Functional test generation for digital circuits described using binary decision diagrams", *IEEE Trans. Comput.*, C-35, pp. 375-379.

Abramovici, M., Y. H. Levendel, and P. R. Menon (1982), "A logic simulation machine", *19th Design Automation Conference*, pp. 65-73.

Agarwal, V. K., and G. M. Masson (1977), "Resolution-oriented fault interrelationships in combinational logic networks", *IEEE Trans. Comput.*, C-26, pp. 1170-1175.

Agarwal, V. K., and G. M. Masson (1979), "A functional form approach to test set coverage in tree networks", *IEEE Trans. Comput.*, C-28, pp. 50-52.

Agarwal, V. K., and G. M. Masson (1979), "Recursive coverage projection of test sets", *IEEE Trans. Comput.*, C-28, pp. 865-870.

Agarwal, V. K., and G. M. Masson (1980), "Generic fault characterizations for table look-up coverage bounding", *IEEE Trans. Comput.*, C-29, pp. 288-299.

Agarwal, V. K. (1980), "Multiple fault detection in programmable logic arrays", *IEEE Trans. Comput.*, C-29, pp. 518-522.

Agarwal, V. K,. and E. Cerny (1981), "Store and Generate Built-In-Testing Approach", *Proc. 11th Int.. Symp. Fault-Tolerant Computing*, pp. 35-40.

Agarwal, V. K., and A. S. F. Fung (1981), "Multiple fault testing of large circuits by single fault test sets", *IEEE Trans. Comput.*, C-30, pp. 855-865.

Agarwal, V. K. (1983), "Increased effectiveness of built-in-testing by output data modification", *Proc. 13th Int. Symp. on Fault Tolerant Computing*, pp. 227-234.

Agrawal, V. D., and P. Agrawal (1972), "An automatic test generation system for Illiac IV logic boards", *IEEE Trans. Comput.*, C-21, pp. 1015-1017.

Agrawal, P., and V. D. Agrawal (1975), "Probabilistic analysis of random test generation method for irredundant combinational logic networks", *IEEE Trans. Comput.*, C-24, pp. 691-695.

Agrawal, P. (1984), "Test generation at switch level", *IEEE Int. Conf. Computer-aided Design*, pp. 128-130.

Akers, S. B. (1973), "Universal test sets for logic networks", *IEEE Trans. Comput.*, C-22, pp. 835-839.

Akers, S. B. (1976), "A Logic System for Fault Test Generation", *IEEE Trans. Comput.*, C-25, pp. 620-630.

Akers, S. B. (1978), "Binary decision diagrams", *IEEE Trans. Comput.*, C-27, pp. 509-516.

Akers, S. B. (1978), "Functional testing with binary decision diagrams", *Dig. Papers, 8th Int. Symp. Fault-Tolerant Comput.*, pp. 82-92.

Akers, S. B. (1985), "On the use of linear sums in exhaustive testing", *15th Int. Symp. Fault Tolerant Comput.*, pp. 148-153.

Akers, S. B. (1980), "Test generation techniques", *Computer*, 13, pp. 9-15.

Anderson, D. A., and G. Metze (1973), "Design of totally self-checking check circuits for m-out-of-n codes", *IEEE Trans. Comput.*, C-22, pp. 263-269.

Ando, H. (1980), "Testing VLSI with random access scan", *Proc. COMPCON Spring*, pp. 50-52.

Armstrong, D. B. (1966), "On finding a nearly minimal set of fault detection tests for combinational nets", *IEEE Trans. Electron. Comput.*, EC-15, pp. 66-73.

Armstrong, D. B. (1972), "A deductive method for simulating faults in logic circuits", *IEEE Trans. Comput.*, C-21, pp. 464-471.

Banerjee, P., and J. A. Abraham (1983), "Generating tests for physical failures in MOS logic circuits", *IEEE Test Conf.*, pp. 554-559.

Bardell, P. H., and W. H. McAnney (1982), "Self-testing of multi-chip logic modules", *Proc, 1982 Int. Test Conf.*, pp. 200-204.

Barzilai, Z., J. Savir, G. Markowsky, and M. G. Smith (1981), "The weighted syndrome sums approach to VLSI testing", *IEEE Trans. Comput.*, C-30, pp. 996-1000.

Barzilai, Z., D. Coppersmith, and A. L. Rosenburg (1983), "Exhaustive generation of bit patterns with applications to VLSI self-testing", *IEEE Trans. Comput.*, C-32, pp. 190-194.

Baschiera, D., and B. Courtois (1984), "Testing CMOS: a challenge", *VLSI Design*, 5, pp. 58-62.

Bassett, J. and C. R. Kime (1972), "Improved procedures for determining diagnostic resolution", *IEEE Trans. Comput.*, C-21, pp. 385-388.

Bate, J. A. and D. M. Miller (1985), "Fault Detection in CMOS Circuits and an Algorithm for Generating Eulerian Cycles in Directed Hypercubes", *Congressus Numerantium*, 47, pp. 107-117.

Batni, R. P., J. D. Russell, and C. R. Kime (1974), "An efficient procedure for finding an irredundant set cover", *J. ACM*, 21, pp. 351-355.

Batni, R. P., and C. R. Kime (1976), "A module-level testing approach for combinational networks", *IEEE Trans. Comput.*, C-25, pp. 594-604.

Bearnson, L. W., and C. C. Carroll (1971), "On the design of minimum length fault tests for combinational circuits", *IEEE Trans. Comput.*, C-20, pp. 1353-1356.

Bennetts, R. G., and R. V. Scott (1976), "Recent developments in the theory and practice of testable logic design", *Computer*, 9, pp. 47-63.

Bennetts, R. G., C. M. Maunder, and G. D. Robinson (1981), "CAMELOT: a computer-aided measure for logic testability", *IEE Proc.*, 128, pp. 177-189.

Benowitz, N., D. F. Calhoun, G. E. Alderson, J. E. Bauer, and C. T. Joeckel (1975), "An advanced fault isolation system for digital logic", *IEEE Trans. Comput.*, C-24, pp. 489-497.

Berg, W. C., and R. D. Hess (1982), COMET: a testability analysis and design modification package", *Proc. IEEE Test Conf.*, pp. 364-378.

Berger, I., and Z. Kohavi (1973), "Fault detection in fanout-free combinational networks", *IEEE Trans. Comput.*, C-22, pp. 908-914.

Betancourt, R. (1973), "Derivation of minimum test sets for unate logical circuits", *IEEE Trans. Comput.*, C-20, pp. 1264-1269.

Bhattacharya, B. B., "On a novel approach of fault detection in an easily testable sequential machine with extra inputs and outputs", *IEEE Trans. Comput.*, to be published.

Bhattacharya, B. B., and B. Gupta (1983), "Syndrome testable design of combinational networks for detecting stuck-at and bridging faults", *Proc. 1983 Int. Test Conf.*, pp. 446-452.

Bhavsar, D. K., and R. W. Heckelman (1981), "Self-testing by polynomial division", *Proc. IEEE Test Conf.*, pp. 208-216.

Bhavsar, D. K., and B. Krishnamurthy (1984), "Can we eliminate fault escape in self testing by polynomial division (signature analysis)?", *Proc. 1984 Int. Test Conf.*, pp. 134-139.

Bose, A. K., *et al* (1982), "A fault simulator for MOS LSI circuits", *Proc. 19th Design Automation Conf.*, pp. 400-409.

Bossen, D. C., and S. J. Hong (1971), "Cause-effect analysis for multiple fault detection in combinational networks", *IEEE Trans. Comput.*, C-20, pp. 1252-1257.

Bouricius, W. G., et al. (1971), "Algorithms for detection of faults in logic circuits", *IEEE Trans. Comput.*, C-20 , pp. 1258-1264.

Boute, R. T. (1974), "Optimal and near-optimal checking experiments for output faults in sequential machines", *IEEE Trans. Comput.* , C-23, pp. 1207-1213.

Bozorgui-Nesbat, S., and E. J. McCluskey (1980), "Structured design for testability to eliminate test pattern generation", *Proc. 10th Int. Symp. Fault-Tolerant Computing*, pp. 158-163.

Bozorgui-Nesbat, S., and E. J. McCluskey (1986), "Lower overhead design for testability of programmable logic arrays", *IEEE Trans. Comput.*, C-35, pp. 379-383.

Braun, R. D., and D. D. Givone (1979), "An improved algorithm for deriving checking experiments", *IEEE Trans. Comput.*, C-28, pp. 153-156.

Breuer, M. A. (1970), "Functional partitioning and simulation of digital circuits", *IEEE Trans. Comput.*, C-19, pp. 1038-1046.

Breuer, M. A. (1971), "A random and an algorithmic technique for fault detection test generation for sequential circuits", *IEEE Trans. Comput.*, C-20, pp. 1364-1370.

Breuer, M. A. (1972), "A note on three-valued logic simulation", *IEEE Trans. Comput.*, C-21, pp. 399-402.

Breuer, M. A. (1972), "Generation of fault tests for linear logic networks", *IEEE Trans. Comput.*, C-21, pp. 79-83.

Breuer, M. A., and A. D. Friedman (1980), "TEST/80 - a proposal for an advanced automatic test generation system", *Proc. IEEE Autotestcon*, pp. 305-312.

Breuer, M. A,. and A. D. Friedman (1980), "Functional level primitives in test generation", *IEEE Trans. Comput.*, C-29, pp. 223-235.

Chang, H. Y., and G. W. Heimbigner (1974), "LAMP: Controllability, observability, and maintenance engineering technique (COMET), *Bell Sys. Tech. J.*, 53, pp. 1505-1534.

Chang, H. Y., S. G. Chappell, C. H. Elmendorf, and L. D. Schmidt (1974), "Comparison of parallel and deductive fault simulation methods", *IEEE Trans. Comput.*, C-23, pp. 1132-1138.

Chappell, S. G., C. H. Elmensdorf, and L. D. Schmidt (1974), "LAMP: logic circuit simulation", *Bell System Technical J.*, 53, pp. 1451-1476.

Chappell, S. G., P. R. Menon, J. F. Pellegrin, and A. M. Schowe (1977), "Functional simulation in the LAMP systems", *J. Design Automat. Fault Tolerant Comput.*, 1, pp. 203-216.

Chiang, A. C. L., I. S. Reed, and A. V. Banes (1972), "Path sensitization, partial Boolean difference, and automated fault diagnosis", *IEEE Trans. Comput.*, C-21, pp. 189-195.

Chiang, K.-W., and Z. G. Vranesic (1982), "Test generation for MOS complex gate networks", *Proc. 19th Design Automation Conf.*, pp. 149-157.

Chin, C. K., and E. J. McCluskey (1987), "Test length for pseudorandom testing", *IEEE Trans. Comput.*, C-36, pp. 252-256.

Coy, W., and A. Vogel (1981), "Inserting test points for simple test generation", *Proc. 4th Int. Conf. Fault-Tolerant Systems and Diagnostics*, pp. 180-186.

Daehn, W., and J. Mucha, (1981), "Hardware test pattern generation for built-in testing", *Proc. IEEE Test Conf.*, pp. 110-113.

Daehn, W., and J. Mucha, (1981), "A hardware approach to self-testing of large programmable logic arrays", *IEEE Trans. Comput.*, C-30, pp. 829-233.

Dandapani, R., and S. M. Reddy (1974), "On the design of logic networks with redundancy and testability considerations", *IEEE Trans. Comput.*, C-23, pp. 1139-1149.

DasGupta, S., C. R. P. Hartmann, and L. D. Rudolph (1980), "Dual-mode logic for function-independent fault testing", *IEEE Trans. Comput.*, C-29, pp. 1025-1029.

Breuer, M. A., and A. C. Parker (1981), "Digital system simulation: current status and future trends or Darwin's theory of simulation", *18th Design Automation Conference*, pp. 269-275.

Breuer, M. A., and X. Zhu (1985), "A knowledge based system for selecting a test methodology for a PLA", *Proc. 22nd Design Automation Conf.*, pp. 259-265.

Brglez, F. (1984), "On testability analysis of combinational networks", *Proc. Int. Symp. Circuits and Systems*, pp. 221-225.

Brown, R. D. (1977), "A recursive algorithm for sequency-ordered fast Walsh transforms", *IEEE Trans. Comput.*, C-26, pp. 819-822.

Bryant, R. E., and M. D. Schuster (1985), "Performance evaluation of FMOSSIM, a concurrent switch-level fault simulator", *Proc. 22nd Design Automation Conf.*, pp. 715-719.

Brzozowski, J. A., and M. Seger (1985), "Design of testable CMOS latches", *Canadian Conf. on VLSI*, Toronto, pp. 225-228.

Brzozowski, J. A. (1986), "On the testability of CMOS latches", *Canadian Conf. on VLSI*, Montreal, pp. 153-158.

Carl, J. W., and R. V. Swartwood (1973), "A hybrid Walsh transform computer", *IEEE Trans. Comput.*, C-22, pp. 669-672.

Carter, J. L. (1982), "The theory of signature testing for VLSI", *Proc. 14th ACM Symp. Theory Comput.*, pp. 289-296.

Carter, W. C. (1982), "Signature testing with guaranteed bound for fault coverage", *Proc. IEEE Test Conf.*, pp. 75-82.

Cha, C. W., W. Donath, and F. Ozguner (1978), "9-V algorithm for test pattern generation of combinational digital circuits", *IEEE Trans. Comput.*, C-27, pp. 193-200.

Cha, C. W. (1978), "A testing strategy for PLAs", *15th ACM/IEEE Design Automation Conf.*, pp. 326-334.

Chandramouli, R. (1981), "On testing stuck-open faults in CMOS combinational cells", *Proc. 19th Allerton Conf. on Commun., Control, and Comput..*, pp. 701-712.

Chandramouli, R. (1983), "On testing stuck-open faults", *IEEE Test Conf.*, pp. 258-265.

DasGupta, S., R. G. Walther, T. W. Williams, E. B. Eichelberger (1981), "An enhancement to LSSD and some applications of LSSD in reliability, availability, and serviceability", *Proc. 11th Symp. Fault-Tolerant Computing*, pp. 32-34.

DasGupta, S., P. Goel, R. G. Walther, and T. W. Williams (1982), "A variation of LSSD and its implication on design and test pattern generation in VLSI", *Proc. IEEE Test Conf.*, pp. 63-66.

Denneau, M. M. (1982), "The Yorktown simulation engine", *19th Design Automation Conference*, pp. 55-59.

Dias, F. J. O. (1975), "Fault masking in combinational logic circuits", *IEEE Trans. Comput.*, C-24, pp. 476-482.

Dias, F. J. O. (1976), "Truth-table verification of an iterative logic array", *IEEE Trans. Comput.*, C-25, pp. 605-613.

Du, M.-W., and C. D. Weiss (1973), "Multiple fault detection in combinational circuits: Algorithms and computational results", *IEEE Trans. Comput.*, C-22, pp. 235-240.

Dussault, J. A. (1978), "A testability measure", *IEEE Semiconductor Test Conf.*, pp. 113-116.

Edwards, C. R. (1975), "The application of the Rademacher-Walsh transform to Boolean function classification and threshold logic synthesis", *IEEE Trans. Comput.*, C-24, pp. 48-62.

Eichelberger, E. B. and T. W. Williams (1978), "A logic design structure for LSI testability", *J. Design Automat. Fault Tolerant Computing*, 2, pp. 165-178.

Eichelberger, E. B., and E. Lindbloom (1983), "Trends in VLSI testing", *VLSI'83*, Elsevier Science Publishers, pp. 339-348.

El-ziq, Y. M., and S. Y. H. Su (1978), "Computer-aided logic design of two-level MOS combinational networks with statistical results", *IEEE Trans. Comput.*, C-27, pp. 911-923.

El-ziq, Y. M., and R. J. Cloutier (1981), "Functional-level test generation for stuck-open faults in CMOS VLSI", *IEEE Test Conf.*, pp. 536-546.

El-ziq, Y. M. (1981), "Automatic test generation for stuck-open faults in CMOS VLSI", *Proc. 18th Design Automation Conf.*, pp. 347-354.

El-ziq, Y. M., and S. Y. H. Su (1982), "Fault diagnosis of MOS combinational networks", *IEEE Trans. Comput.*, C-31, pp. 129-139.

El-ziq, Y. M., and H. H. Butt (1983), "A mixed-mode built-in self-test technique using scan path and signature snalysis", *Proc. 1983 Int. Test Conf.*, pp. 269-274.

Eldred, R. D. (1959), "Test routines based on symbolic logic statements", *J. Assoc. Comput. Mach.*, 6, pp. 33-36.

Eris, E., and D. M. Miller (1983), "Syndrome-testable internally unate combinational networks", *Electron. Lett.*, 19, p. 16.

Fantauzzi, G., and A. Marsella (1974), "Multiple-fault detection and location in fan-out free combinational circuits", *IEEE Trans. Comput.*, C-23, pp. 48-55.

Fantauzzi, G. (1974), "An algebraic model for the analysis of logical circuits", *IEEE Trans. Comput.*, C-23, pp. 576-581.

Farmer, D. E. (1973), "Algorithms for designing fault-detection experiments for sequential machines", *IEEE Trans. Comput.*, C-22, pp. 159-167.

Fridrich, M., and W. A. Davis (1974), "Minimal fault tests for combinational networks", *IEEE Trans. Comput.*, C-23, pp. 850-859.

Friedman, A. D., and P. R. Menon (1973), "Restricted checking sequences for sequential machines", *IEEE Trans. Comput.*, C-22, pp. 397-399.

Friedman, A. D. (1967), "Fault detection in redundant circuits", *IEEE Trans. Electron. Comput.*, EC-16, pp. 99-100.

Friedman, A. D. (1973), "Easily testable iterative systems", *IEEE Trans. Comput.*, C-22, pp. 1061-1064.

Frohwerk, R. A. (1977), "Signature analysis: A new digital field service method", *Hewlett-Packard J.*, pp. 2-8.

Fujiwara, E. (1983), "A self-testing group-parity prediction checker and its use for built-in testing", *Proc. 13th Int. Symp. Fault Tolerant Computing*, pp. 146-153.

Fujiwara, E., and K. Matsuoka (1987), "A self-checking generalized prediction checker and its use for built-in testing", *IEEE Trans. Comput.*, C-36, pp. 86-93.

Fujiwara, H., and K. Kinoshita (1974), "Design of diagnosable sequential machines utilizing extra outputs", *IEEE Trans. Comput.*, C-23, pp. 138-145.

Fujiwara, H., Y. Nagao, T. Sasao, and K. Kinoshita (1975), "Easily testable sequential machines with extra inputs", *IEEE Trans. Comput.*, C-24, pp. 821-826.

Fujiwara, H., and K. Kinoshita (1978), "Testing logic circuits with compressed data", *Proc. 8th Int. Symp. on Fault Tolerant Computing*, pp. 108-113.

Fujiwara H., K. Kinoshita, and H. Ozaki (1980), "Universal test sets for programmable logic arrays", *Proc. 10th Int. Symp. Fault-Tolerant Computing*, pp. 137-142.

Fujiwara, H. (1981), "On closedness and test complexity of logic circuits", *IEEE Trans. Comput.*, C-30, pp. 556-562.

Fujiwara, H., and K. Kinoshita (1981), "A design of programmable logic arrays with universal tests", *IEEE Trans. Comput.*, C-30, pp. 823-828.

Fujiwara, H., and S. Toida (1982), "The Complexity of Fault Detection: An Approach to Design For Testability", *Proc. 12th Int. Symp. Fault-Tolerant Computing*, pp. 101-108.

Fujiwara, H., and S. Toida (1982), "The complexity of fault detection problems for combinational circuits", *IEEE Trans. Comput.*, C-31, pp. 555-560.

Fujiwara, H., and T. Shimono (1983), "On the acceleration of test generation algorithms", *IEEE Trans. Comput.*, C-32, pp. 98-105.

Fujiwara, H. (1984), "A new PLA design for universal testability", *IEEE Trans. Comput.*, C-33, pp. 745-750.

Funatsu, S., N. Wakatsuki, and T. Arima (1975), "Test generation in Japan", *Proc. 12th Design Automation Symp.*, pp. 114-122.

Galiay, J., Y. Crouzet, and M. Vergniault (1980), "Physical versus logical fault models for MOS LSI circuits: impact on their testability", *IEEE Trans. Comput.*, C-29, pp. 527-531.

Gault, J. W., J. P. Robinson, and S. M. Reddy (1972), "Multiple fault detection in combinational networks", *IEEE Trans. Comput.*, C-21, pp. 31-36.

Goel, P. (1981), "An implicit enumeration algorithm to generate tests for combinational logic circuits", *IEEE Trans. Comput.*, C-30, pp. 215-222.

Goel, P., and B. C. Rosales, (1981), "PODEM-X: An automatic test generation system for VLSI logic structures", *Proc.18th Design Automation Conf.*, pp. 260-268.

Goldstein, L. H. (1979), "Controllability/observability analysis of digital circuits", *IEEE Trans. Circuits and Systems*, CAS-26, pp. 685-693.

Gonenc, G. (1970), "A method for the design of fault detection experiments", *IEEE Trans. Comput.*, C-19, pp. 551-558.

Grason, J., and A. W. Nagle (1981), "Digital test generation and design for testability", *J. of Digital Systems*, pp. 319-359.

Hardie, F. H., and R. J. Suhocki (1967), "Design and use of fault simulation for Saturn computer design", *IEEE Trans. Electron. Comput.*, EC-16, pp. 412-429.

Hassan, S. Z. (1984), "Signature testing of sequential machines", *IEEE Trans. Comput.*, C-33, pp. 762-764.

Hassan, S. Z., and E. J. McCluskey (1983), "Testing PLAs using multiple parallel signature analyzers", *Proc. 13th Int. Symp. on Fault Tolerant Computing*, pp. 422-425.

Hassan, S. Z., and E. J. McCluskey (1984), "Increased fault coverage through multiple signatures", *Proc. 14th Int. Symp. on Fault Tolerant Computing*, pp. 354-359.

Hayes, J. P. (1971), "On realizations of Boolean functions requiring a minimal or near-minimal number of tests", *IEEE Trans. Comput.*, C-20, pp. 1506-1513.

Hayes, J. P. (1971), "A NAND model for fault diagnosis in combinational logic networks", *IEEE Trans. Comput.*, C-20, pp. 1496-1506.

Hayes, J. P. (1974), "On modifying logic networks to improve their diagnosability", *IEEE Trans. Comput.*, C-23, pp. 56-62.

Hayes, J. P., and A. D. Friedman (1974), "Test point placement to simplify fault detection", *IEEE Trans. Comput.*, C-23, pp. 727-735.

Hayes, J. P. (1975), "Detection of pattern-sensitive faults in random-access memories", *IEEE Trans. Comput.*, C-24, pp. 150-157.

Hayes, J. P. (1975), "The fanout structure of switching functions", *J. Assoc. Comput. Mach.*, 22, pp. 551-571.

Hayes, J. P. (1976), "Transition count testing of combinational logic circuits", *IEEE Trans. Comput.*, C-25, pp. 613-620.

Hayes, J. P. (1976), "Check sum methods for test data compression", *J. Design Automation and Fault Tolerant Comput.*, 1, pp. 3-17.

Hayes, J. P. (1978), "Generation of optimal transition count tests", *IEEE Trans. Comput.*", C-27, pp. 36-41.

Hayes, J. P. (1985), "Fault modeling", *IEEE Design and Test*, 2, pp. 88-95.

Hochwald W., and J. D. Bastian (1979), "A DC approach for analog fault dictionary determination", *IEEE Trans. on Circuits and Systems*, CAS-26, pp. 523-529.

Hong, S. J., and D. L. Ostapko (1980), "FITPLA: a programmable logic array for function independent testing", *Proc. 10th Int. Symp. Fault-Tolerant Computing*, pp. 131-136.

Hsiao, M. Y., and D. K. Chia (1971), "Boolean difference for fault detection in asynchronous sequential machines", *IEEE Trans. Comput.*, C-20, pp. 1356-1361.

Hsaio, M. Y., A. M. Patel, and D .K. Pradhan (1977), "Store address generator with on-line fault-detection capability", *IEEE Trans. Comput.*, C-26, pp. 1144-1147.

Hsiao, T.-C., and S. C. Seth (1984), "An analysis of the use of Rademacher-Walsh spectrum in compact testing", *IEEE Trans. Comput.*, C-33, pp. 934-937.

Hsieh, E. P. (1971), "Checking experiments for sequential machines", *IEEE Trans. Comput.*, C-20, pp. 1152-1166.

Hua, K. A., J.-Y. Jou, and J. A. Abraham (1984), "Built-in tests for VLSI finite state machines", *14th Int. Symp. Fault Tolerant Comput.*, pp. 292-297.

Hurst, S. L. (1986), "The use of linerisation in the input and output compaction testing of digital networks", *Proc. 2nd Int. Workshop on Spectral Techniques*, Montreal.

Jain, S. K., and V. K. Agarwal (1983), "Test generation for MOS circuits using the D algorithm", *Proc. 20th Design Automation Conf.*, pp. 64-70.

Jain, S. K., and V. K. Agarwal (1985), "Modeling and test generation algorithm for MOS circuits", *IEEE Trans. Comput.*, C-34, pp. 426-433.

Jha, N. (1986), "Detecting multiple faults in CMOS circuits", *IEEE Test Conf.*, pp. 514-519.

Kamal, S., and C. V. Page (1974), "Intermittent faults: A model and detection procedure", *IEEE Trans. Comput.*, C-23, pp. 713-719.

Karpovsky, M. G., and S. Y. H. Su (1980), "Detection and location of input and feedback bridging faults among input and output lines", *IEEE Trans. Comput.*, C-29, pp. 523-527.

Kautz, W. H. (1968), "Fault testing and diagnosis in combinational digital circuits", *IEEE Trans. Comput.*, C-17, pp. 352-366.

Kautz, W. H. (1971), "Testing faults in combinational cellular logic arrays", *Proc. 8th Ann. Symp. Switching and Automata Theory*, pp. 161-174.

Keiner, W., and R. West (1977), "Testability measures", *Proc. IEEE Autotestcon*, pp. 49-55.

Khakbaz, J., and E. J. McCluskey (1982), "Concurrent error detection and testing for large PLAs", *IEEE J. Solid-State Circuits*, SC-17, pp. 386-394.

Khakbaz, J. (1984), "A testable PLA design with low overhead and high fault coverage", *IEEE Trans. Comput.*, C-33, pp. 743-745.

Kime, C. R. (1970), "An analysis model for digital system diagnosis", *IEEE Trans. Comput.*, C-19, pp. 1063-1073.

Kime, C. R. (1975), "Fault-tolerant computing: An introduction and a perspective", *IEEE Trans. Comput.*, C-24, pp. 457-460.

Kinoshita, K., Y. Takamatsu, and M. Shibata (1980), "Test generation for combinational circuits by structure description functions", *Proc. 10th Int. Symp. Fault-Tolerant Computing*, pp. 152-154.

Kodandapani, K. L. (1974), "A note on easily testable realizations for logic functions", *IEEE Trans. Comput.*, C-23, pp. 332-333.

Kohavi, Z., and P. Lavallee (1967), "Design of sequential machines with fault detection capabilities", *IEEE Trans. Electron. Comput.*, EC-16, pp. 473-484.

Kohavi, I., and Z. Kohavi (1972), "Detection of multiple faults in combinational networks", *IEEE Trans. Comput.*, C-21, pp. 556-568.

Komonytsky, D. (1982), "LSI level-sensitive scan design using LSSD and signature analysis", *Proc. IEEE Test Conf.*, pp. 414-424.

Konemann, B., J. Mucha, and G. Zwiehoff (1980), "Built-in test for complex digital integrated circuits", *IEEE J. Solid State Circuits*, SC-15, pp. 315-318.

Kovijanic, E. K., and G. Pfister (1979), "Testability analysis", *Proc. IEEE Test Conf.*, pp. 310-316.

Kronstadt, E., and G. Pfister (1982), "Software support for the Yorktown simulation engine", *19th Design Automation Conf.*, pp. 60-64.

Ku, C.-T., and G. M. Masson (1975), "The Boolean difference and multiple fault analysis", *IEEE Trans. Comput.*, C-24, pp. 62-71.

Kuban, J., and W. Bruce (1983), "The MC6804P2 Built-In Self-Test", *Proc. 1983 Int. Test Conf.*, pp. 295-300.

Kuhl, J. G., and S. M. Reddy (1978), "On the detection of terminal stuck faults", *IEEE Trans. Comput.*, C-27, pp. 467-469.

Lackey, R. B., and D. Meltzer (1971), "A simplified definition of Walsh functions", *IEEE Trans. Comput.*, C-20, pp. 211-213.

Landgraff, R. W., and S. S. Yau (1971), "Design of diagnosable iterative arrays", *IEEE Trans. Comput.*, C-20, pp. 867-877.

Lee, H.-P. S., and E. S. Davidson (1974), "Redundancy testing in combinational networks", *IEEE Trans. Comput.*, C-23, pp. 1029-1047.

Levendel, Y. H., and P. R. Menon (1982), "Test generation algorithms for computer hardware description languages, *IEEE Trans. Comput.*, C-31, pp. 577-588.

Lineback, J. R. (1982), "Logic simulation speeded with new special hardware", *Electronics*, June, pp. 45-46.

Losq, J. (1978), "Efficiency of random compact testing", *IEEE Trans. Comput.*, C-27, pp. 516-525.

McCluskey, E. J., and F.W. Clegg (1971), "Fault equivalence in combinational logic networks", *IEEE Trans. Comput.*, C-20, pp. 1286-1293.

McCluskey, E. J., and S. Bozorgui-Nesbat (1981), "Design for autonomous test", *IEEE Trans. Comput.*, C-30, pp. 866-875.

McCluskey, E. J. (1982), "Verification Testing", *Proc. 19th Design Automation Conf.*, pp. 495-500.

McCluskey, E. J. (1984), "Verification testing -- a pseudoexhaustive test technique", *IEEE Trans. Comput.*, C-33, pp. 541-546.

McCluskey, E. J. (1985), "Built-in self-test techniques", *IEEE Design and Test of Comput.*, pp. 21-28.

Malaiya, Y. K., and S. Y. H. Su (1982), "A new fault model and testing technique for CMOS devices", *IEEE Test Conf.*, pp. 25-34.

Mandl, K. D. (1984), "CMOS VLSI challenges for test", *IEEE Test Conf.*, pp. 642-648.

Markowsky, G. (1981), "Syndrome-testability can be achieved by circuit modification", *IEEE Trans. Comput.*, C-30, pp. 604-606.

Maunder, C. (1985), "Built-in test: a review", *IEE Electronics & Power*, March, pp. 204-208.

Mei, K. C. Y. (1974), "Bridging and stuck-at faults", *IEEE Trans. Comput.*, C-23, pp. 720-727.

Menon, P. R., and A. D. Friedman (1971), "Fault detection in iterative logic arrays", *IEEE Trans. Comput.*, C-20, pp. 524-535.

Menon, P. R., and S. G. Chappell (1978), "Deductive fault simulation with functional blocks", *IEEE Trans. Comput.*, C-27, pp. 689-695.

Mercer, M. R., V. D. Agarwal, and C. M. Roman (1981), "Test generation for highly sequential scan-testable circuits through logic transformation", *Proc. IEEE Test Conf.*, pp. 561-565.

Miller, D. M., and J. C. Muzio (1983), "Spectral fault signatures for internally unate combinational networks", *IEEE Trans. Comput.*, C-32, pp. 1058-1062.

Miller, D. M., and J. C. Muzio (1984), "Spectral fault signatures for single stuck-at faults in combinational networks", *IEEE Trans. Comput.*, C-33, pp. 765-769 .

Mine, H., and Y. Koga (1967), "Basic properties and a construction method for fail-safe logical systems", *IEEE Trans. Elec. Comp.*, EC-16, pp. 282-289.

Mor, M. (1976), "On the three-valued simulation of digital systems", *IEEE Trans. Comput.*, C-25, pp. 1152-1156.

Moritz, P. S., and L. M. Thorsen (1985), "CMOS circuit testability", *IEEE Custom Integrated Circuits Conf.*, pp. 311-314.

Muehldorf, E. I., and A. D. Savkar (1981), "LSI logic testing -- An overview", *IEEE Trans. Comput.*, C-30, pp. 1-17.

Muehldorf, E. I. (1973), "Theory of a switching calculus for the diagnosis of ninary logic", *Elektron. Rechenanlagen*, 15, pp. 215-222.

Murakami, S.-I., K. Kinoshita, and H. Ozaki (1970), "Sequential machines capable of fault diagnosis", *IEEE Trans. Comput.*, C-19, pp. 1079-1085.

Muth, P. (1976), "A nine-valued circuit model for test generation", *IEEE Trans. Comput.*, C-25, pp. 630-636.

Muzio, J. C., and S.L. Hurst (1978), "The computation of complete and reduced sets of orthogonal spectral coefficients for logic design and pattern recognition puposes", *Comput. Elec. Eng.*, 5, pp. 231-249.

Muzio, J. C. (1980), "Composite spectra and the analysis of switching circuits", *IEEE Trans. Comput.*, C-29, pp. 750-753.

Muzio, J. C., and M. Serra (1985), "Built-in testing and design for testability of stuck-at neighbour faults", *Canadian Conf. on VLSI*, Toronto, pp. 229-233.

Muzio, J. C., and M. Serra (1986), "Spectral criteria for the detection of bridging faults", *2nd Int. Workshop on Spectral Techniques*, Montreal.

Ogus, R. C. (1975), "The probability of a correct output from a combinational circuit", *IEEE Trans. Comput.*, C-24, pp. 534-544.

Oklobdzija, V. G., and M. D. Ercegovac (1982), "Testability of VLSI using circuit structures", *Proc. IEEE ICCC'82*, pp. 198-201.

Ostapko, D. L. and S. J. Hong (1979), "Fault analysis and test generation for programmable logic arrays (PLA's)", *IEEE Trans. Comput.*, C-28, pp. 617-627.

Page, E. W. (1980), "Minimally testable Reed-Muller canonical forms", *IEEE Trans. Comput.*, C-29, pp. 746-750.

Paige, M. R. (1973), "Synthesis of diagnosable FET networks", *IEEE Trans. Comput.*, C-22, pp. 513-516

Papaioannou, S. G. (1977), "Optimal test generation in combinational networks by pseudo-Boolean programming", *IEEE Trans. Comput.*, C-26, pp. 553-560.

Parker, K. P., and E. J. McCluskey (1975), "Analysis of logic circuits with faults using input signal probabilities", *IEEE Trans. Comput.*, C-24, pp. 573-578.

Parker, K. P., and E. J. McCluskey (1975), "Probabilistic treatment of general combinational networks", *IEEE Trans. Comput.*, C-24, pp. 668-670.

Pfister, G. F. (1982), "The Yorktown simulation engine: Introduction", *19th Design Automation Conf.*, pp. 51-54.

Podaima, B., P. D. Hortensius, R. D. Schneider, H. C. Card, and R. D. McLeod (1986), "CALBO - cellular automata logic block observation", *Proc. Canadian VLSI Conf.*, pp. 171-176.

Pradham, D. K. (1978), "Universal test sets for multiple-fault detection in AND-EXOR arrays", *IEEE Trans. Comput.*, C-27, pp. 181-187.

Pradham, D. K., and K. Son (1980), "The effect of untestable faults in PLA and a design for testability", *Proc. IEEE Test Conf.*, pp. 359-367.

Pradhan, D. K. (1983), "Sequential network design using extra inputs for fault detection", *IEEE Trans. Comput.*, C-32, pp. 319-323.

Putzolu, G. R., and J. P. Roth (1971), "A heuristic algorithm for testing of asynchronous circuits", *IEEE Trans. Comput.*, C-20, pp. 639-647.

Ramaoorthy, C. V., and W. Mayeda (1971), "Computer diagnosis using the blocking gate approach", *IEEE Trans. Comput.*, C-20, pp. 1294-1299.

Ramantha, K. S., and N. N. Biswas, (1982), "A design for complete testability of programmable logic arrays", *IEEE Test Conf.*, pp. 67-74.

Ramantha, K. S., and N. N. Biswas, (1983), "A design for testability of undetectable crosspoint faults in programmable logic arrays", *IEEE Trans. Comput.*, C-32, pp. 551-557.

Reddy, K. M., S. M. Reddy, and V. K. Agarwal (1984), "Robust test for stuck-open faults in CMOS combinational logic circuits", *Proc. 14th Int. Symp. Fault Tolerant Comput.*, pp. 44-49.

Reddy, M. K., and S. M. Reddy (1985), "On FET stuck-open fault detectable CMOS memory elements", *IEEE Test Conf.*, pp. 424-429.

Reddy, S. M. (1972), "Easily testable realization for logic functions", *IEEE Trans. Comput.*, C-21, pp. 1183-1188.

Reddy, S. M. (1972), "A design procedure for fault-locatable switching circuits", *IEEE Trans. Comput.*, C-21, pp. 1421-1426.

Reddy, S. M. (1973), "Complete test sets for logic functions", *IEEE Trans. Comput.*, C-22, pp. 1016-1020.

Reddy, S. M. (1977), "A note on testing logic circuits by transition counting", *IEEE Trans. Comput.*, C-26, pp. 313-314.

Reddy, S. M. (1977), "Comments on "Minimal fault tests for combinational networks", *IEEE Trans. Comput.*, C-26, pp. 318-319.

Reddy, S. M., M. K. Reddy, and J. G. Kuhl (1983), "On testable design for CMOS logic circuits", *IEEE Test Conf.*, pp. 435-441.

Reddy, S. M., and D. S. Ha (1987), "A new approach to the design of testable PLAs", *IEEE Trans. Comput.*, C-36, pp. 201-211.

Renovell, M, and G. Cambon (1986), "Topology dependence of floating gate faults in MOS integrated circuits", *Electronics Letters*, 22, pp. 152-157.

Roberts, M. W., and P. K. Lala (1984), "An algorithm for the partitioning of logic circuits", *IEE Proc. Pt. E*, 131, pp. 113-118.

Robinson, J. P., and N. R. Saxena (1987), "A unified view of test compression methods", *IEEE Trans. Comput.*, C-36, pp. 94-99.

Rosenbluth, W. (1976), "Design automation architecture and applications", *Computer*, 9, pp. 12-17.

Roth, J. P. (1966), "Diagnosis of automata failures: a calculus and a method", *IBM J. Res. Dev.*, 10, pp. 278-281.

Roth, J. P., W. G. Bouricius, and P. R. Schneider (1967), "Programmed algorithms to compute tests to detect and distinguish between failures in logic circuits", *IEEE Trans. Electron. Comput.*, EC-16, pp. 567-580.

Russell, J. D., and C. R. Kime (1971), "Structural factors in the fault diagnosis of combinational networks", *IEEE Trans. Comput.*, C-20, pp. 1276-1285.

Saluja, K. K., and S. M. Reddy (1974), "On minimally testable logic networks", *IEEE Trans. Comput.*, C-23, pp. 552-554.

Saluja, K. K,. and S. M. Reddy (1974), "Easily testable two-dimensional cellular logic arrays", *IEEE Trans. Comput.*, C-23, pp. 1204-1207.

Saluja, K. K., and S. M. Reddy (1975), "Fault detecting test sets for Reed-Muller canonic networks", *IEEE Trans. Comput.*, C-24, pp. 995-998.

Saluja, K. K., K. Kinoshita, and H. Fujiwara (1981), "A multiple fault testable design of programmable logic arrays", *Proc. 11th Symp. Fault-Tolerant Computing*, pp. 44-46.

Saluja, K. K. (1982), "An enhancement of LSSD to reduce test pattern generation effort and increase fault coverage", *19th Design Automation Conference*, pp. 489-494.

Saluja, K. K., and M. G. Karpovsky (1983), "Testing computer hardware through data compression in space and time", *Proc. 1983 Int. Test Conf.*, pp. 83-88.

Saluja, K. K., K. Kinoshita, and H. Fujiwara (1983), "An easily testable design of programmable logic arrays for multiple faults", *IEEE Trans. Comput.*, C-32, pp. 1038-1046.

Saluja, K. K., and R. Dandapani (1986) "An alternative to scan design methods for sequential machines", *IEEE Trans. Comput.*, C-35, pp. 384-388.

Sasaki, T., N. Koike, K. Ohmori, and K. Tomita (1983), "HAL: A block level hardware logic simulator", *Proc. 20th Design Automation Conf.*, pp. 150-156.

Savir, J. (1980), "Syndrome-testable design of combinational circuits", *IEEE Trans. Comput.*, C-29, pp. 442-451, correction: pp. 1012-1013.

Savir, J. (1981), "Syndrome-testing of 'syndrome-untestable' combinational circuits", *IEEE Trans. Comput.*, C-30, pp. 606-608.

Savir, J., and P. H. Bardell (1983), "Random pattern testability", *13th Int. Symp. Fault-Tolerant Comput.*, pp. 80-89.

Savir, J., and W. H. McAnney (1985), "On the masking probability with one's count and transistion count", *IEEE Int. Conf. Computer-Aided Design*, pp. 111-113.

Saxena, N. R., and J. P. Robinson (1986), "Accumulator compression testing", *IEEE Trans. Comput.*, C-35, pp. 317-321.

Schertz, D. R., and G. Metze (1971), "On the design of multiple fault diagnosable networks", *IEEE Trans. Comput.*, C-20, pp. 1361-1364.

Schertz, D. R., and G. Metze (1972), "A new representation for faults in combinational digital circuits", *IEEE Trans. Comput.*, C-21, pp. 858-866.

Schnurmann, H. D., E. Lindbloom, and R. G. Carpenter (1975), "The weighted random test-pattern generator", *IEEE Trans. Comput.*, C-24, pp. 695-700.

Sedmak, R. M. (1980), "Design for self-verification: an approach for dealing with testability problems in VLSI-based designs", *Proc. IEEE Test Conf.*, pp. 112-120.

Segers, M. T. M. (1981), "A self-test method for digital circuits", *Proc. IEEE Test Conf.*, pp. 79-85.

Sellers, E. F., M. Y. Hsiao, and L. W. Bearnson (1968), "Analyzing errors with the Boolean difference", *IEEE Trans. Electron. Comput.*, EC-17, pp. 676-683.

Serra, M., and J. C. Muzio (1987), "Testing programmable logic arrays by sum of syndromes", *IEEE Trans. Comput.*, forthcoming.

Seth, S. C. and K. L. Kodandapani (1977), "Diagnosis of faults in linear tree networks", *IEEE Trans. Comput.*, C-26, pp. 29-33.

Shedletsky, J. J., and E. J. McCluskey (1976), "The error latency of a fault in a sequential digital circuit", *IEEE Trans. Comput.*, C-25, pp. 655-659.

Smith, J. E. (1979), "Detection of faults in programmable logic arrays", *IEEE Trans. Comput.*, C-28, pp. 845-853.

Smith, J. E. (1980), "Measures of effectiveness of fault signature analysis", *IEEE Trans. Comput.*, C-29, pp. 510-514.

Somenzi, F., S. Gai, M. Mezzalama, and P. Prinetto (1984), "PART: programmable array testing based on a partitioning algorithm", *IEEE Trans. Computer-Aided Design*, CAD-3, pp. 142-149.

Son, K., and D. K. Pradham (1980), "Design of programmable logic arrays for testability", *Proc. IEEE Test Conf.*, pp. 163-166.

Sridhar, T. and J. P. Hayes, (1981), "A functional approach to testing bit-sliced microprocessors", *IEEE Trans. Comput.*, C-30, pp. 563-571.

Sridhar, T., D. S. Ho, T. J. Powell, and S. M. Thatte (1982), "Analysis and simulation of parallel signature analysis", *Proc. IEEE Test Conf.*, pp. 656-661.

Srini, V. P. (1978), "Fault location in a semiconductor random-access memory unit", *IEEE Trans. Comput.*, C-27, pp. 349-358.

Stephenson, J. E., and J. Grason (1976), "A testability measure for register transfer level digital circuits", *Proc. 6th Int. Symp. Fault Tolerant Computing*, pp. 101-107.

Susskind, A. K. (1973), "Diagnostics for logic networks", *IEEE Spectrum*, 10, pp. 40-47.

Susskind, A. K. (1983), "Testing by verifying Walsh coefficients", *IEEE Trans. Comput.*, C-32 , pp. 198-201.

Sutton, J. C., and J. G. Bredeson (1980), "Minimal redundant logic for high reliability and irredundant testability", *IEEE Trans. Comput.*, C-29, pp. 648-656.

Swartzlander, E. E., J. A. Eldon, and D. D. Hsu (1985), "VLSI testing: a decade of experience", *IEEE Spring COMPCON*, pp. 392-395.

Szygenda, S. A., and E. W. Thompson (1976), "Modeling and digital simulation for design verification and diagnosis", *IEEE Trans. Comput.*, C-25, pp. 1242-1253.

Tang, D. T., and L. S. Woo (1983), "Exhaustive test pattern generation with constant weight vectors", *IEEE Trans. Comput.*, C-32, pp. 1145-1150.

Tang, D. T., and C.-L. Chen, (1984), "Logic test pattern generation using linear codes", *IEEE Trans. Comput.*, C-33, pp. 845-850.

Thatte, S. M., D. S. Ho, H.-T. Yuan, T. Sridhar, and T. J. Powell (1982), "An architecture for testable VLSI processors", *Proc. IEEE Test Conf.*, pp. 484-492.

Thayse, A., and M. Davio (1973), "Boolean differential calculus and its applications to switching theory", *IEEE Trans. Comput.*, C-22, pp. 409-420.

Thomas, J. J. (1971), "Automated diagnostic test programs for digital networks", *Computer Des.*, pp. 63-67.

Timoc, C., *et al* (1983), "Logic models of physical failures", *IEEE Test Conf.*, pp. 546-553.

To, K. (1973), "Fault folding for irredundant and redundant combinational circuits", *IEEE Trans. Comput.*, C-22, pp. 1008-1015.

Toth, A., and C. Holt (1974), "Automated database-driven digital testing", *Computer*, 7, pp. 13-19.

Totton, K. A. E. (1985), "Review of built-in test techniques for gate arrays", *IEE Proc. Pt. I*, 132, pp. 121-129.

Trischler, E. (1980), "Incomplete scan path with an automatic test-generation methodology", *IEEE Test Conf.*, pp. 153-162.

Tzidon, A., I. Berger, and M. Yoeli, (1978), "A practical approach to fault detection in combinational networks", *IEEE Trans. Comput.*, C-27, pp. 968-971.

Ulrich, E. G., and T. Baker (1974), "Concurrent simulation of nearly identical digital networks", *Computer*, 7, pp. 39-44.

van Bokhoven, W. M. G. (1982), "Mixed level and mixed mode simulation by a piecewise linear approach", *IEEE Int.. Symposium on Circuits and Systems*, pp. 1256-8.

Varma, P., A. P. Ambler, and K. Barker (1985), "On-chip testing of embedded PLAs", *IEEE Custom Integrated Circuits Conf.*, pp. 306-310.

Wadsack, R. L. (1978), "Fault Modelling and Logic Simulation of CMOS and MOS Integrated Circuits", *Bell System Technical Journal*, 57, pp. 1449-1473.

Wang, D. T. (1975), "An algorithm for the generation of test sets for combinational logic networks", *IEEE Trans. Comput.*, C-24, pp. 742-746.

Wang, D. T. (1975), "Properties of faults and criticalities of values under tests for combinational networks", *IEEE Trans. Comput.*, C-24, pp. 746-750.

Wang, L.-T., and E. J. McCluskey (1986), "Condensed linear feedback shift register (LFSR) testing -- a pseudoexhaustive test technique", *IEEE Trans. Comput.*, C-35, pp. 367-369.

Want, F., and H. H. Schrieber (1979), "A pragmatic approach to automatic test generation and failure isolation of analog systems", *IEEE Trans. on Circuits and Systems*, CAS-26, pp.584-585.

Williams, M. J. Y., and J. B. Angell (1973), "Enhancing testability of large-scale integrated circuits via test points and additional logic", *IEEE Trans. Comput.*, C-22, pp. 46-60.

Williams, T. W., and K. P. Parker (1979), "Testing logic networks and design for testability", *IEEE Computer*, 12, pp. 9-21.

Williams, T. W., and K. P. Parker (1982), "Design for testability -- a survey", *IEEE Trans. Comput.*, C-31, pp. 2-15.

Williams, T. W. (1984), "VLSI testing", *IEEE Computer*, 17, pp. 126-136.

Wood, C. T. (1979), "The quantitative measure of testability", *Proc. IEEE Autotestcon*, pp. 286-291.

Yajima, S., and T. Aramaki (1981), "Autonomously testable programmable logic arrays", *Proc. 11th Symp. Fault-Tolerant Computing*, pp. 41-43.

Yamada, T. (1983), "Syndrome-testable design of programmable logic arrays", *Proc. 1983 Int. Test Conf.*, pp. 453-458.

Yau, S. S., and Y. S. Tang (1971), "An efficient algorithm for generating complete test sets for combinational logic circuits", *IEEE Trans. Comput.*, C-20, pp. 1245-1251.

Yau, S. S., and S.-C. Yang (1975), "Multiple fault detection for combinational logic circuits", *IEEE Trans. Comput.*, C-24, pp. 233-242.

Zasao, J. J. (1985), "MOS stuck fault testing of CMOS VLSI", *IEEE COMPCON Spring*, pp. 388-391.

Zorian, Y., and V. K. Agarwal (1984), "Higher certainty of error coverage by output data modification", *Proc. 1984 Int. Test Conf.*, pp. 140-147.

Index

A

accumulator compression testing (ACT), 187
accumulator testing, 172, 186, 200
adjacency matrix, 64
aliasing, 8, 116, 172, 257
  average case, 174
  empirical results for, 188
  heuristics for, 266
  probability function, 260
  signature analysis, 267
  syndrome testing, 267
  validation tests of, 269
  worst case, 173
analogue
  circuits, 391
    digital modeling of, 394
    failures in, 395
    test sets for, 398, 404
  fault
    diagnosis, 398
    models, 397
  simulation, 392, 405
arithmetic
  expansion, canonic, 90
  coefficient testing, 98
  spectrum, 91
augmented gate, 62
available sequences (AS), 239

B

ß-distance, 258, 263
Boolean testing, 17, 22
Boyle models, 396
bridging fault model, (*see* fault models, bridging)
built-in logic block observation (BILBO), 292
built-in self-test (BIST), 2, 7, 219, 292, 360
  by output data modification, 242
  general model of, 225

C

canonic arithmetic expansion, 90
canonic minterm expansion, 90
cellular automata
  for random sequences, 310
  logic block observation (CALBO), 310
circuit database, 405
clean fault, 315

complementary logic redundancy, 367
complementary metal oxide semi-conductor (CMOS)
    cells, 283, 319
    control transistors, 334, 341
    dynamic faults in, 327
    failure modes in, 285, 325
    fault classification for, 329, 335
    full complementary, 321
    graph representation of, 319
    inverter, 283
    latches, 348
    nand, two-input, 284
    technology, 281, 319
    transistors, (*see* transistors, CMOS)
complex redundancy, 58, 70
concurrent checking (*see* concurrent testing)
concurrent circuit testing (*see* concurrent testing)
concurrent testing, 2, 17, 361
    for ALU, 376
    for data path, 384
    for multiplier, 375, 383
    for PLA, 377
    for rotate and shift operations, 382
    techniques, 366
constrained dependence fault model, 35
counting
    edge, 223
    generalized techniques, 184, 200
    ones, (*see* syndrome)
    syndrome, (*see* syndrome)

transition, 97, 223
covering condition
    strong, 177
    weak, 176
cross point fault model, 135
cut-critical transistor, 324
cut-essential transistor, 324

D

data
    compaction, 9, 40, 88
    compression, 116, 170
        parallel signature and syndrome, 148
        guidelines for, 166
deception volume, 221
delay failures, 363
dependent error
    model, 258
    sequences, 259
dependent fanout, 78
design
    errors, 15
    for testability, 2, 139, 288, 360
    verification, 1, 23
    verification vectors, 16, 23
dictionary, fault, 40
diagnostic matrix, 64
digital modeling of analogue, 394
distance distribution, 262
documentation, 15
dominant logic value, 50
dynamic faults in CMOS, 327

E

easily generated sequences (EGS), 226, 241
edge counting, 223
equal size segments (ESS), 227
  modifier sequence generation, 239
equally likely errors (ELE), 257
equivalent length register, 268
error
  control circuit, 32
  coverage, 221
    improving, 222
  dependent model of, 258
  dependent sequences, 259
  design, 15
  equally likely (ELE), 257
  masking, 220
  masking probability, 221, 257
  unequally likely (ULE), 257
Eulerian cycle, 257
  circuit to generate, 302
  non-recursive generation of, 299
  recursive generation of, 298
  test generator, 295
exhaustive
  test set, 33
  testing, 86, 291
expert gate analysis (EGA), 273
  distance distribution for, 275
explicit testing, 17
exponentially increasing segments (EIS), 227, 238
  double input method, 235
  multi-seed possibility, 232
  single input method, 227

F

failures, 13, 362
  delay, 363
  functional, 1
  in analogue circuits, 395
  in CMOS cells, 285, 325
  intermittent, 2, 364
  permanent, 2
fan-out free circuit, 51
fault
  clean, in CMOS, 315
  CMOS, 329
    classification, 335
    latches, 348
  collapsing, 193
  coverage, 18, 21, 40, 272
  diagnosis, analogue, 398
  dictionary, 40
  grading, 18
  models, 3, 29, 279, 393
    analogue, 397
    bridging, 4, 29, 34, 136, 280
    cross point, 135
    constrained dependence, 35
    input bridging, 138
    multiple stuck-at, 4, 48, 280
    programmable logic array (PLA), 134
    single stuck-at, 3, 48, 188, 280, 316, 362
  stuck-at neighbour, 4
  stuck-on, 5, 317
  stuck-open, 4, 287, 316
  static, 315

simulation, 19, 269, 351
   statistical, 20
   unclean, in CMOS, 327
   untestable, 30
full redundancy, 48
functional
   failures, 1
   test set, 38
   testing, 22

G
Gallois field, 9
generalized counting techniques, 184, 200
generating function, 178
grading, fault, 18

H
Hadamard spectrum, 94, 118
Hamming code, 149, 369
hardware redundancy, 366
hazards, 32, 344
   avoidance of, 346
hybrid integrated circuits, 391

I
implicit testing (*see* concurrent testing)
improved compressor, 226, 247
independence condition, 177
independence, in CMOS cells, 325
independent fanout, 52
information redundancy, 368
input bridging fault model, 138
integrated circuit
   analogue, 391
   hybrid, 391
   testing of, (*see* testing)
   yield, 21
intermittent failures, 2, 364
internally unate lines, 141
inverter, CMOS, 283
irredundant circuit, 48, 54

L
level sensitive scan design (LSSD), 6
linear feedback shift register (LFSR), 7, 116
   testing, 9, 170, 173, 200
   HP model, 171
linearity condition, 177
low cost residue code, 371

M
macro modeling, 395
masking (*see* aliasing)
matrix
   adjacency, 64
   diagnostic, 64
   multiplication, diagonal method, 66
   parity check, 369
   representation of digital circuits, 61
   structural, 63
minterm expansion, canonic, 90
modifier block, 246
   realization of, 251

modulo 3 low cost residue code, 361
  anomalies, 379
  comparator, 374
  encoder, 375
  generator, 374
  multiplier, 375
Monte Carlo simulation, 163
multiple-input LFSR (MISR), 10, 219, 245, 293
multiple-output
  circuits, 74, 115
  redundancy, 75
multiple stuck-at fault, (*see* fault models, multiple stuck-at)

N

N-channel CMOS transistor, 281
nand
  two-input CMOS, 284
negative function, 322
non-exhaustive testing, 86
number system
  one's complement, 380
  two's complement, 381

O

off-line testing, 2
one's complement number system, 380
ones counting (*see* syndrome)
operating surface, 401
output compaction (*see* data compaction)

output data modification, 219, 222
  as a BIST scheme, 242
  implementation of, 247

P

P-channel CMOS transistor, 282
parallel / serial converter, 244
parity
  check matrix, 369
  code, 369
partial redundancy, 48
permanent failures, 2
pin test set, 27
positive
  canonic Reed-Müller expansion, 92
  function, 322
primitive polynomial, 149, 159
probability function
  aliasing, 260
programmable logic array (PLA)
  concurrent testing of, 377
  testing by WSPS, 132
pseudo-exhaustive test set, 23
pseudo-random test set, 36

R

Radamacher-Walsh spectrum, 96, 118
random test generation, 170
reconvergent
  fanout, 51
  path, 52
redundancy, 30, 48
  complementary logic, 367

redundancy (*cont.*)
  complex, 58, 70
  full, 48
  hardware, 366
  information, 368
  multiple-output, 75
  partial, 48
  simple, 56, 67
  trivial, 49
redundant fanout branch, 56
Reed-Müller
  coefficient testing, 98
  expansion, positive canonic, 92
residue codes, 371
  for logical operations, 373
  low cost, 371
  modulo 3, (*see* modulo 3 low cost residue code)
response compaction (*see* data compaction)

S

scan path testing, 6, 68
separable code, 371
sequences
  available (AS), 239
  easily generated (EGS), 226, 241
  equal size segment (ESS), 227, 239
signature, 7, 89, 106, 116, 148, 170
  analysis, 99, 258, 267
    aliasing in, 267
simple redundancy, 56, 67
simulation

analogue, 392, 405
  fault, 19, 269, 351
  statistical, 20
  Monte Carlo, 163
  vectors (*see* design verification vectors)
single stuck-at fault (*see* fault model, single stuck-at)
spectral coefficients, 95
  Hadamard, 100
  Rademacher-Walsh, 96, 118
  Reed-Müller, 92
  relationships between, 100
spectral coefficient testing, 9, 116, 293
spectrum
  arithmetic, 91
  Hadamard, 94, 118
  Rademacher-Walsh, 96, 118
standard logic symbols, 16
static fault, 315
structural testing, 28
structural matrix, 63
strong covering condition, 177
stuck-at
  fault, (*see* fault models, single *and* multiple stuck-at)
  fault test set, 36
  neighbour fault model, 4
stuck-on, fault model, 5, 317
stuck-open fault
  detection by test vectors, 288, 326, 350
  exhaustive testing of, 291, 294
  memory cells, 290

model, 4, 287, 316
syndrome
  counter, 171
  counting, 182, 223
  testing, 9, 97, 109, 116, 119, 148, 171, 198, 258, 267, 293, 305
    aliasing in, 267

T

test set, 5, 116, 170, 288, 359
  analogue circuit, 398, 404
  effectiveness of, 38
  exhaustive, 33
  functional, 38
  PLA, 132
  pin, 27
  pseudo-exhaustive, 23
  pseudo-random, 36
  stuck-at fault, 36
test vectors, 5, 24
  random generation of, 170
  toggle, 27
testing
  built-in self- (BIST), (*see* built-in self-test)
  Boolean, 17, 22
  concurrent (*see* concurrent testing)
  exhaustive, 86, 291
  explicit, 17
  functional, 22
  LFSR, (*see* LFSR testing)
  non-exhaustive, 86
  off-line, 2
  Reed-Müller coefficient, 98

scan path, 6, 68
spectral coefficient, 9, 116, 293
structural, 28
weighted spectral sum, (*see* weighted spectral sum)
weighted syndrome sum, 117
tie-critical transistor, 323
tie-essential transistor, 323
toggle test vectors, 27
transistors, CMOS
  control, 334, 341
  cut-critical, 324
  cut-essential, 324
  N-channel, 281
  P-channel, 282
  tie-critical, 323
  tie-essential, 323
transition counting, 97, 223
trivial redundancy, 49
two-level modification scheme, 239
two's complement number system, 381

U

unclean faults in CMOS, 327
undetectable faults, 30
unequally likely errors (ULE), 257
unimodal, 187
untestable faults, 30

V

verification
  design, 1, 23
  vectors, 16, 23

W

weak covering condition, 176
weight of function, 119, 171
weighted spectral sum (WSPS),
    117, 121, 123
  testability
    complexity of, 142
    of input lines, 124, 126
    of internal lines, 127, 129
weighted syndrome sum (WSS),
    117

Y

yield for integrated circuits, 21

PERSPECTIVES IN COMPUTING

Vol. 1 John R. Bourne, *Laboratory Minicomputing*

Vol. 2 Carl Tropper, *Local Computer Network Technologies*

Vol. 3 Kendall Preston, Jr., and Leonard Uhr, editors, *Multicomputers and Image Processing: Algorithms and Programs*

Vol. 4 Stephen S. Lavenberg, editor, *Computer Performance Modeling Handbook*

Vol. 5 R. Michael Hord, *Digital Image Processing of Remotely Sensed Data*

Vol. 6 Sakti P. Ghosh, Y. Kambayashi, and W. Lipski, editors, *Data Base File Organization: Theory and Applications of the Consecutive Retrieval Property*

Vol. 7 Ulrich W. Kulisch and Willard L. Miranker, editors, *A New Approach to Scientific Computation*

Vol. 8 Jacob Beck, Barbara Hope, and Azriel Rosenfeld, editors, *Human and Machine Vision*

Vol. 9 Edgar W. Kaucher and Willard L. Miranker, *Self-Validating Numerics for Function Space Problems: Computation with Guarantees for Differential and Integral Equations*

Vol. 10 Mohamed S. Abdel-Hameed, Erhan Çınlar, and Joseph Quinn, editors, *Reliability Theory and Models: Stochastic Failure Models, Optimal Maintenance Policies, Life Testing, and Structures*

Vol. 11 Mark G. Karpovsky, editor, *Spectral Techniques and Fault Detection*

Vol. 12 Selim G. Akl, *Parallel Sorting Algorithms*

Vol. 13 Azriel Rosenfeld, editor, *Human and Machine Vision II*

Vol. 14 Y. C. Tay, *Locking Performance in Centralized Databases*

Vol. 15 David S. Johnson, Takao Nishizeki, Akihiro Nozaki, Herbert S. Wilf, editors, *Discrete Algorithms and Complexity Theory: Proceedings of the Japan-US Joint Seminar, June 4–6, 1986, Kyoto, Japan*

Vol. 16   J. Duane Northcutt, *Mechanisms for Reliable Distributed Real-Time Operating Systems: The Alpha Kernel*

Vol. 17   Gerd Bohlender, Christian Ullrich, Jürgen Wolff von Gudenberg, and Louis B. Rall, *Pascal-SC: A Computer Language for Scientific Computation*

Vol. 18   D. M. Miller, editor, *Developments in Integrated Circuit Testing*

Volumes 1–12 were published as **Notes and Reports in Computer Science and Applied Mathematics.**